SUPPORT VECTOR MACHINE
in Chemistry

SUPPORT VECTOR MACHINE
in Chemistry

Nianyi Chen
Shanghai University, China

Wencong Lu
Shanghai University, China

Jie Yang
Shanghai Jiao Tong University, China

Guozheng Li
Shanghai Jiao Tong University, China

World Scientific

NEW JERSEY · LONDON · SINGAPORE · BEIJING · SHANGHAI · HONG KONG · TAIPEI · CHENNAI

Published by

World Scientific Publishing Co. Pte. Ltd.

5 Toh Tuck Link, Singapore 596224

USA office: 27 Warren Street, Suite 401–402, Hackensack, NJ 07601

UK office: 57 Shelton Street, Covent Garden, London WC2H 9HE

British Library Cataloguing-in-Publication Data
A catalogue record for this book is available from the British Library.

SUPPORT VECTOR MACHINE IN CHEMISTRY

ISBN 981-238-922-9

Printed in Singapore by World Scientific Printers (S) Pte Ltd

Preface

In recent 25 years, my colleagues and I have been dealing with research work of computer chemistry, especially in the field of materials design, phase diagram prediction and optimal control in petrochemical factories. Although our work has some good results, we often meet two kinds of difficulties: overfitting and underfitting. About 5 years ago, I began to cooperate with professor Jie Yang and his student, doctor Chenzhou Ye. In scientific cooperation, I realized that a newly proposed method of machine learning, support vector machine, could be used to overcome the difficulties induced by overfitting. Later, I found that this new method had been used in QSAR work, with good results. Then I decided to organize an interdisplinary research project in this field: to apply this new method in various fields in chemistry and chemical technology. In our research work, we have collected the relevant literatures from international journals. This book is just the product of our four years research work and the result of our collection work about relevant international literatures.

As one of the authors of this book, I wish to emphasize that this book is a product of collective cooperation. Professor Wencong Lu, one of my chief cooperators, has excellent work about the software of SVM and its applications. He is the writer of two chapters of this book. Professor Jie Yang and Doctor Guozheng Li, are the writers of three chapters of this book. Professor Xinhua Bao has integrated SVM into an intelligent data base for the assessment of phase diagrams of molten salt systems. Professor Yimin Ding has finished a series of experimental measurements of the phase diagrams to confirm our results of computerized prediction by SVM method. Doctor Zhirong Lu, the chief manager of Seawall Data Technology Co., Ltd, helps us to apply SVM algorithm to the optimal control in petrochemical industry. Yonggang Wang, Dong Liang, Guopin Liu also take part in the preparation of the manuscript of this book. I am grateful to Xiaobo Ji, Ning Dong and Shansheng Yang for their hard work in the preparation of the manuscript of this book. Professor Yuh-Kang Pan, a professor of Boston College, has given good advice and great help to us in our research work described in this book. We wish to express heartful thanks to him.

Professor Nianyi Chen
Laboratory of Chemical Data Mining
Shanghai University
24 March 2004, Shanghai, China

Contents

Chapter 1

Introduction

1.1 Support Vector Machine: Data Processing Method for Problems of Small Sample Size[*]

Carrying out experimental work, finding the regularities of the data obtained, and making prediction for some unknown phenomena, are the chief mode of the research work in the fields of chemistry and related disciplines, including chemical engineering, materials science and environmental science. Since the progress and achievement of computer science and technology, computerized data processing, or so-called *machine learning*, has been widely used in chemical research work and chemical industrial optimal control. Up to now, the statistical methods used in chemistry are almost all based on the classical statistical theory. It is well known that one of the basic principles in classical statistics is the *law of large numbers.* According to this principle, *when the number of observations tends to infinity, the empirical distribution function* $F_\ell(x)$ *converges to the actual distribution function* $F(x)$. In other words, for getting a reliable mathematical model by using machine learning, we have to provide the data set including *infinite number* of samples used as training set. In any practical problem-solving work, including the machine learning tasks in chemistry, however, it is impossible to have so many samples for training and mathematical model

---

[*]*Problem of small sample size* is a technical term used by Vapnik. It means a *problem of small data set.*

1

building. On the contrary, in most of the chemical data processing work the number of training samples is usually *quite small*. For example, QSAR study is one of the most important steps in molecular design. In QSAR work, the known data of some similar compounds are used as training samples, and the number of training samples is usually not more than several tens. Another example is the work of experimental design. People wish to find the clue for searching the best experimental results on the basis of data processing of a small set of known data. So it is quite natural to ask a question: Does the contradiction between *law of large numbers* and the *small number of training samples* have any significant influence on the reliability of the mathematical models built by using machine learning?

In recent years, a widely recognized theory of statistical science, the *statistical learning theory (SLT),* has been proposed to find the answer of the above-mentioned question [133; 135]. And several newly proposed methods of machine learning, including *support vector machine (SVM)* and *weight-decay artificial neural network (WD-ANN)*, have been proposed based on the spirit of statistical learning theory [69; 127]. These new methods of computation have been used in many fields of application, including image recognition, text categorization and DNA research, with rather good results. Now these powerful data processing techniques have been also used in the fields of chemistry and related disciplines. As compared with other algorithms used in computer chemistry, SVM has some outstanding advantages: it can be used for both classification (support vector classification, abbreviated as SVC) and regression (support vector regression, abbreviated as SVR); it is suitable for both linear and nonlinear data processing; it has special generalization ability, especially for problems of small sample size; SVM has no trouble of local minimum problem. As a newly proposed algorithm, SVM has bright future as a powerful tool for chemistry and related fields owing to these advantages.

In this chapter, the basic principles of statistical learning theory will be introduced. And the possibility of application of support vector machine to various fields in chemistry and chemical technology will be discussed.

1.2 Support Vector Machine: Data Processing Method for Complicated Data Sets in Chemistry

Although the classical methods of statistics have been successfully applied in many fields of chemistry and chemical technology, there are still some difficult problems unsolved in these fields. The principal origin of these difficulties is that most of the data sets in chemistry and chemical technology are *complicated data sets*. It is often difficult to extract the useful information completely and efficiently from such kind of data sets by using classical statistical methods, because of the following characteristics of these complicated data sets:

(1) Nonlinearity: Many classical statistical methods are especially suitable for the linear data processing problems. But most of the data processing problems in chemistry and chemical technology are nonlinear problems. Of course, if some data sets indeed exhibit linear relationships or even nearly linear relationships, the data processing process will be greatly simplified, and the results of machine learning will be more reliable. As a matter of fact, however, only a small part of data sets in practical problems can be considered as linear or nearly linear data sets. It is reasonable to use the multiple correlation coefficient or the PRESS of PLS as the criterion for the linearity of the relationship of a data set. Using these criteria, it can be shown that most of the data sets in real chemical problems exhibit more or less degree of nonlinearity.

Table 1.1 shows the nonlinearities of some data sets treated in our previous work. In our research work, we usually use the multiple correlation coefficient larger than 0.9 or the PRESS of PLS regression lower than 0.2 as the criterion for deciding a data set suitable to be treated by linear regression. This is, of course, a very rough criterion. And it should be emphasized that different standards for the justification of usability of linear methods of data processing may be necessary for different problems.

Strictly speaking, among the examples of real chemical problems listed above, only one or two problems can be considered as nearly linear problems. And the nonlinear nature of all other practical problems cannot be ignored. Otherwise, the results of data processing will be unreliable.

Table 1.1   The nonlinearity of some data processing problems.

Data processing problems	PRESS of normalized data	Multiple Correlation coefficient
Data for high T_c superconductor exploration	0.830	0.469
Data for new phosphor materials exploration	0.307	0.847
Data of leaching rate in alumina production	0.930	0.282
Data of magnetic property of some alloys	0.511	0.718
Data for optimal control of butadiene rubber production	0.215	0.894
Data of electrochemical capacity of Ni/H battery	0.956	0.257
Data of carbon content change in steel making	0.270	0.862
Data for VPTC materials research work	0.138	0.938
Melting points modeling of complex halides of A_3BX_6 type	0.356	0.816
Modeling enthalpy of mixing of MX-MX' systems (M=alkali metals, X=halogens)	0.804	0.622
Modeling condition of Cr electroplating	0.454	0.753
Breast cancer mortality and trace element intake relationship	0.226	0.901

Before the development of SVM techniques, there are two usually used techniques for the data processing of nonlinear data sets. One is nonlinear regression with polynomials, and the other is artificial neural network. It is well known that the former often needs too many terms and too many adjustable parameters in regression. This is so-called "curse of dimensionality". And the latter often suffers from overfitting, i.e., having low reliability of the prediction results.

The development of SVM has provided a new way of data processing for solving nonlinear problems. And in many cases the adaptability of SVM is better than other techniques. Therefore SVM should be considered as a new powerful tool for the data processing in the field of chemistry and chemical technology.

(2) Multivariate problems: chemical reactions are usually influenced by many factors, such as temperature, pressure, concentration, the presence and activity of catalysts, the kind of solvents, etc. The physical or chemical behaviors of materials also depend on many factors, such as their chemical composition, phase composition, particle size, presence of impurities, etc. The production processes in chemical or metallurgical industry usually involve heat transfer, mass transfer, fluid flow and a series of chemical reactions, so that there are always many factors

influencing the technical situations of a production process. According to our experience, there are usually more than five or six chief factors must be considered for solving a practical optimization problem in chemical or metallurgical industry, and these five or six chief factors must be selected from several dozens of factors via feature selection procedures in data processing work.

Sometimes the situation is even more complicated. In many cases it is even difficult to decide whether a selected feature set is a complete feature set for the exact description of the relationship between the target and the affecting factors.

In classical methods, it is difficult to treat a data set with too many affecting factors, because in this case the high dimension of feature space shall induce uncertainty of the results of data processing. But Vapnik and his coworkers have found that high dimension can be made less harmful in SVM computation by using the *principle* of *large margin* and *kernel functions*. So the development of SVM has provided an effective way to overcome the *curse of dimensionality* for solving multivariate problems in chemistry and chemical technology.

(3) High noise: One of the requirements of classical statistical methods is that the noise in data set should be low enough, but this requirement cannot be satisfied in many cases of the data processing in chemistry and chemical technology. Chemical processes are usually affected by many factors. It is usually very difficult to confirm accurately how many factors should be considered in the solution of a practical problem. Therefore, the influence of the neglected factors shall be considered as noise. In the production processes of chemical or metallurgical industry, the uncertainty problems are more serious. For example, the composition of raw materials in a large petroleum refinery often changes since the crude oil composition in every batch from the tanker is usually not the same, and the activities of the catalysts in many chemical processes always changes in their life of use. Besides, many exothermic chemical processes in chemical plants may induce chaotic phenomena. All these factors are the origin of uncertainty or noise in production processes.

The presence of noise gives rise to many difficulties in the data processing work in chemistry and chemical technology, especially in the problems with small sample size.

The use of SVM cannot solve all problems of noise in data processing, but it is possible to use SVM technique to improve noisy data processing in many ways. For example, it can provide some ways for outlier deleting: By leave-one-out (LOO) cross-validation method, we can delete the data samples with large error in prediction, and make the improvement of data files. Besides, the adoption of ε-insensitive loss function in support vector regression makes it more robust to noisy data sets.

1.3   Underfitting and Overfitting: Problems of Machine Learning

According to statistical learning theory, the machine learning is a process of choosing an appropriate function from a given set of functions to correlate the data set. The set of functions used is called hypothesis functions or indicator functions[*]. For example, in the process of linear regression or linear separation for different classes of samples, all linear functions are used as hypothesis functions. Since the appropriate function has to be chosen from the hypothesis functions only, the mathematical model built by using machine learning is always constrained within the scope of hypothesis functions used. For example, if a linear regression method is used as learning process, the mathematical model found shall be surely linear one, even if the actual data set exhibits some nonlinearity, because this nonlinearity has been treated as noise or residue and eliminated in the process of machine learning. Since most of the chemical data sets exhibit more or less nonlinearity nature, the results of linear modeling of many chemical data sets usually undergo some degree of *underfitting*. Underfitting is obviously a source of the inaccuracy of the mathematical model obtained. As an example, Fig. 1.1 illustrates the result of linear regression of the data of the measured thickness of the thin films of indium oxide by PLS method (the background of this data

---

[*]Indicator functions denote the hypothesis function set in classification problems.

will be described in chapter 8 in this book). From the comparison of the experimental data and the data calculated by using PLS method, it can be seen that the result somewhat deviates from linearity, and perhaps an equation with quadratic terms may be more suitable to describe the regularity. Figure 1.2 shows another example. This data set expresses the relationship between the rate of recovery of Al_2O_3 and the raw materials composition in alumina production (the background will be described in chapter 14). In this case, the result of regression by using PLS method is not satisfactory since the data set exhibits strong nonlinearity. It is evident that the use of linear functions in data processing cannot give satisfactory mathematical model when the nature of data set is strongly nonlinear. So it is clear that *too narrow scope* of function set used in data processing will give rise to underfitting problems.

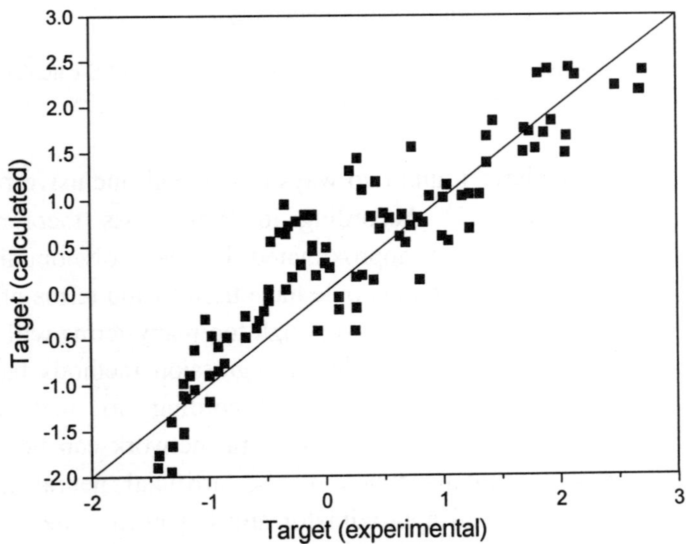

Fig. 1.1   Comparison between the experimental data and the PLS calculated data of the thickness of the semiconductor thin films of indium oxide.

Since too narrow scope of function set used cannot give good result, it is natural to think that the use of wider scope of function set in data processing may give better results of machine learning, and that an all-inclusive function set can be used to avoid underfitting problem.

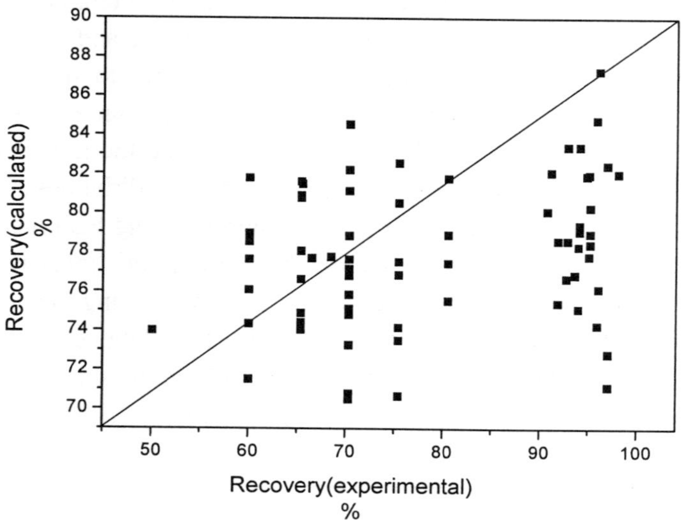

Fig. 1.2   Comparison between the experimental data and the PLS calculated data of the recovery of Al_2O_3 in alumina production.

Actually, people have found two ways to make all-inclusive function sets for this purpose: (1) According to Weierstrass theorem, any continuous function can be approximated by the polynomials with infinite number of terms, so that people have tried to add terms of higher degree to improve the data fitting. Although too many terms will lead to "curse of dimensionality", the nonlinear regression methods based on this strategy are widely used; (2) According to mathematical argumentation, a three-layered artificial neural network can be used to approximate any continuous function.[*] So artificial neural network appears to be the universal approximator and has been widely used in chemical data processing. Although the use of artificial neural network can avoid underfitting problems for some nonlinear data processing work, unfortunately, another serious problem, *overfitting* problem, has been found in the application of artificial neural network. In data processing

---

[*]Since the invention of support vector machine, we have a new method to imitate nonlinear data set: kernel functions.

practice with artificial neural network, it can be found that sometimes the fitting of known data in training set is rather good, but the results of prediction for unknown data or test data are not so good or even completely unacceptable. This is so-called *overfitting problem.*

Here we will demonstrate the problem of overfitting with some examples of chemical data processing. Table 1.2 demonstrates a set of data about the preparation of bismuth-based high-temperature superconductors. And Table 1.3 demonstrates a set of data about the preparation of VPTC ceramic semiconductors. In these tables, the samples of class "1" are those with "good" properties, and those of class "2" with unsatisfactory properties.

The purpose of the data processing work here is to find some clues for searching the conditions (composition and technological conditions) of preparation for good samples of superconductor or semiconductor. Figure 1.3 illustrates the rate of correctness in training (influence of underfitting) and that of prediction in LOO cross-validation test (influence of overfitting) as functions of the number of iteration in the computation of ANN and support vector machine. It can be seen that the rate of correctness in training process of ANN increases monotonically with the increase of number of iteration steps, and after 250000 steps the rate of correctness approaches to 100% (no underfitting). While the rate of correctness of prediction (in LOO cross-validation test) changes in a quite different manner: it firstly increases and then decreases after 50000 iteration steps (due to overfitting). From this example, it can be seen that the errors due to underfitting and overfitting are not the same thing, and that the good training results cannot guarantee to minimize overfitting or good ability of prediction. Besides, it is meaningful to see that the best result of prediction of ANN is still not so good as the prediction result of support vector machine, as demonstrated in Fig. 1.3. It means that the early-stopping (at 10000 steps) cannot avoid overfitting of ANN completely in this case (In the real computation of early stopping ANN, the result may be even worse than this case because the number of test samples may be more than one, so the number of training samples should be less than the LOO cross validation test).

Table 1.2   Data of preparation of Bismuth-based high temperature superconductors.

Sample No.	Class	Bi[*]	O	t (minute)	T °C
1	1	1.65	9.3	200	840
2	1	1.65	9.3	240	835
3	1	1.60	9.3	190	835
4	1	1.64	9.3	190	830
5	1	1.65	9.0	200	835
6	1	1.50	9.0	180	835
7	2	1.50	9.4	240	835
8	2	1.60	9.4	200	840
9	2	1.60	9.4	160	835
10	2	1.60	9.3	200	840
11	2	1.70	9.5	160	835
12	2	1.80	9.3	200	840
13	2	1.90	9.3	200	835
14	2	1.60	10.0	160	890
15	2	1.65	10.0	140	890
16	2	1.70	10.0	130	890
17	2	1.80	10.0	125	895
18	2	1.40	9.0	240	840
19	2	1.80	9.6	140	840
20	2	1.90	10.0	120	895
21	2	1.40	9.0	220	840
22	2	1.80	9.8	120	840
23	1	1.60	9.3	200	835

[*]Bi and O denote the stoichiometrical ratio of bismuth and oxygen in the empirical formula of samples respectively; t and T denote the time and temperature of sintering process of sample preparation respectively.

Table 1.3   Data of preparation of VPTC ceramic semiconductors.

Sample No.	Class	Tb_2O_3%	ExcessTiO$_2$%	Sintering time (hr)	Relative cooling rate
1	1	0.4	1	4	0.5
2	1	0.3	1	4	0.5
3	1	0.4	0	0.25	0.5
4	1	0.4	1	0.25	0.5
5	1	0.4	1	2	0.8
6	1	0.14	0	0.25	0.1
7	2	0.15	1	1	0.5
8	2	0.13	1	0.25	0.5
9	2	0.11	1	0.25	0.5
10	2	0.15	1	0.25	0.5
11	2	0.13	0	0.25	0.9
12	2	0.11	1	0.25	0.9
13	2	0.15	0	0.25	0.1

The result of data processing for the data set in Table 1.3 is quite similar. Although the rate of correctness in training process increases very quickly (it means that the structure of the data set is relatively simple and can be imitated by using ANN very easily), the minimum number of errors in prediction test (by LOO cross-validation method) of ANN is still higher than that of support vector machine, as shown in Fig. 1.4.

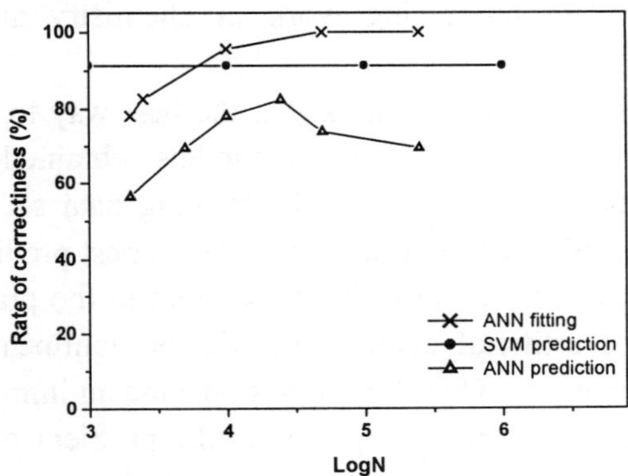

Fig. 1.3   Rate of correctness via iteration steps (N) of SVM and ANN for T_c data processing.

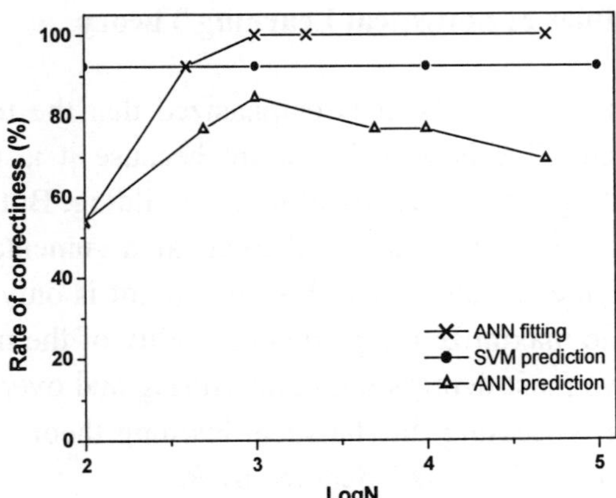

Fig. 1.4   Rate of correctness via iteration steps (N) of SVM and ANN for VPTC data processing.

Therefore, in machine learning work, we have two "enemies": underfitting and overfitting. The enlargement of the scope of hypothesis functions can only avoid the underfitting problem. However, it often makes overfitting becoming more serious problem.

So, what is the *origin of overfitting*? How to *avoid, depress or control overfitting*? In other words, how to improve the prediction ability (in statistical learning theory, generalization ability is used as the measure of prediction performance)? The answer of these questions is doubtless very important for the data processing work in chemistry and chemical technology.

In the past, people was apt to think that the best way to increase the prediction ability of the mathematical models obtained from data processing is to find a function to fit the training data set *as close as possible*. In other words, best training could assure best prediction result. But this concept has been found to be not correct in the practice of the application work of artificial neural networks or nonlinear regression with polynomial equations. Therefore, it has become an imminent task to find a strict mathematical theory for solving the problem of overfitting [68].

1.4   Theory of Overfitting and Underfitting Control, ERM and SRM Principles of Statistical Learning Theory

In classical statistical methods, it is emphasized that the training error must be minimized. This is very important because it is necessary to minimize underfitting for mathematical model building. But the practice of the application of ANN and the theoretical argumentation of the statistical learning theory tell us that this view-point is only a *one-sided concept*. In order to maximize the prediction ability of the mathematical model obtained, we must depress both underfitting and overfitting at the same time in data processing. In statistical learning theory, the error of training is called "empirical risk", denoted by R_{emp}.

According to the principle of empirical risk minimization (ERM) it is necessary to depress the training error. But this is not enough, since the risk of prediction still contains another term for risk due to overfitting:

$$R_{pred} \leq R_{emp} + \sqrt{\dfrac{h(\ln\dfrac{2\ell}{h}+1)-\ln(\dfrac{\eta}{4})}{\ell}} \qquad (1.1)$$

where R_{pred} is the total risk of prediction, ℓ is the number of samples in training set, and $1-\eta$ is the probability for the equation to be true. h is a very important concept: *VC dimension of the indicator function. The use of indicator function set with small h (VC dimension) is the method to depress the overfitting in data processing.*

VC dimension is one of the most basic concepts in statistical learning theory. It can be defined as follows: *In the feature space, the largest number of data points which can be shattered by a set of indicator functions is equal to the VC dimension of this set of indicator functions.* The meaning of *shatter* is to separate the set of points in all ways for the classification of these points into any two classes. For example, a set of sample points with three points on a 2-dimensional plane can be shattered by straight line (as shown in Fig.1.5a), but cannot separate four points in all possible ways of separation with one straight line. So the VC dimension of straight lines on a 2-dimensional plane is 3. It is easily understood that the set of indicator functions having large VC dimension is more powerful in data processing work. In statistical learning theory, VC dimension is defined as a measure of the capacity of a set of indicator functions.

So, in order to get good prediction reliability, we have to minimize

$$\min\left\{R_{emp} + \sqrt{\dfrac{h(\ln\dfrac{2\ell}{h}+1)-\ln(\dfrac{\eta}{4})}{\ell}}\right\} \qquad (1.2)$$

instead of minimizing R_{emp} only. This requirement of minimizing is called *principle of structure risk minimization (SRM)*. This principle requires us to trade off two somewhat contradictory requirements, because these two requirements (to minimize the value of training error and to use a set of indicator functions with small VC dimension) are somewhat contradictory. To minimize the number of training errors, one need to choose a function from a wide set of functions, while a narrow set of indicator functions has small VC dimension. Therefore, to find the best guaranteed solution, one has to make a compromise between the accuracy of approximation of the training set and the capacity (the VC dimension) of the set of indicator functions, in order to minimize the errors of prediction. This is the basic concept of the SRM principle.

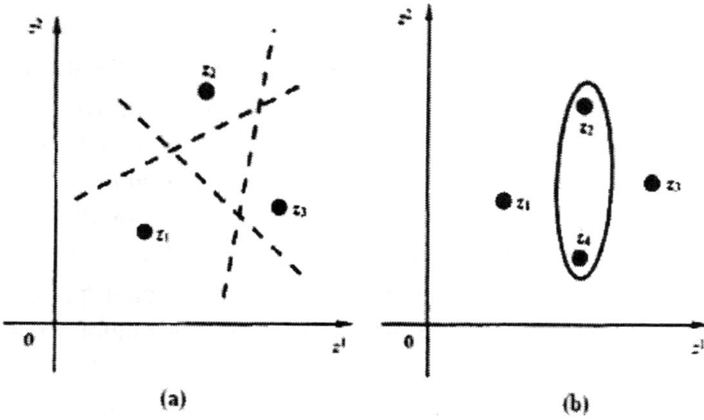

(a)　　　　(b)

Fig. 1.5 A straight line on a plane can shatter three points (a), but cannot shatter four points in all possible ways (b).

Based on the concept of classical statistical mathematics, people are apt to think that the number of free parameters (or the dimension of feature space) is the controlling factor deciding the reliability of mathematical model for prediction. But Vapnik and his coworkers have proved that the controlling factor for the reliability in classification problems is VC dimension. And VC dimension exhibits no one-to-one correspondence to the number of free parameters. One of the most important achievements of Vapnik and his coworkers is the *large margin concept*. It is found that the VC dimension can be greatly depressed if the sample points of different classes can be mapped into another feature space to make a *wide margin* between the points of two classes. It is just this achievement that makes the success of support vector machine.

1.5 Concept of Large Margin—A Basic Concept of SVM

Vapnik and his coworkers found that if there existed a *wide margin* between the regions of distribution of sample points of different kinds, the VC dimension of the indicator functions could be greatly decreased, and the mathematical model obtained as an *optimal hyperplane* exhibited very good prediction ability, even if the dimension of the feature space was very high and the equation of this optimal hyperplane had to be expressed by many adjustable parameters. So they decided to propose a novel strategy of computation: firstly the sample points in *input space* were mapped into a *feature space* with higher dimension by linear or nonlinear transformation, to make the distribution of sample points form a wide margin, and then an optimal hyperplane was used to describe the criterion of classification of different classes. By this way, the mathematical models of the linearly or non-linearly separable data sets can be obtained with good prediction ability. So not only the linearly separable data set, but also nonlinearly separable data set can be classified with mathematical models obtained. Figure 1.6 illustrates this concept. Here the optimal hyperplane denotes the unique hyperplane having largest distances with the sample points of different classes. And the sample points located on the border of the margin are called *support vectors*. It can be seen that the position of the optimal hyperplane is decided by the support vectors only. And the sample points other than support vectors have no influence on the position of this hyperplane.

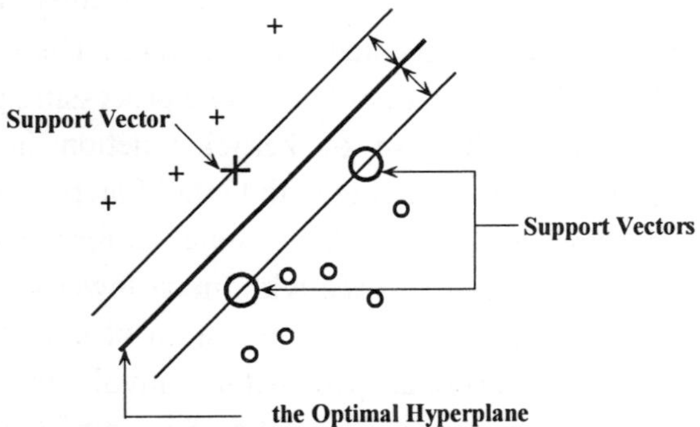

Fig. 1.6   Optimal hyperplane and large margin in SVM Method.

It can be proved that the optimal hyperplane for separation can be described by a linear equation:

$$\langle \mathbf{w} \cdot \mathbf{x} \rangle + b = 0 \qquad (1.3)$$

with minimum value of $\|\mathbf{w}\|^2$.

From the standpoint of statistical learning theory, the principle of SVM is much different from those of other commonly used methods in chemical data processing. It is well known that the preprocessing step of commonly used data processing methods is dimension reduction, in order to overcome the curse of dimensionality and to make the mathematical model more reliable, but in SVM algorithm the most important task is *not dimension reduction* but *dimension elevation*: to use *kernel function* to map the sample points of *input space* into a *feature space* with higher dimensionality by nonlinear transformation. In these high dimensional feature spaces, nonlinear separable sample points in input space can become linearly separable with wide margin in feature space, and a linear separation algorithm can be used to make the mathematical model with good prediction ability.

1.6 Kernel Functions: Technique for Nonlinear Data Processing by Linear Algorithm

Kernel function was originally a kind of functions used in integral operator research. But Vapnik and his coworkers cleverly implemented this function into their newly invented SVM method. The use of kernel function makes SVM able to treat nonlinear data processing problems by using linear algorithms. The use of kernel function has also two additional advantages: (1) By mapping with kernel function we can make classification and regression in high dimensional *feature space* but only via the computation within *input space* with much lower dimension; (2) The use of kernel function can make computation of nonlinear data set problems in very high-dimensional space without involving large number of free parameters in computation. Therefore, the use of kernel function is one of the chief factors to make SVM become an effective tool of machine learning.

Strictly speaking, Vapnik was not the first people to use kernel functions in data processing. As early as the years of sixty in 20-th century, Aizermann and his coworkers invented potential function method, which was very familiar to computer chemists. Suppose we have a linearly nonseparable data point set in some hyperspace, Aizermann proposed a method very similar to KNN method. In this method, every sample point of the training set is considered as a point charge: the sample points of the first class are considered to be positively charged, and the sample points of the second class are considered as negatively charged. So, the electric field at the place of a test sample point should be the difference of the field strength induced by all sample points of training set:

$$\Phi = \Phi_+ - \Phi_- \qquad (1.4)$$

where Φ_+ and Φ_- are the electric field induced by the positively charged and negatively charged sample points respectively. The values of Φ_+ and Φ_- can be calculated by Coulomb law as the functions of the distances between the test sample points and all sample points of the training set.

So the class of a point of test sample in the hyperspace can be determined by the sign of Φ: it will belong to class "1" if $\Phi_+ - \Phi_- > 0$, or belong to class "2" if $\Phi_+ - \Phi_- < 0$. It means that through the calculation of Φ_+ and Φ_- the classification problem can be solved by a very simple linear function.

By this way, Aizermann successfully converted a linearly nonseparable problem to a very simple linearly separable problem: to take the difference of electric fields at every point. It is easy to understand that potential functions other than Coulomb potential function are also applicable in this method. Aizermann also suggested the following function for the evaluation of the field strength around every sample point instead of Coulomb potential function:

$$\text{potential function} = \exp(-\|\mathbf{x}\text{-}\mathbf{z}\|^2 / \sigma^2) \qquad (1.5)$$

This is Gaussian kernel function. It is just one of the kernel functions used in SVM.

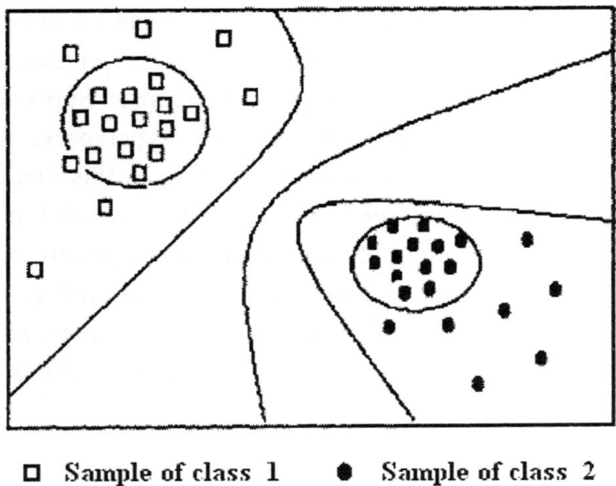

Fig. 1.7　By potential function method, a nonlinear problem can be converted to a linear problem for solution.

The work of Aizermann actually demonstrated the function of kernels: kernels could be used to convert a nonlinear classification problem into a linear one. But at that time Aizermann was limited by the physical models of his method, so he was not aware of the general usability of kernel function in machine learning.

It is Vapnik and his coworkers who found the trick of using kernel function as a tool to convert nonlinear data processing problems into linearly solvable problems in about 25 years later after Aizermann's work about the potential function method.

The definition of kernel function can be stated as follows: *a kernel is an inner product*

$$K(\mathbf{x}, \mathbf{z}) = \langle \phi(\mathbf{x}) \cdot \phi(\mathbf{z}) \rangle \qquad (1.6)$$

where ϕ *is a mapping from input space to feature space.*

1.7 Support Vector Regression: Regression Based on Principle of Statistical Learning Theory

The statistical learning theory can be also applied to regression problems by the introduction of a new loss function: the ε-insensitive loss function. The physical meaning of this function can be understood through the following example:

Suppose the yield of some compound by organic synthesis is dependent on the composition of raw materials (C), the temperature of reaction (T), the quantity of catalyst (K) and the time of reaction (t). If we want to find some empirical relationships between these variables by regression method, some functions should be used to fit the experimental data:

$$\text{Yield} = w_1 C + w_2 T + w_3 K + w_4\, t + ... \tag{1.7}$$

Based on the minimization of the least square loss function by traditional regression method, unique solution can be obtained. This selection is actually based on the assumption that all experimental data are *infinitely accurate*. But strictly speaking this is not true. Actually all experimental data on both side of the above-mentioned equation have some experimental errors. If we minimize the loss function very accurately, the mathematical model should contain not only the true relationships between these variables, but also the errors in the measurements. So although the fitting of the data of training set may be rather good, the results of prediction of unknown sample will be not so good. Moreover, since the influence of noise is usually not very regular, it is quite necessary to use more complicated functions to fit such kind of irregularities. This is what we called the problem of overfitting.

In order to avoid fitting the values of experimental error into the mathematical model, Vapnik and his coworkers proposed a new kind of loss function:

$$L(y) = \begin{cases} 0 & \text{if } |y\text{-}f(\mathbf{x})| < \varepsilon \\ |y\text{-}f(\mathbf{x})| - \varepsilon & \text{otherwise} \end{cases} \tag{1.8}$$

In other words, the loss function will be zero after the residue less than some small value ε. By this loss function, we will obtain infinite number of solutions, not a unique solution. In order to get a unique solution having the smallest overfitting, the sum of the square of coefficients of regression, $\|\mathbf{w}\|^2$, should be minimized. Since the relationship between the error of the calculated value of a dependent variable (Δy) and the experimental errors of the independent variables (Δx_i) in the regression equation can be expressed as follows:

$$\Delta y = w_1 \Delta x_1 + w_2 \Delta x_2 + w_3 \Delta x_3 ... \tag{1.9}$$

The error of the calculated value should be reduced when the values of coefficients become smaller. So the error of the calculated value can be depressed by minimization of $\|\mathbf{w}\|^2$ in the equation obtained. Using ε-insensitive loss function and minimizing the $\|\mathbf{w}\|^2$ value are two principles of support vector regression.

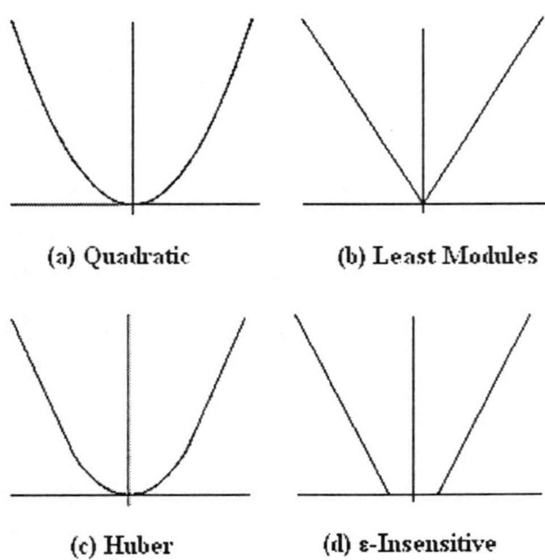

Fig. 1.8   Comparison between ε-insensitive loss function and other type of loss functions.

1.8 Other Machine Learning Methods Related to Statistical Learning Theory

The success of support vector machine stimulates many computer scientists to search various new methods of machine learning on the basis of the spirit of statistical learning theory. In order to control or depress the overfitting of artificial neural networks, an effective method is to minimize the weights of ANN, just as the minimization of $\|w\|^2$ in support vector regression. Based on this idea, we can have weight decay ANN (WD-ANN).

PCA, PLS and Fisher methods are the commonly used algorithms in computer chemistry. They are all linear projection methods. It is easy to understand that a linear projection may be inappropriate to reveal some nonlinear regularities of the data set processed. Since kernel function provides a very effective way for the data processing of nonlinear data sets by linear algorithm, the coordination of kernel function and these widely used linear projection methods soon lead to the invention of three new algorithms. These are kernel PCA (abbreviated as KPCA), kernel PLS (KPLS) and kernel Fisher method (KFD). These methods can provide large number of nonlinear projection maps from a data set processed. When the nonlinear mapping exhibits some relationships with the physical background of the chemical phenomena investigated, these methods can be applied to reveal the useful information for problem-solving.

1.9 Some Comments on the Application of SVM in Chemistry

As a newly developed method for chemical data processing, SVM has following obvious advantages in comparison with classical chemometrical methods: (1) It can treat both linear and nonlinear data sets so the trouble of underfitting can be depressed or controlled in some problems; (2) It is so designed that the overfitting can be depressed or controlled by the capacity control of the indicator functions used so that the prediction results are often more reliable; (3) As compared with ANN, SVM has no local minimum problem and the solution is unique. As mentioned above, SVM is based on the strict mathematical theory: statistical learning theory, so it is relatively more believable. SVM is especially suitable for the data processing to solve the problems of small sample size. According to Vapnik's concept of the problems of small sample size, the data set in machine learning with $\ell/h < 20$ can be roughly considered as the data set of this category (Here ℓ is the number of training samples, and h is the VC dimension of indicator functions used). For example, if the number of influencing factors of a chemical problem is six, and the sample points of different classes can be linearly separated in the six-dimensional hyperspace (so the VC dimension h should be 6+1=7), then any data file with the number of samples less than 140 will belong to this category. And if the sample points of different classes cannot be linearly separated, the VC dimension will be still higher and the data sets with much more than 140 will also belong to this category. It is obvious that most of chemical data processing problems should be included into the scope suitable for the application of SVM. And as will be shown in the following chapters of this book, the results of applications of SVM in many fields of chemistry and related sciences are indeed rather satisfactory [31].

On the other hand, we should not forget that every new method or theory has its limitations. Although the strict system of statistical learning theory has been established on the basis of more than 30 years research work of Vapnik and others, the practical calculation of VC dimension is still not successful for many indicator functions and machine learning methods (for example, the estimation of the VC dimension of early stopping ANN is still not successful).

Before the end of this chapter, we wish to emphasize that the different methods of data processing have different fields of applications. They are not competitive, but complementary to each other. The application of SVM does not mean that the traditional methods are useless. On the contrary, the traditional methods and SVM should be considered as mutually complementary to each other, since many traditional methods also have their advantages as compared with SVM. For example, the traditional methods, including PCA, PLS and Fisher methods can give many linear projection figures. These figures contain plentiful information. Domain experts, including chemists and chemical engineers, can find very useful inspiration from these projection figures. Although the separability of different kinds of sample points is not always good by such kind of linear projection, the simple relationships of linear projection make these projection figures easier to be understood based on domain knowledge. This is an indisputable advantage of linear projection method, since human beings have good ability to understand the structure of high dimensional space by looking at its two-dimensional projections. Therefore, a clever strategy is to build a data processing system integrating different algorithms comprehensively.

At last, it should be emphasized that the results of SVC or SVR obtained by using normalized data sets are usually better than that without normalization in data processing. In this book, all mathematical models are expressed by normalized data sets.

Chapter 2

Support Vector Machine

In this chapter, we will give a comprehensive introduction to support vector machine (SVM) in an accessible and self-contained way. The organization of this chapter is as follows: We start from the central concepts about margin, from which the support vector methods are developed; Second, the SVM for classification problems are introduced and the derivation in both linear and nonlinear cases will be described in detail; Third, we discuss the support vector regression, i.e. the SVM in regression problems. At last, a variant of SVM, ν-SVM is briefly introduced.

2.1 Margin and Optimal Separating Plane

2.1.1 *Linear classification problem*

Suppose that we are given a set of training data

$$(\mathbf{x}_1, y_1), \ldots, (\mathbf{x}_i, y_i), \ldots, (\mathbf{x}_\ell, y_\ell), \qquad \mathbf{x}_i \in \mathbb{R}^n, \quad y_i \in \{+1, -1\}$$

The goal is to find some decision function $g : \mathbb{R}^n \to \{+1, -1\}$ that can accurately predict the labels of unseen data $(\mathbf{x}, \mathbf{y})$. That is, the binary classification is performed by using a real-valued function,

$$f : \mathbb{R}^n \to \mathbb{R}$$

whose output is filtered by a threshold function to yield the final classification $g(\mathbf{x}) = \mathrm{sgn}(f(\mathbf{x}))$. We consider a simple example: linear decision function. In this case the linear classification decision function can be written as

$$g(\mathbf{x}) = \mathrm{sgn}(f(\mathbf{x})) \qquad (2.1)$$

$$= \mathrm{sgn}(\langle \mathbf{w} \cdot \mathbf{x} \rangle + b) \qquad (2.2)$$

$$= \mathrm{sgn}(\sum_{i=1}^{\ell} w_i x_i + b) \qquad (2.3)$$

where $\mathbf{w} \in \mathbb{R}^n$ and $b \in \mathbb{R}$. A geometric interpretation of this kind of hypothesis is that the input space X is split into two parts by the $n-1$ dimensional hyperplane defined by the decision boundary $\langle \mathbf{w} \cdot \mathbf{x} \rangle + b = 0$. This situation is illustrated in Fig. 2.1 where the vector $\mathbf{w}$ defines a direction perpendicular to the hyperplane, while varying the value of b moves the hyperplane parallel to itself.

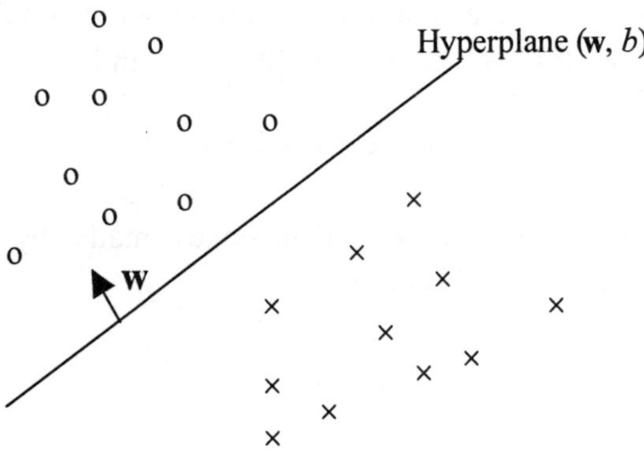

Fig. 2.1   A hyperplane (**w**, b) for a linear classification problem.

The problem of the classification can be transformed into finding a set of parameters $\mathbf{w}$ and b, the so called the *weight vector* and *bias* respectively in some literatures. Several simple iterative algorithms with different cost functions were introduced in the 1960s for separating points of two kinds by means of a hyperplane. One of the famous examples is *perceptron*. In such a simple system, there exist most of the central concepts that are needed for the theory of support vector machine.

2.1.2   *An important theorem in perceptron algorithm*

The perceptron algorithm was proposed by Frank Rosenblatt in 1956 and has created a great deal of interest since then. It starts with an initial weight vector $\mathbf{w}_0$ and adapts it each time to a training point which is misclassified by the current weights. The algorithm is a 'mistake-driven' procedure [42], i.e. the weight vector and bias are only updated on the misclassified examples.

The following theorem shows that if the training sample is consistent with some simple perceptron, then this algorithm converges after a finite number of iterations. In this theorem, $\mathbf{w}^*$ and b^* define a decision boundary that correctly classifies all training examples, and every training sample point is at least having distance γ from the decision boundary.

Theorem 2.1 Let S be a non-trivial training set. Suppose that there exists a $\gamma > 0$, a vector $\mathbf{w}^*$ such that $\left\| \mathbf{w}^* \right\| = 1$ and

$$y_i(\left\langle \mathbf{w}^* \cdot \mathbf{x}_i \right\rangle + b^*) \geq \gamma \tag{2.4}$$

for $1 \leq i \leq \ell$. Then the number of mistakes made by the on-line perceptron algorithm on S is at most

$$\left(\frac{2R}{\gamma} \right)^2 \tag{2.5}$$

where $R = \max_{1 \leq i \leq \ell} \left\| \mathbf{x}_i \right\|$.

The proof of this theorem is in Ref. [42]. Obviously, the critical quantity in the convergence bound is the distance γ of the training sample from the decision boundary.

2.1.3 *Margin*

The quantity γ in Theorem 2.1 determines how well the two classes can be separated and consequently how fast the perceptron learning algorithm converges. We call the quantity a *margin* and give a more formal definition as follows.

Definition 2.1   Denote $(\mathbf{w}, b)$ a hyperplane determined by $f : \mathbb{R}^n \to \mathbb{R}$, a real valued linear function used for classification. We define the margin of a sample point $(\mathbf{x}_i, y_i)$ with respect to this hyperplane to be the quantity

$$\gamma_i = y_i \left(\langle \mathbf{w} \cdot \mathbf{x}_i \rangle + b \right). \tag{2.6}$$

Note that $\gamma_i > 0$ implies correct classification of $(\mathbf{x}_i, y_i)$. According to Definition 2.1, furthermore, a few of other relative definitions are derived as follows.

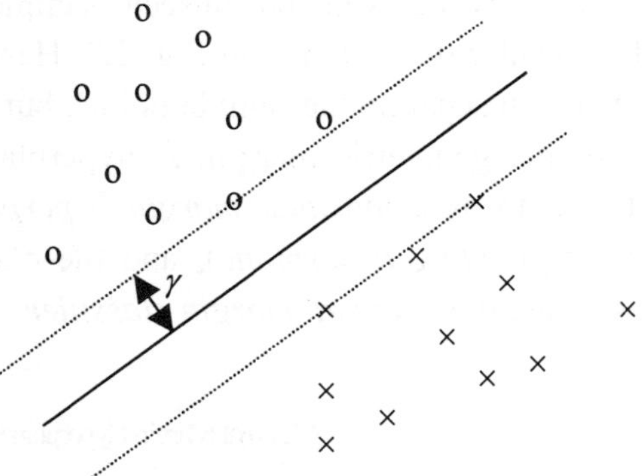

Fig. 2.2   The margin of a training set in two dimensions.

The *margin distribution of a hyperplane* $(\mathbf{w}, b)$ on a training set S is the distribution of the margins of all sample points in S.

Usually, the minimum of the margin distribution is referred to as the *margin of a hyperplane* $(\mathbf{w}, b)$ *on a training set* S.

Geometric margin: the margin in Definition 2.1 sometimes is named as the *functional margin*. In order to understand it more clearly, we introduce the *geometric margin,* which is the ratio of the functional margin to the normalized weight vector:

$$\gamma_{geom,i} = \frac{y_i\left(\langle \mathbf{w} \cdot \mathbf{x}_i \rangle + b\right)}{\|\mathbf{w}\|} \tag{2.7}$$

The geometric margin will equal the functional margin if the weight vector is a unit vector [42].

The Margin of a training set S is the maximum geometric margin over all hyperplanes.

Figure 2.2 shows the margin of a training set in two dimensions.

2.1.4 *Maximal margin hyperplane*

Intuitively, it appears desirable to have classifiers that achieve a large margin since one might expect that an estimate that is "reliable" on the training set will also work well on unseen sample points, i.e. it generalizes well. Consider the example in Fig. 2.3. Here there are many linear classifiers that can separate the sample points, but there exists only one that maximizes the geometric margin. A hyperplane realizing this maximum is referred to as a *maximal margin hyperplane* (sometimes known as *optimal separating hyperplane*), and the classifier based on this hyperplane is called the *maximal margin classifier.*

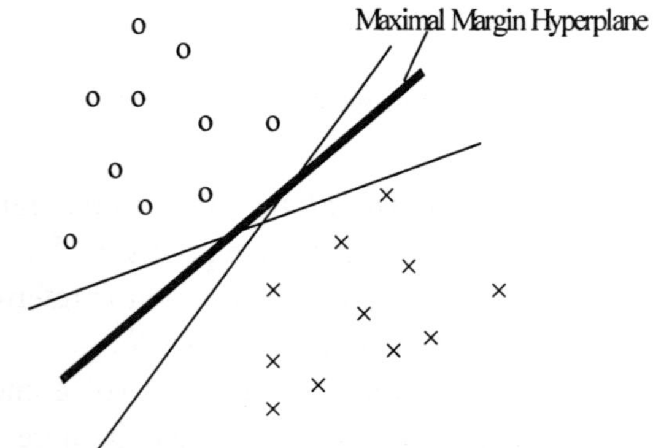

Fig. 2.3   The maximal margin hyperplane.

We sometimes refer to the minimum of the margin distribution as the *(functional) margin of a hyperplane* $(\mathbf{w}, b)$ on a training set S [42]. Hence, the problem of seeking the maximal margin classifier is to determine whether there exists some $(\mathbf{w}^*, b^*)$ by solving the following problem:

$$\text{maximize} \quad \gamma_{\mathbf{w},b} = \min_{i=1}^{\ell} y_i f_{\mathbf{w},b}(\mathbf{x}_i). \tag{2.8}$$

A slight modification in the formula should be carried out before the next steps. Note that in the definition of the functional margin there is an inconsistent problem in use. The linear function $f(\mathbf{x})$ associated with the hyperplane $(\mathbf{w}, b)$ does not change if we rescale the hyperplane to $(\lambda \mathbf{w}, \lambda b)$, for $\lambda \in \mathbb{R}^+$. There will, however, be a change in the functional margin as opposed to the geometric margin. Hence, the geometric margin is maximized by the weight vector $\mathbf{w}^*$ and bias b^* that satisfying the following optimization problem [126]:

$$\text{maximize}_{\mathbf{w},b} \min_{i=1}^{\ell} \frac{y_i \left(\langle \mathbf{w} \cdot \mathbf{x}_i \rangle + b \right)}{\|\mathbf{w}\|} \tag{2.9}$$

$$= \min_{i=1}^{\ell} y_i \, \text{sgn} \left(\langle \mathbf{w} \cdot \mathbf{x}_i \rangle + b \right) \left\| \frac{\langle \mathbf{w} \cdot \mathbf{x}_i \rangle + b}{\|\mathbf{w}\|} \right\| \tag{2.10}$$

$$= \min_{i=1}^{\ell} y_i \, \text{sgn} \left(\langle \mathbf{w} \cdot \mathbf{x}_i \rangle + b \right) \left\| \frac{\langle \mathbf{w} \cdot \mathbf{x}_i \rangle + b}{\|\mathbf{w}\|} \frac{\mathbf{w}}{\|\mathbf{w}\|} \right\| \tag{2.11}$$

$$= \min_{i=1}^{\ell} y_i \, \text{sgn} \left(\langle \mathbf{w} \cdot \mathbf{x}_i \rangle + b \right) \left\| \frac{\langle \mathbf{w} \cdot \mathbf{x}_i \rangle}{\|\mathbf{w}\|^2} \mathbf{w} + \frac{b}{\|\mathbf{w}\|^2} \mathbf{w} \right\| \tag{2.12}$$

In the derivation of the above formulas, note that the output $f(\mathbf{x})$ is a scalar, therefore the multiplication by the unit weight vector $\mathbf{w}/\|\mathbf{w}\|$ should not change the norm. The formula (2.12) has a simple geometric interpretation (see Fig. 2.4): since $\langle \mathbf{w} \cdot \left(-b\mathbf{w}/\|\mathbf{w}\|^2 \right) \rangle + b = 0$, $-b\mathbf{w}/\|\mathbf{w}\|^2$ is the vector in direction $\mathbf{w}$ that ends right on the decision hyperplane ($\overrightarrow{OA}$ in Fig. 2.4), and for a sample vector $\mathbf{x}_i$, $\langle \mathbf{w} \cdot \mathbf{x}_i \rangle \mathbf{w}/\|\mathbf{w}\|^2$ is the projection of $\mathbf{x}_i$ onto $\mathbf{w}$ ($\overrightarrow{OB}$ in Fig. 2.4). Thus, the problem of the maximal margin classifier is converted into seeking appropriate $(\mathbf{w}^*, b^*)$ maximizing $\overrightarrow{BA}$, the norm of the vector differences $\langle \mathbf{w} \cdot \mathbf{x}_i \rangle \mathbf{w}/\|\mathbf{w}\|^2 - (-b\mathbf{w}/\|\mathbf{w}\|^2)$ signed by $y_i g(\mathbf{x}_i)$.

Define $\gamma = \min\limits_{i=1}^{\ell} y_i \left(\langle \mathbf{w} \cdot \mathbf{x}_i \rangle + b \right)/\|\mathbf{w}\|$ to be the lower bound on the margin. Here, the max-min problem (2.9) can be transformed into the following constrained optimization task:

Maximize $\gamma_{\mathbf{w},b}$,

subject to
$$\frac{y_i \left(\langle \mathbf{w} \cdot \mathbf{x}_i \rangle + b \right)}{\|\mathbf{w}\|} \geq \gamma \tag{2.13}$$

which is in turn equivalent to the following problem:

maximize $\gamma_{\mathbf{w},b}$

subject to $\|\mathbf{w}\| = 1$ and $y_i \left(\langle \mathbf{w} \cdot \mathbf{x}_i \rangle + b \right) \geq \gamma$ \hfill (2.14)

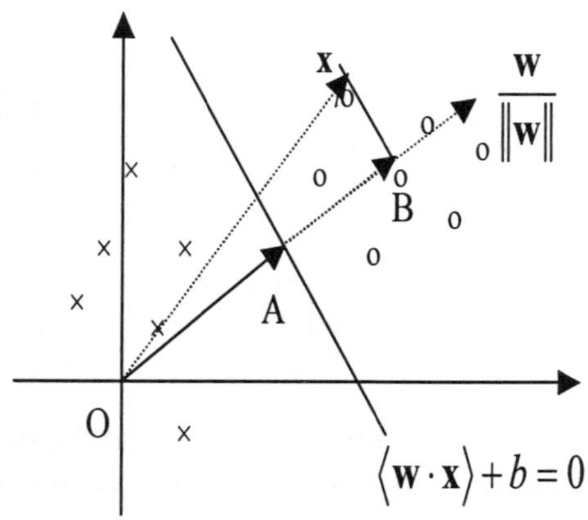

Fig. 2.4   The maximal hyperplane and the geometric interpretation of (2.12).

The last formula (2.14) is in an intuitive form [126]. We are seeking a weight vector $\mathbf{w}$ that obtains large dot products $y_i \langle \mathbf{w} \cdot \mathbf{x}_i \rangle$, but constrain the weight vector to lie on the unit sphere to prevent obtaining such large dot products "for free" by scaling up $\mathbf{w}$. The hyperplanes with functional margin 1 are sometimes known as *canonical hyperplanes* (see Fig. 2.5). If $\mathbf{w}$ is the weight vector realizing a functional margin of 1 on the positive point $\mathbf{x}_1$ and the negative point $\mathbf{x}_2$ (see Fig. 2.5), we can compute its geometric margin as follows. Recall that a functional margin of 1 implies

$$\langle \mathbf{w} \cdot \mathbf{x}_1 \rangle + b = +1,$$
$$\langle \mathbf{w} \cdot \mathbf{x}_2 \rangle + b = -1,$$

while to compute the geometric margin we must normalize $\mathbf{w}$. The geometric margin γ is then the functional margin of the resulting classifier

$$\gamma = \frac{1}{2}\left(\left\langle \frac{\mathbf{w}}{\|\mathbf{w}\|} \cdot \mathbf{x}_1 \right\rangle - \left\langle \frac{\mathbf{w}}{\|\mathbf{w}\|} \cdot \mathbf{x}_2 \right\rangle\right)$$

$$= \frac{1}{2\|\mathbf{w}\|}\left(\langle \mathbf{w} \cdot \mathbf{x}_1 \rangle - \langle \mathbf{w} \cdot \mathbf{x}_2 \rangle\right)$$

$$= \frac{1}{\|\mathbf{w}\|} \tag{2.15}$$

Hence, the resulting geometric margin will be equal to $1/\|\mathbf{w}\|$ and we can get the $\left(\mathbf{w}^*, b^*\right)$ by solving the following task from (2.14):

Minimize $\|\mathbf{w}\|$

subject to $\qquad\qquad y_i\left(\langle \mathbf{w} \cdot \mathbf{x}_i \rangle + b\right) \geq 1, \tag{2.16}$

i.e., the maximal margin is equal to the minimal norm of the weight vector and the optimal separating hyperplane will be obtained by minimizing the latter. In the next section, we shall see why the size of $\mathbf{w}$ is a good measure of the complexity of the classifier.

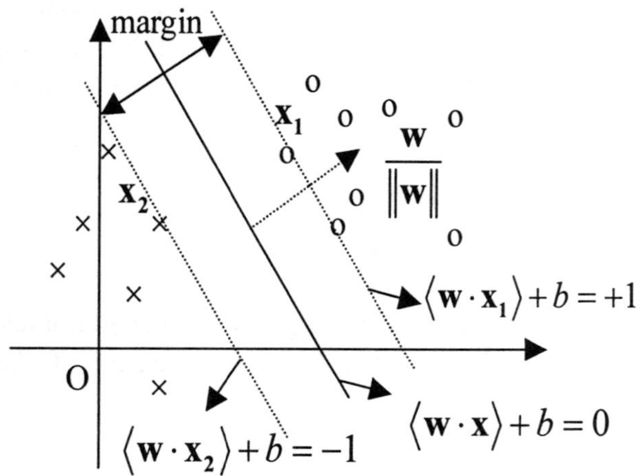

Fig. 2.5 The canonical hyperplanes.

2.2 Interpretation by Statistical Learning Therory

Section 2.1 has attempted to determine the maximal margin hyperplane in an intuitive way. Support Vector Machine, the successful implementation of statistical learning theory (SLT), are built on the basis of the maximal margin hyperplane described above. It is important to reveal the relationship between the formula (2.16) and SLT.

Given a set of functions, the relationship between the empirical risk and the actual risk for this set of functions is one of most important research directions, which is known as the *bounds on generalization ability of learning machines*. As for binary classification problems, the following basic bounds describing the generalization ability of a threshold real-valued function (also known as indicator function) that minimize the empirical risk functional [131]:

Theorem 2.2 [132] For some η such that $0 < \eta \leq 1$, with probability $1-\eta$ any of the bounds

$$R(\alpha) \leq R_{emp}(\alpha) + \sqrt{\frac{h\big(\ln(2\ell/h)+1\big)-\ln(\eta/4)}{\ell}} \qquad (2.17)$$

hold true, where h is a positive integer called the *Vapnik Chervonenkis (VC) dimension* of the set of indicator functions $\{Q(\mathbf{x},\alpha),\alpha \in \Lambda\}$ and α is the parameters of the functions.

The inequality (2.17) shows that the actual risk of the learning machine consists of two parts: the first term on the right hand side of the inequality is the empirical risk (corresponding to the training errors) and the second term is called the "*VC confidence* ", which depends on the VC dimension of the set of functions (h) and the number of sample points (ℓ). Obviously, the VC confidence is a monotonic increasing function of h, which is true for any value of ℓ.

The bound in (2.17) gives a principled method for choosing a learning machine for a given task, and is the essential idea of structural risk minimization. Here we will only consider functions that correspond to the binary classification case, so that $f(\mathbf{x},\alpha) \in \{+1,-1\}\ \forall \mathbf{x},\alpha$. We will give a definition of a more general separating margin hyperplane as follows.

Definition 2.2 We call a hyperplane

$$\langle \mathbf{w}^{*} \cdot \mathbf{x} \rangle + b = 0, \qquad \left\| \mathbf{w}^{*} \right\| = 1 \qquad (2.18)$$

the *γ-margin separating hyperplane* if it classifies vector $\mathbf{x}$ as follows:

$$y = \begin{cases} +1, & if\ \langle \mathbf{w}^{*} \cdot \mathbf{x} \rangle + b \geq \gamma \\ -1, & if\ \langle \mathbf{w}^{*} \cdot \mathbf{x} \rangle + b \leq -\gamma \end{cases}. \qquad (2.19)$$

According to Theorem 2.2, given some selection of learning machines whose empirical risk is zero, one wants to choose that learning machine whose associated set of functions has minimal VC dimension. At present, for the γ-margin separating hyperplane, we quote an important theorem without proof as follows. For more details, see [132].

Theorem 2.3 Let vectors $\mathbf{x} \in X$ belong to a sphere of radius R. Then the set of γ-margin separating hyperplanes has the VC dimension h bounded by the inequality

$$h \leq \min\left(\left[\frac{R^2}{\gamma^2}\right], n\right) + 1 \tag{2.20}$$

In general the VC dimension of the set of hyperplanes is equal to $n+1$, where n is dimensionality of input space. However, the VC dimension of the set of γ-margin separating hyperplanes (with a large value of margin γ) can be less than $n+1$. Hence, according to (2.20), minimizing an upper bound on the VC dimension h is equivalent to maximizing the margin (see (2.16)). This will lead to a better upper bound on the actual error in the inequality (2.17).

We have now laid the groundwork necessary to begin our exploration of support vector machine.

2.3 Support Vector Classification

2.3.1 *Linearly separable case*

Recall that the optimal separating hyperplane is given by minimizing the norm of the weight vector according to the formula (2.16). Hence, it is easy to find that the optimal hyperplane with geometric margin $\gamma = 1/\|\mathbf{w}\|$ can be realized by solving the following optimization problem

Minimize $\langle \mathbf{w} \cdot \mathbf{w} \rangle$

subject to $\quad y_i\left(\langle \mathbf{w} \cdot \mathbf{x}_i \rangle + b\right) \geq 1, \quad i = 1, \ldots, \ell$.

Note that the minimization is taken with respect to both weight vector $\mathbf{w}$ and bias b. The solution to the constrained optimization problem of the formula (2.20) is given by the saddle point of the Lagrange functional (Lagrangian) [100],

$$L(\mathbf{w},b,\boldsymbol{\alpha}) = \frac{1}{2}\langle \mathbf{w}\cdot\mathbf{w}\rangle - \sum_{i=1}^{\ell}\alpha_i\Big[y_i\big(\langle \mathbf{w}\cdot\mathbf{x}_i\rangle + b\big)-1\Big] \qquad (2.21)$$

where $\alpha_i \geq 0$ are the Lagrange multipliers. The Lagrangian has to be minimized with respect to $\mathbf{w}$, b and maximized with respect to $\boldsymbol{\alpha}$. One can solve this problem in the *primal space* – the space of parameters $\mathbf{w}$ and b. However, the deeper results can be obtained by solving this quadratic optimization problem in the *dual space* – the space of Lagrange multipliers $\boldsymbol{\alpha}$. Below we consider this type of solution.

The corresponding dual is found by differentiating with respect to $\mathbf{w}$ and b, imposing stationarity,

$$\frac{\partial L(\mathbf{w},b,\boldsymbol{\alpha})}{\partial \mathbf{w}} = \mathbf{w} - \sum_{i=1}^{\ell}y_i\alpha_i\mathbf{x}_i = \mathbf{0}, \qquad (2.22)$$

$$\frac{\partial L(\mathbf{w},b,\boldsymbol{\alpha})}{\partial b} = \sum_{i=1}^{\ell}y_i\alpha_i = 0, \qquad (2.23)$$

and resubstituting the relations obtained,

$$\mathbf{w} = \sum_{i=1}^{\ell}y_i\alpha_i\mathbf{x}_i, \qquad (2.24)$$

$$\sum_{i=1}^{\ell}y_i\alpha_i = 0, \qquad (2.25)$$

into the primal to obtain

$$W(\boldsymbol{\alpha}) = \frac{1}{2}\sum_{i,j=1}^{\ell}y_iy_j\alpha_i\alpha_j\langle \mathbf{x}_i\cdot\mathbf{x}_j\rangle - \sum_{i,j=1}^{\ell}y_iy_j\alpha_i\alpha_j\langle \mathbf{x}_i\cdot\mathbf{x}_j\rangle + \sum_{i=1}^{\ell}\alpha_i$$

$$= \sum_{i=1}^{\ell} \alpha_i - \frac{1}{2} \sum_{i,j=1}^{\ell} y_i y_j \alpha_i \alpha_j \langle \mathbf{x}_i \cdot \mathbf{x}_j \rangle \qquad (2.26)$$

Note that the primal (2.21) and the corresponding dual (2.26) arise from the same objective function but with different constraints; and the solution is found by minimizing the prime or by maximizing the dual. Now to construct the optimal hyperplane one has to find the coefficients $\boldsymbol{\alpha}^*$ that maximize the function $W(\boldsymbol{\alpha})$, subject to constraints (2.25) and positivity of the α_i, with solution $\mathbf{w}^*$ given by (2.24), i.e.,

$$\mathbf{w}^* = \sum_{i=1}^{\ell} y_i \alpha^*_i \mathbf{x}_i .$$

As an immediate application, note that, while $\mathbf{w}^*$ is explicitly determined by (2.24), the bias b^* is not, although it is implicitly determined. However b^* is easily found by using the Karush-Kuhn-Tucker (KKT) complementarity condition, which will be described as follows.

The *Kuhn-Tucker theorem* plays a central role in giving conditions for an optimum solution to a general constrained optimization problem. For the primal problem mentioned above, these conditions may be stated [53]:

$$\frac{\partial L(\mathbf{w}^*, b^*, \boldsymbol{\alpha}^*)}{\partial \mathbf{w}} = \mathbf{0} \qquad (2.27)$$

$$\frac{\partial L(\mathbf{w}^*, b^*, \boldsymbol{\alpha}^*)}{\partial b} = 0, \qquad (2.28)$$

$$y_i \left(\langle \mathbf{w}^* \cdot \mathbf{x}_i \rangle + b^* \right) - 1 \geq 0 \qquad i = 1, \ldots \ell, \qquad (2.29)$$

$$\alpha_i^* \geq 0 \qquad i = 1, \ldots, \ell, \qquad (2.30)$$

$$\alpha_i^* \left(y_i \left(\left\langle \mathbf{w}^* \cdot \mathbf{x}_i \right\rangle + b^* \right) - 1 \right) = 0 \qquad i = 1, \ldots, \ell. \qquad (2.31)$$

The last relation (2.31) is known as *Karush-Kuhn-Tucker complementarity condition*. From this condition one can not only compute b^* by choosing any i for which $\alpha_i \neq 0$, but also conclude that nonzero coefficients α_i^* correspond only to the vectors $\mathbf{x}_i$ that satisfy the equality

$$y_i \left(\left\langle \mathbf{w}^* \cdot \mathbf{x}_i \right\rangle + b^* \right) = 1. \qquad (2.32)$$

Geometrically, these vectors are the closest to the optimal hyperplane (see Fig. 2.6). They are called *support vectors*. The support vectors play a crucial role in constructing the learning algorithms of *support vector machine* (SVM) since the weight vector $\mathbf{w}$ of the optimal hyperplane are linear combination of the support vectors of the training set (see the formula (2.24)); if all other training vectors are removed, and training is repeated, the separating hyperplane found should be the same one. In this case, therefore, the KKT condition implies sparseness which is one of the fundamental properties of SVM.

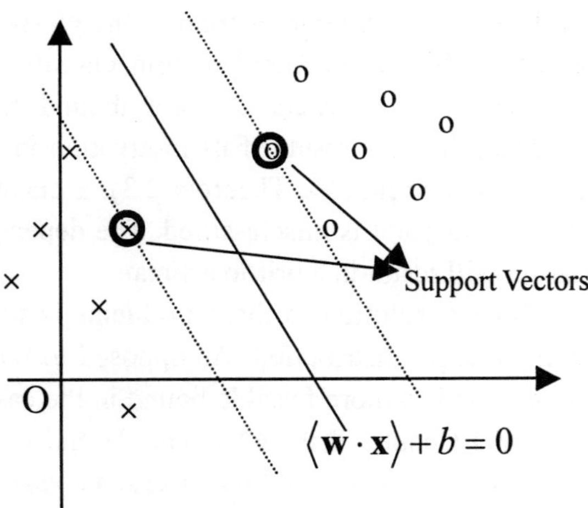

Fig. 2.6   The support vectors.

Finally, the optimal separating hyperplane decision function can thus be written as

$$g(\mathbf{x}) = \mathrm{sgn}\left(\sum_{i=1}^{\ell} y_i \alpha_i^* \langle \mathbf{x} \cdot \mathbf{x}_i \rangle + b^* \right)$$

$$= \mathrm{sgn}\left(\sum_{i \in SV} y_i \alpha_i^* \langle \mathbf{x} \cdot \mathbf{x}_i \rangle + b^* \right) \tag{2.33}$$

Note that both the separation hyperplane in (2.33) and the objective function of our optimization problem (2.26) do not depend explicitly on the dimensionality of the vector $\mathbf{x}$ but depend only on the inner product of two vectors. This fact will allow us later to construct separating hyperplanes in high-dimensional spaces.

2.3.2 *Linearly non-separable case*

So far the discussion has been restricted to the case that the training data is linearly separable. However, in general this will not be the case. An example of non-separable cases is that a separating hyperplane may not exist if a high noise level causes a large overlap of the classes (see Fig. 2.7). The main problem with the maximal margin classifier is that it always produces perfectly a consistent decision boundary with no training error. In essence, this is a result of its motivation in terms of a bound that depends on the margin (see Theorem 2.3), a quantity that is negative only when the data point is misclassified. The dependence on a quantity like the margin will result in a brittle estimator.

A powerful and efficient solution to these problems is to use more robust measures of the margin distribution. As opposed to the maximal bound, such measures provide a more feasible bound in the case of noise and outliers (see [41] and Chapter 4 in [42]). This bound is associated with non-negative variables, $\xi_i \geq 0$, also known as *slack variables*.

According to Cortes and Vapnik [41], slack variables ξ_i and a penalty function as follows are introduced to construct the optimal hyperplane in the case when data are linearly non-separable:

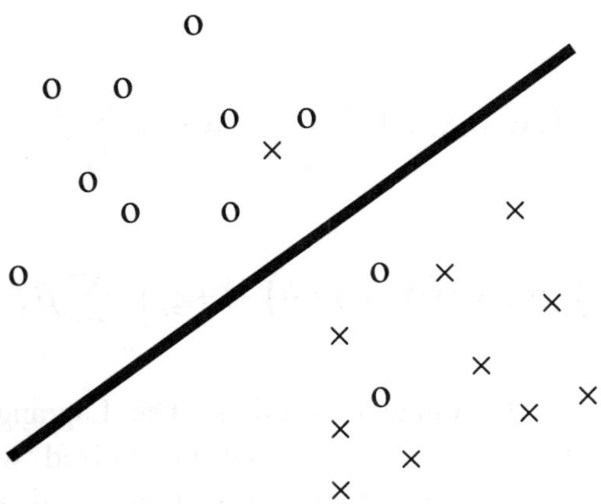

Fig. 2.7   The non-separable case.

$$F_\sigma(\xi) = \sum_{i=1}^{\ell} \xi_i^\sigma, \quad \sigma > 0.$$

(2.34)

Since ξ_i is a measure of the classification errors, the optimization problem is now posed so as to minimize the classification error as well as to minimiz the bound on the VC dimension of the classifier. The constraints discussed in the maximal margin classifier are modified for the non-separable case to

$$y_i\left(\langle \mathbf{w} \cdot \mathbf{x}_i \rangle + b\right) \geq 1 - \xi_i, \quad i = 1, \ldots, \ell .$$

(2.35)

In a natural way, therefore, the generalized optimal separating hyperplane is determined by solving the following functional:

Minimize
$$\frac{1}{2}\langle \mathbf{w} \cdot \mathbf{w} \rangle + C\sum_{i=1}^{\ell} \xi_i$$

(2.36)

subject to the constraints of the formula (2.35), where C is a parameter to be chosen by the user, a larger C corresponding to assigning a higher penalty to errors. As it stands, this is a convex quadratic programming problem and its solution is given by the saddle point of the Lagrangian,

$$L(\mathbf{w},b,\xi,\boldsymbol{\alpha},\boldsymbol{\beta}) = \frac{1}{2}\langle \mathbf{w}\cdot\mathbf{w}\rangle + C\sum_{i=1}^{\ell}\xi_i$$

$$-\sum_{i=1}^{\ell}\alpha_i\left[y_i\left(\langle\mathbf{w}\cdot\mathbf{x}_i\rangle+b\right)-1+\xi_i\right]-\sum_{i=1}^{\ell}\beta_i\xi_i \qquad (2.37)$$

where $\boldsymbol{\alpha},\boldsymbol{\beta}$ are the Lagrange multipliers. The Lagrangian has to be minimized with respect to $\mathbf{w},b,\mathbf{x}$ and maximized with respect to $\alpha_i \geq 0$ and $\beta_i \geq 0$. As before, the corresponding dual is found by differentiating with respect to $\mathbf{w},b$ and ξ, imposing stationarity,

$$\frac{\partial L(\mathbf{w},b,\xi,\boldsymbol{\alpha},\boldsymbol{\beta})}{\partial \mathbf{w}} = \mathbf{w} - \sum_{i=1}^{\ell} y_i\alpha_i\mathbf{x}_i = \mathbf{0}, \qquad (2.38)$$

$$\frac{\partial L(\mathbf{w},b,\xi,\boldsymbol{\alpha},\boldsymbol{\beta})}{\partial b} = \sum_{i=1}^{\ell} y_i\alpha_i = 0, \qquad (2.39)$$

$$\frac{\partial L(\mathbf{w},b,\xi,\boldsymbol{\alpha},\boldsymbol{\beta})}{\partial \xi_i} = C - \alpha_i - \beta_i = 0, \qquad (2.40)$$

and resubstituting the relations obtained into the primal; we obtain the following adaptation of the dual objective function:

$$W(\boldsymbol{\alpha}) = \sum_{i=1}^{\ell}\alpha_i - \frac{1}{2}\sum_{i,j=1}^{\ell} y_i y_j \alpha_i \alpha_j \langle\mathbf{x}_i\cdot\mathbf{x}_j\rangle, \qquad (2.41)$$

which curiously is identical to that for the maximal margin. The only difference is that the constraint $C - \alpha_i - \beta_i = 0$, together with $\beta_i \geq 0$, enforces $\alpha_i \leq C$, while $\xi_i \neq 0$ only if $\beta_i = 0$ and therefore $\alpha_i = C$. The KKT conditions for the primal problem (2.37) are therefore

$$\frac{\partial L(\mathbf{w}, b, \xi, \alpha, \beta)}{\partial \mathbf{w}} = \mathbf{0}, \tag{2.42}$$

$$\frac{\partial L(\mathbf{w}, b, \xi, \alpha, \beta)}{\partial b} = \mathbf{0}, \tag{2.43}$$

$$\frac{\partial L(\mathbf{w}, b, \xi, \alpha, \beta)}{\partial \xi} = \mathbf{0}, \tag{2.44}$$

$$y_i \left(\langle \mathbf{w} \cdot \mathbf{x}_i \rangle + b \right) - 1 + \xi_i \geq 0, \tag{2.45}$$

$$\xi_i \geq 0, \tag{2.46}$$

$$\alpha_i \geq 0, \tag{2.47}$$

$$\beta_i \geq 0, \tag{2.48}$$

$$\alpha_i \left[y_i \left(\langle \mathbf{w} \cdot \mathbf{x}_i \rangle + b \right) - 1 + \xi_i \right] = 0, \tag{2.49}$$

$$\beta_i \xi_i = 0. \tag{2.50}$$

As before, we can use the KKT complementarity conditions, (2.49) and (2.50), to determine the bias b. Note that the formula (2.40) combined with (2.50) shows that $\xi_i = 0$ if $\alpha_i < C$. Thus we can simply take any training data for which $0 < \alpha_i < C$ to use the formula (2.49) (with $\xi_i = 0$) to compute b.

Optimizing the norms of the margin slack vector has a diffuse effect on the margin. For this reason, it is referred to as a *soft margin* in contrast to the maximal margin, which depends critically on a small subset of points and is therefore often called a *hard margin*. In addition, the decision function corresponding to the soft margin is called the *soft margin classifier*.

2.3.3　*Non-linear case*

In general, complex real-world applications require more expressive
decision functions than linear functions. There exist two approaches to
constructing non-linear classifiers to solve these problems: One is to
create a net of simple linear classifiers, e.g. a neural network with lots of
neurons simulating the human's brains. Some problems have appeared in
this category, such as local minima, many parameters not easy to adjust,
heuristics needed to train, etc. Alternatively a more attractive solution is
to map data into a feature space including non-linear features, and then
use a linear classifier. Figure 2.8 illustrates the procedure of this method.
Working in a high dimensional feature space can solve the problem of
expressing complex functions, although, some other problems occur:
There are a computational problem and a generalization theory problem
(known as *curse of dimensionality* [12]).

In the literature [15], a rather flexible trick [12] was introduced. First
note that the only way in which the data appeard in the training problem,
the formulas (2.26), (2.33) and (2.41), is in the form of dot products,
$\langle \mathbf{x}_i \cdot \mathbf{x}_j \rangle$. Now suppose we first mapped the data to some other (possibly
infinite dimensional) feature space F, using mapping ϕ (see Fig. 2.8):
$$\phi : X \rightarrow F.$$

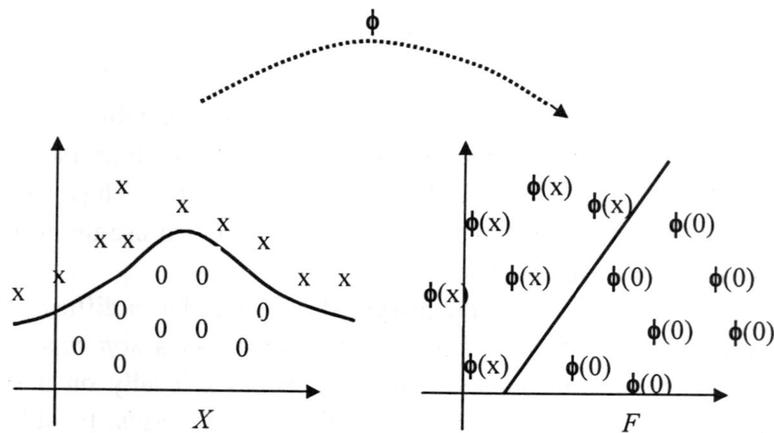

Fig. 2.8　Mapping the input space into a feature space where they are linearly
separable.

Then of course the training algorithm will only depend on the data through dot products in F, i.e. on functions of the form $\langle \phi(\mathbf{x}_i) \cdot \phi(\mathbf{x}_j) \rangle$. If one replaces $\langle \mathbf{x}_i \cdot \mathbf{x}_j \rangle$ by $\langle \phi(\mathbf{x}_i) \cdot \phi(\mathbf{x}_j) \rangle$ everywhere in the training algorithm of above linear cases, the algorithm will happily produce a support vector machine which lives in a high-dimensional space. Thus, the dual representation (2.41) will be turned into

$$W(\boldsymbol{\alpha}) = \sum_{i=1}^{\ell} \alpha_i - \frac{1}{2} \sum_{i,j=1}^{\ell} y_i y_j \alpha_i \alpha_j \langle \phi(\mathbf{x}_i) \cdot \phi(\mathbf{x}_j) \rangle \qquad (2.51)$$

Now if there is a "kernel function" K such that

$$K(\mathbf{x}_i, \mathbf{x}_j) = \langle \phi(\mathbf{x}_i) \cdot \phi(\mathbf{x}_j) \rangle, \qquad (2.52)$$

we will only need to use K in the training algorithm and the dimensionality of space F is not necessarily important. We may not even know the concrete function after mapping by ϕ. In this case, the decision function in SVM is in the following form:

$$g(\mathbf{x}) = \mathrm{sgn}\left(f(\mathbf{x})\right) = \mathrm{sgn}\left(\sum_{i \in SV} a_i y_i \langle \phi(\mathbf{x}) \cdot \phi(\mathbf{x}_i) \rangle + b\right)$$

$$= \mathrm{sgn}\left(\sum_{i \in SV} a_i y_i K(\mathbf{x}, \mathbf{x}_i) + b\right) \qquad (2.53)$$

where the $\mathbf{x}_i$ is the support vectors. Thus, we produce the SVM algorithm with the roughly same amount of computational time it will take to train on the un-mapping data. In addition, all the considerations in the previous subsections still hold, since we are still doing the linear separation, but in a different space. One example of kernel function is a 2-order polynomial kernel of the form,

$$K(\mathbf{x}, \mathbf{x}') = \left(\langle \mathbf{x} \cdot \mathbf{x}' \rangle + 1\right)^2,$$

which maps a two dimensional input vector into a six dimensional feature space.

Now there is a further problem about the kernel function, i.e. given a function K, how to verify that it is a kernel. The answer is given by Mercer's condition [132], which will be discussed in next chapter.

2.4   Support Vector Regression

SVM were firstly developed to solve the classification problems, but later they have been extended to the domain of regression problems, retaining all the main properties that characterize the maximal margin algorithm, such as duality, sparseness, kernel and convexity. As a difference, support vector regression algorithms introduce a loss function that ignores errors that are within a certain distance of the true value. This type of function is referred to as an ε-insensitive loss function and can control a parameter that is equivalent to the margin parameter for separating hyperplanes [133]. Another motivation for considering the ε-insensitive loss function is that it will ensure sparseness of the dual variables, just as the case in SVM for classification.

In this section we first introduce the definition of the ε-insensitive loss function, then show that the same quadratic optimization technique that was used in Section 2.3 for constructing approximations to indicator functions provides an approximation to real-valued functions, involving the linear case and nonlinear case.

2.4.1   *ε-insensitive loss functions*

Suppose we are given training data $\{(\mathbf{x}_1, y_1), \ldots, (\mathbf{x}_\ell, y_\ell)\} \subset \mathbb{R}^n \times \mathbb{R}$. In support vector regression [132], our goal is to find a function $f(\mathbf{x})$ that has at most ε deviation from the actually obtained targets y_i for all the training data, and at the same time is as flat as possible. In other words, we do not care about errors as long as they are less than ε, but will not accept any deviation larger than this.

Definition 2.3   The (linear) ε-insensitive loss function $L(\mathbf{x}, \mathbf{y}, f)$ is defined by

$$L^{\varepsilon}(\mathbf{x},\mathbf{y},f)=\left|\mathbf{y}-f(\mathbf{x})\right|_{\varepsilon}=\begin{cases}0 & if\ \left|\mathbf{y}-f(\mathbf{x})\right|\leq\varepsilon \\ \left|\mathbf{y}-f(\mathbf{x})\right|-\varepsilon & otherwise\end{cases}, \qquad (2.54)$$

where f is a real-valued function on a X. Similarly the quadratic ε-insensitive loss is given by

$$L_2^{\varepsilon}(\mathbf{x},\mathbf{y},f)=\left|\mathbf{y}-f(\mathbf{x})\right|_{\varepsilon}^2. \qquad (2.55)$$

Figure 2.9 shows the form of the linear and quadratic ε-insensitive loss function for zero and non-zero ε.

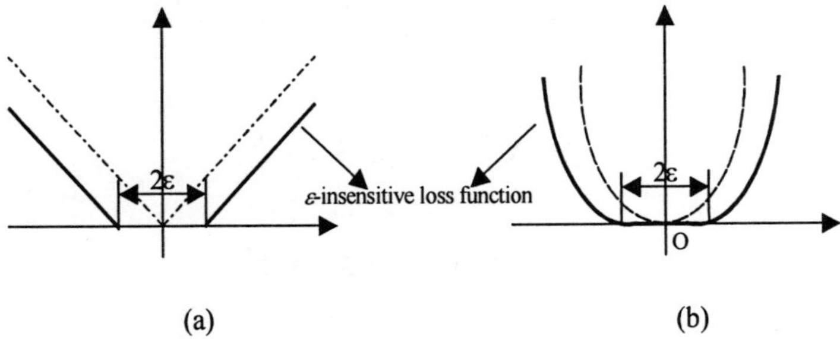

Fig. 2.9   The linear (a) and quadratic (b) ε-insensitive loss function.

2.4.2   *Linear regression*

We begin by describing the case of linear functions f, taking the form

$$f(\mathbf{x})=\langle\mathbf{w}\cdot\mathbf{x}\rangle+b. \qquad (2.56)$$

Flatness in the case of (2.56) means that one seeks a small $\mathbf{w}$ [4]. One way to ensure this is to minimize the norm, i.e. $\|\mathbf{w}\|^2=\langle\mathbf{w}\cdot\mathbf{w}\rangle$. We can write this problem as a convex optimization problem:

Minimize
$$\frac{1}{2}\|\mathbf{w}\|^2$$

subject to
$$\langle\mathbf{w}\cdot\mathbf{x}_i\rangle+b-y_i\leq\varepsilon \qquad (2.57)$$

and
$$y_i - \langle \mathbf{w} \cdot \mathbf{x}_i \rangle - b \leq \varepsilon.$$

In general, the assumption in (2.57) does not hold in many real-world problems: if the data is much noisy, such a function f may not exist that approximates all pairs $(\mathbf{x}_i, y_i)$ with ε precision, or in other words, that convex optimization problem becomes infeasible. Analogously to the soft margin classifier mentioned in Section 2.3, one can introduce slack variables ξ_i, ξ_i^* to seek the optimal regression function by the minimum of the functional,

$$\Phi(\mathbf{w}, \xi, \xi^*) = \frac{1}{2}\|\mathbf{w}\|^2 + C\sum_{i=1}^{\ell}\left(\xi_i + \xi_i^*\right), \tag{2.58}$$

subject to
$$\langle \mathbf{w} \cdot \mathbf{x}_i \rangle + b - y_i \leq \varepsilon + \xi_i$$

$$y_i - \langle \mathbf{w} \cdot \mathbf{x}_i \rangle - b \leq \varepsilon + \xi_i^*$$

and
$$\xi_i, \xi_i^* \geq 0$$

where the constant $C > 0$ determines the trade-off between the flatness of f and the amount up to which deviations larger than ε are tolerated [4].

(1) Linear ε-insensitive loss

Using a linear ε-insensitive loss function (2.54), the solution of (2.58) is given by the saddle point of the Lagrangian,

$$L(\mathbf{w}, b, \xi, \xi^*, \alpha, \alpha^*, \beta, \beta^*) = \frac{1}{2}\|\mathbf{w}\|^2 + C\sum_{i=1}^{\ell}\left(\xi_i + \xi_i^*\right) - \sum_{i=1}^{\ell}\left(\beta_i\xi_i + \beta_i^*\xi_i^*\right)$$

$$-\sum_{i=1}^{\ell}\alpha_i\left(\varepsilon + \xi_i - y_i + \langle \mathbf{w} \cdot \mathbf{x}_i \rangle + b\right)$$

$$-\sum_{i=1}^{\ell} \alpha_i^* \left(\varepsilon + \xi_i^* + y_i - \langle \mathbf{w} \cdot \mathbf{x}_i \rangle - b \right) \tag{2.59}$$

where $\alpha_i, \alpha_i^*, \beta_i, \beta_i^*$ are Lagrange multipliers and hence

$$\alpha_i^{(*)}, \beta_i^{(*)} \geq 0 \tag{2.60}$$

Note that by $\alpha_i^{(*)}$, we refer to α_i and α_i^*.

As before, the corresponding dual formula is found by differentiating L with respect to the primal variables $\left(\mathbf{w}, b, \xi_i, \xi_i^* \right)$, imposing stationarity,

$$\frac{\partial L}{\partial \mathbf{w}} = \mathbf{w} - \sum_{i=1}^{\ell} \left(\alpha_i - \alpha_i^* \right) \mathbf{x}_i = \mathbf{0}, \tag{2.61}$$

$$\frac{\partial L}{\partial b} = \sum_{i=1}^{\ell} \left(\alpha_i^* - \alpha_i \right) = 0, \tag{2.62}$$

$$\frac{\partial L}{\partial \xi_i^{(*)}} = C - \alpha_i^{(*)} - \beta_i^{(*)} = 0, \tag{2.63}$$

and resubstituting the relations obtained into the primal (2.59); we obtain the following dual optimization problem:

$$\text{maximize} \quad W(\boldsymbol{\alpha}) = -\frac{1}{2} \sum_{i,j=1}^{\ell} \left(\alpha_i - \alpha_i^* \right) \left(\alpha_j - \alpha_j^* \right) \langle \mathbf{x}_i \cdot \mathbf{x}_j \rangle \tag{2.64}$$

$$-\varepsilon \sum_{i=1}^{\ell} \left(\alpha_i + \alpha_i^* \right) + \sum_{i=1}^{\ell} y_i \left(\alpha_i - \alpha_i^* \right)$$

$$\text{subject to} \quad \sum_{i=1}^{\ell} \left(\alpha_i - \alpha_i^* \right) = 0 \ \text{ and } \ 0 < \alpha_i^{(*)} \leq C.$$

In deriving (2.64) we already eliminated the dual variables β_i, β_i^* through condition (2.63) which can be reformulated as $\eta_i^{(*)} = C - \alpha_i^{(*)}$. The formula (2.62) can be rewritten as follows

$$\mathbf{w} = \sum_{i=1}^{\ell} \left(\alpha_i - \alpha_i^* \right) \mathbf{x}_i \,, \tag{2.65}$$

thus

$$f(\mathbf{x}) = \sum_{i=1}^{\ell} \left(\alpha_i - \alpha_i^* \right) \langle \mathbf{x} \cdot \mathbf{x}_i \rangle + b \,. \tag{2.66}$$

Note that in (2.65) $\mathbf{w}$ is described as a linear combination of the training vector $\mathbf{x}_i$. In a sense, the complexity of a function's representation by SVs is independent of the dimensionality of the input space X, and depends only on the number of SVs.

As before, the value of b can be determined by exploiting the KKT conditions. They state that at the point of the solution the product between dual variables and constraints has to vanish.

$$\alpha_i \left(\varepsilon + \xi_i - y_i + \langle \mathbf{w} \cdot \mathbf{x}_i \rangle + b \right) = 0$$

$$\alpha_i^* \left(\varepsilon + \xi_i^* + y_i - \langle \mathbf{w} \cdot \mathbf{x}_i \rangle - b \right) = 0$$

$$\xi_i \xi_i^* = 0, \alpha_i \alpha_i^* = 0 \tag{2.67}$$

$$\left(C - \alpha_i \right) \xi_i = 0, \quad \left(C - \alpha_i^* \right) \xi_i^* = 0$$

From these conditions, we can conclude: i) only samples $\left(\mathbf{x}_i, y_i \right)$ with corresponding $\alpha_i^{(*)} = C$ lie outside the ε-insensitive tube; ii) we have

$$\max \left\{ -\varepsilon + y_i - \langle \mathbf{w} \cdot \mathbf{x}_i \rangle \middle| \alpha_i < C \text{ or } \alpha_i^* > 0 \right\} \le b \le$$

$$\min \left\{ -\varepsilon + y_i - \langle \mathbf{w} \cdot \mathbf{x}_i \rangle \middle| \alpha_i > 0 \text{ and } \alpha_i^* < C \right\} \tag{2.68}$$

If some $\alpha_i^{(*)} \in (0, C)$ the inequalities become equalities [4].

(2) Quadratic ε-insensitive loss

Using a quadratic loss function, we have the primal problem defined as follows:

Minimize
$$\|\mathbf{w}\|^2 + C\sum_{i=1}^{\ell}\left(\xi_i^2 + \xi_i^{*2}\right)$$

subject to
$$\langle \mathbf{w} \cdot \mathbf{x}_i \rangle + b - y_i \leq \varepsilon + \xi_i$$

$$y_i - \langle \mathbf{w} \cdot \mathbf{x}_i \rangle - b \leq \varepsilon + \xi_i^*$$

and
$$\xi_i, \xi_i^* \geq 0$$

The corresponding dual problem can be derived using the standard method and taking into account that $\xi_i \xi_i^* = 0$ and therefore that the same relation $\alpha_i \alpha_i^* = 0$ holds for the corresponding Lagrange multipliers:

Maximize $$W(\boldsymbol{\alpha}) = -\frac{1}{2}\sum_{i,j=1}^{\ell}\left(\alpha_i - \alpha_i^*\right)\left(\alpha_j - \alpha_j^*\right)\left(\langle \mathbf{x}_i \cdot \mathbf{x}_j \rangle + \frac{1}{C}\delta_{ij}\right) \quad (2.69)$$

$$-\varepsilon\sum_{i=1}^{\ell}\left(\alpha_i + \alpha_i^*\right) + \sum_{i=1}^{\ell}y_i\left(\alpha_i - \alpha_i^*\right)$$

subject to
$$\sum_{i=1}^{\ell}\left(\alpha_i - \alpha_i^*\right) = 0 \quad \text{and} \quad \alpha_i^{(*)} \geq 0.$$

The optimization problem can be simplified by the corresponding KKT conditions:

$$\alpha_i\left(\langle \mathbf{w} \cdot \mathbf{x}_i \rangle + b - y - \varepsilon - \xi_i\right) = 0$$

$$\alpha_i^* \left(y_i - \langle \mathbf{w} \cdot \mathbf{x}_i \rangle - b - \varepsilon - \xi_i^* \right) = 0$$

$$\xi_i \xi_i^* = 0, \; \alpha_i \alpha_i^* = 0$$

2.4.3 *Nonlinear regression*

In practice, a non-linear model is often required for adequate data fitting. In the same manner as the non-linear support vector classification approach, a non-linear mapping can be used to map the data into a high dimensional feature space where linear regression can be used (see Fig. 2.10). As noted in the previous subsection, the complete SVM can be described in terms of dot products between the data. The nonlinear SVR solution, using an ε-insensitive loss function (2.54) is given by solving the problem:

$$\text{Maximize} \quad W(\boldsymbol{\alpha}) = -\frac{1}{2} \sum_{i,j=1}^{\ell} \left(\alpha_i - \alpha_i^* \right) \left(\alpha_j - \alpha_j^* \right) K\left(\mathbf{x}_i, \mathbf{x}_j \right) \tag{2.70}$$

$$-\varepsilon \sum_{i=1}^{\ell} \left(\alpha_i + \alpha_i^* \right) + \sum_{i=1}^{\ell} y_i \left(\alpha_i - \alpha_i^* \right)$$

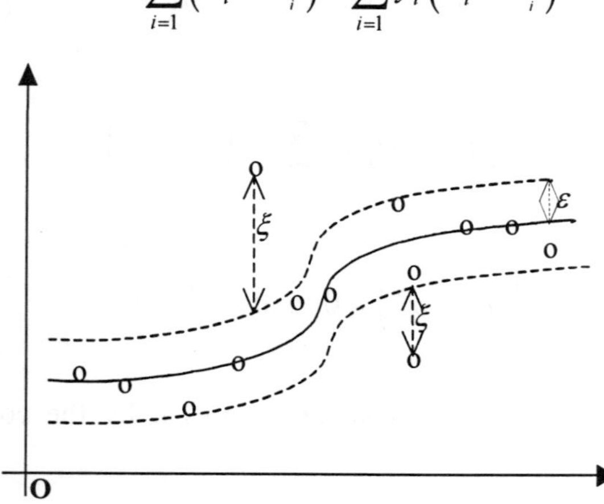

Fig. 2.10   The ε-tube for a non-linear regression function.

subject to
$$\sum_{i=1}^{\ell}\left(\alpha_i - \alpha_i^*\right) = 0 \quad \text{and} \quad 0 < \alpha_i^{(*)} \le C.$$

Likewise the formulas (2.65) and (2.66) can be written as

$$\mathbf{w} = \sum_{i=1}^{\ell}\left(\alpha_i - \alpha_i^*\right)\boldsymbol{\phi}\left(\mathbf{x}_i\right) \quad \text{and} \quad f(\mathbf{x}) = \sum_{i=1}^{\ell}\left(\alpha_i - \alpha_i^*\right)K\left(\mathbf{x}, \mathbf{x}_i\right) + b.$$

2.5 $\nu - $ SVM

In practice, the soft margin versions of the standard SVM (also known as C-SVM) described in the previous sections often suffer from the following problems. Firstly, there is a problem of how to determine the error penalty parameter C. Although the cross-validation technique can be used to determine this parameter, it is still hard to explain. Secondly, the time taken for a support vector classifier to compute the class of a new sample is proportional to the number of support vectors, so if that number is large, the computation is time-consuming.

The ν-support vector machine for both classification and regression problems introduced by Schölkopf et al. [121] attempt to overcome the above mentioned disadvantages of C-SVM. The formulation of ν-SVM removes the constant C, and introduces a new parameter ν. As a primal problem for ν-SVM, the following optimization problem is considered:

Minimize
$$\Phi(\mathbf{w}, \xi, \rho) = \frac{1}{2}\langle \mathbf{w} \cdot \mathbf{w} \rangle - \nu\rho + \frac{1}{\ell}\sum_{i=1}^{\ell}\xi_i \qquad (2.71)$$

subject to
$$y_i\left(\langle \mathbf{w} \cdot \mathbf{x}_i \rangle + b\right) \ge \rho - \xi_i, \qquad i = 1, \ldots, \ell$$

$$\xi_i \ge 0, \quad \rho \ge 0.$$

For the dual form of (2.75), we have

Maximize
$$W(\boldsymbol{\alpha}) = -\frac{1}{2}\sum_{i,j=1}^{\ell} y_i y_j \alpha_i \alpha_j K(\mathbf{x}_i, \mathbf{x}_j) \qquad (2.72)$$

subject to
$$0 \leq \alpha_i \leq 1/\ell$$

$$\sum_{i=1}^{\ell} \alpha_i y_i = 0$$

$$\sum_{i=1}^{\ell} \alpha_i \geq \nu . \qquad (2.73)$$

It can be shown that ν gives an upper bound on the fraction of the training set that are margin errors and provides a lower bound on the total number of support vectors. Accordingly, when the sample size goes to infinity, both fractions tend almost surely to ν under rather general assumptions on the learning problem and the used kernel.

Chapter 3

Kernel Functions

3.1 Introduction

As we have seen in chapter 1 and chapter 2, most of the data sets in chemistry and chemical engineering are nonlinear, while SVM is based on a linear learning machine with large margin. So it is necessary to have some nonlinear mapping techniques to make most of the data sets in chemistry or chemical engineering suitable to be treated by SVM. Kernel function is just an effective tool for this purpose.

3.2 Mercer Kernel

In order to learn nonlinear relationships with a linear machine, we can apply a fixed nonlinear mapping of the data in input space to a feature space, and that the decision function is

$$f(\mathbf{x}) = \sum_{i=1}^{\ell} w_i \phi_i(\mathbf{x}) + b \tag{3.1}$$

where $\boldsymbol{\phi} : X \to F$ is a nonlinear mapping from the input space X to feature space. Figure 3.1 shows that a nonlinear machine can be built in two steps: Firstly, a fixed nonlinear mapping transforms the data into a feature space F. Secondly, a linear machine is used to classify them in the feature space.

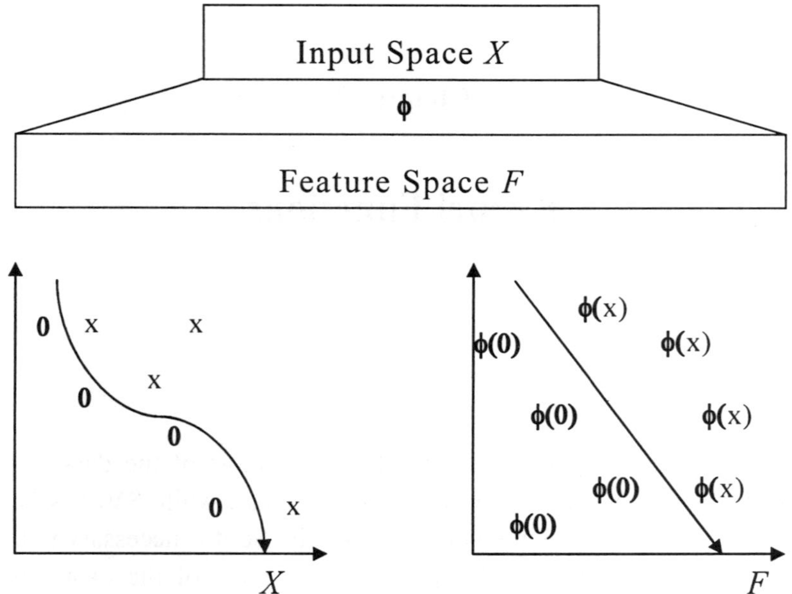

Fig. 3.1   The nonlinear mapping can make a linearly inseparable data set to be linearly separable.

As we know, one important property of the linear learning machine is that they can be expressed in a dual representation. It means that the hypothesis function can be expressed as a linear combination of the training points, so that the decision rule can be evaluated using just inner products between the test point and the training points (see equation (2.24) and (2.53)).

$$f(\mathbf{x}) = \sum_{i=1}^{\ell} \alpha_i y_i \left\langle \phi(\mathbf{x}_i) \cdot \phi(\mathbf{x}) \right\rangle + b \qquad (3.2)$$

If we have a way of computing the inner product $\left\langle \phi(\mathbf{x}_i) \cdot \phi(\mathbf{x}) \right\rangle$ in feature space directly as a function of the original input points, it becomes possible to merge the two steps to build a nonlinear learning machine. Such a direct computation method is called *kernel function* method.

Definition 3.1   *A kernel is a function K, such that for all* $\mathbf{x}, \mathbf{z} \in X$

$$K(\mathbf{x},\mathbf{z}) = \langle \phi(\mathbf{x}) \cdot \phi(\mathbf{z}) \rangle \tag{3.3}$$

where ϕ *is the mapping from X to the feature space F.*

As one does not express the feature vectors explicitly, the number of operations required to compute the inner product is not necessarily proportional to the number of features by employing the kernel function. The use of kernel makes it possible to map the data implicitly into a feature space and to train a linear machine in such a space, the only information used about the training samples is the kernel matrix, and the key is to find a kernel function that can be evaluated efficiently. Once we have such a function, the decision rule can be evaluated by the computation of the kernel. In accordance with (2.21) to (2.33):

$$f(\mathbf{x}) = \sum_{i=1}^{\ell} \alpha_i y_i K(\mathbf{x}_i,\mathbf{x}) + b \tag{3.4}$$

The use of the kernel function is an attractive computational *short-cut*. A curious fact about using a kernel is that we do not need to know the underlying feature map which can learn in the feature space. In practice the approach taken is to define a kernel function directly, hence implicitly to define the feature space. In this way, we avoid the feature space not only in the computation of inner product, but also in the design of the learning machine itself.

Now we determine what properties of a kernel function $K(\mathbf{x},\mathbf{z})$ are necessary to ensure that it is a kernel for some feature space. Clearly, the function must be symmetric,

$$K(\mathbf{x},\mathbf{z}) = \langle \phi(\mathbf{x}) \cdot \phi(\mathbf{z}) \rangle = \langle \phi(\mathbf{z}) \cdot \phi(\mathbf{x}) \rangle = K(\mathbf{z},\mathbf{x}) \tag{3.5}$$

and satisfy the inequality that follows from the Cauchy-Schwarz inequalities.

$$K(\mathbf{x},\mathbf{z})^2 = \langle \phi(\mathbf{x}) \cdot \phi(\mathbf{z}) \rangle^2 \leq \|\phi(\mathbf{x})\|^2 \|\phi(\mathbf{z})\|^2$$
$$= \langle \phi(\mathbf{x}) \cdot \phi(\mathbf{x}) \rangle \langle \phi(\mathbf{z}) \cdot \phi(\mathbf{z}) \rangle = K(\mathbf{x},\mathbf{x}) K(\mathbf{z},\mathbf{z}) \tag{3.6}$$

But these conditions are still not sufficient to guarantee the existence of a feature space.

Then what is the sufficient condition for a function to be a kernel function? The theoretical derivation can prove that if a function satisfies Mercer condition, it can be regard as a kernel function.

Theorem 2.1 (Mercer) *Let X be a compact subset of $\mathbb{R}^n$. Suppose K is a continuous symmetric function such that the integral operator $T_K : L_2(X) \to L_2(X)$.*

$$(T_K f)(\cdot) = \int_X K(\cdot, \mathbf{x}) f(\mathbf{x}) d\mathbf{x} \tag{3.7}$$

is positive, that is

$$\int_{X \times X} K(\mathbf{x}, \mathbf{z}) f(\mathbf{x}) f(\mathbf{z}) d\mathbf{x} d\mathbf{z} \geq 0 \tag{3.8}$$

for all $f \in L_2(X)$. Then we can expand $K(\mathbf{x}, \mathbf{z})$ in a uniformly convergent series (on $X \times X$) in terms of T_K's eigen-functions $\phi_j \in L_2(X)$, normalised in such a way that $\left\| \phi_j \right\|_{L_2} = 1$, and positive associated eigenvalues $\lambda_j \geq 0$.

$$K(\mathbf{x}, \mathbf{z}) = \sum_{j=1}^{\infty} \lambda_j \phi_j(\mathbf{x}) \phi_j(\mathbf{z}) \tag{3.9}$$

Let us observe this theorem, the positivity condition $\int_{X \times X} K(\mathbf{x}, \mathbf{z}) f(\mathbf{x}) f(\mathbf{z}) d\mathbf{x} d\mathbf{z} \geq 0$, $\forall f \in L_2(X)$, corresponds to the positive semi-definite condition in the finite case, this gives the second characterization of a kernel function that is proved to be most useful when we come to constructing kernels. We use the term *kernel* to refer to function satisfying this property, but in the literature these are often called Mercer kernel.

Here we give a simple example of kernel functions [101].

In Fig. 3.2, let $\mathbf{x}_i$ $i = 1, \ldots, \ell$, be vectors, $\mathbf{x}_i \in \mathbb{R}^2$, kernel function $K(\mathbf{x}_i, \mathbf{x}_j) = \langle \mathbf{x}_i \cdot \mathbf{x}_j \rangle^2$. It is simple to find a space H which satisfies the condition: define $\boldsymbol{\phi}$ as a function that takes samples from $\mathbb{R}^2$ and maps them into H, and $\langle \mathbf{x} \cdot \mathbf{y} \rangle^2 = \langle \boldsymbol{\phi}(\mathbf{x}) \cdot \boldsymbol{\phi}(\mathbf{y}) \rangle$, $H = \mathbb{R}^3$, and

$$\phi(\mathbf{x}) = \begin{pmatrix} x_1^2 \\ \sqrt{2}x_1x_2 \\ x_2^2 \end{pmatrix} \tag{3.10}$$

is well to satisfy the requirement. For a fixed kernel function, it is necessary to point that mapping ϕ and space H is not unique. For example we also can select H is $\mathbb{R}^3$ and

$$\phi(\mathbf{x}) = \frac{1}{\sqrt{2}} \begin{pmatrix} \left(x_1^2 - x_2^2\right) \\ 2x_1x_2 \\ \left(x_1^2 + x_2^2\right) \end{pmatrix} \tag{3.11}$$

satisfies the requirement well, also.

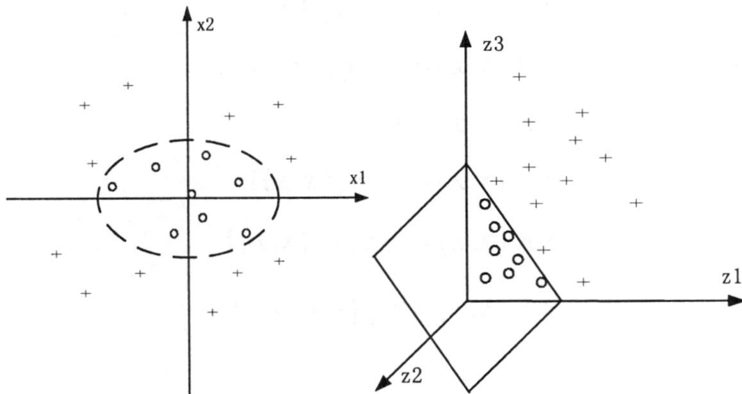

Fig. 3.2 A two dimensional classification example. Using the second order monomials x_1^2, $\sqrt{2}x_1x_2$ and x_2^2 as features a separation in feature space can be found using a linear hyperplane (right). In input space this construction corresponds to a nonlinear ellipsoidal decision boundary (left).

3.3 Properties of Kernel

The key to verifying a new symmetric function as a kernel is the condition stated in Mercer Theorem, that is, the requirement that the matrix defined by restricting the function to any finite set of points is positive semi-definite. Then a number of properties of kernel can be given based on this criterion [120].

Let K_1 and K_2 be kernels over $X \times X$, $X \subseteq \mathbb{R}^n$, $\mathbf{x}, \mathbf{z} \in X$, $a \in \mathbb{R}^+$, $f(\cdot)$ a real-valued function on X,

$$\phi : X \rightarrow \mathbb{R}^m$$

with K_3 a kernel over $\mathbb{R}^m \times \mathbb{R}^m$, and $\mathbf{B}$ a symmetric positive semi-definite $n \times n$ matrix, $p(\mathbf{x})$ a polynomial with positive coefficients. Then the following functions are kernels:

$$
\begin{aligned}
&1. K(\mathbf{x}, \mathbf{z}) = K_1(\mathbf{x}, \mathbf{z}) + K_2(\mathbf{x}, \mathbf{z}) \\
&2. K(\mathbf{x}, \mathbf{z}) = a K_1(\mathbf{x}, \mathbf{z}) \\
&3. K(\mathbf{x}, \mathbf{z}) = K_1(\mathbf{x}, \mathbf{z}) K_2(\mathbf{x}, \mathbf{z}) \\
&4. K(\mathbf{x}, \mathbf{z}) = f(\mathbf{x}) f(\mathbf{z}) \\
&5. K(\mathbf{x}, \mathbf{z}) = K_3 \langle \phi(\mathbf{x}), \phi(\mathbf{z}) \rangle \\
&6. K(\mathbf{x}, \mathbf{z}) = \mathbf{x}' \mathbf{B} \mathbf{z} \\
&7. K(\mathbf{x}, \mathbf{z}) = p\left(K_1(\mathbf{x}, \mathbf{z})\right) \\
&8. K(\mathbf{x}, \mathbf{z}) = \exp\left(K_1(\mathbf{x}, \mathbf{z})\right) \\
&9. K(\mathbf{x}, \mathbf{z}) = \exp\left(-\|\mathbf{x} - \mathbf{z}\|^2 / \sigma^2\right)
\end{aligned}
\tag{3.12}
$$

3.4   Kernel Selection

There are four commonly used kernel functions:

Linear Kernel
$$K(\mathbf{x}, \mathbf{y}) = \langle \mathbf{x} \cdot \mathbf{y} \rangle + \theta$$

Gaussian (RBF) Kernel
$$K(\mathbf{x}, \mathbf{y}) = \exp\left(\frac{-\|\mathbf{x} - \mathbf{y}\|^2}{\sigma^2}\right)$$

Polynomial Kernel $\quad K(\mathbf{x}, \mathbf{y}) = (\langle \mathbf{x} \cdot \mathbf{y} \rangle + \theta)^d$

Sigmoid Kernel $\quad K(\mathbf{x}, \mathbf{y}) = \tanh(v \langle \mathbf{x} \cdot \mathbf{y} \rangle + r)$

If the data set is linear or nearly linear one, the linear kernel should be tried at first for computation, because a linear kernel usually exhibits better generalization ability. For nonlinear data set, the Gaussian kernel and polynomial kernel should be tried in computation at first [75; 39; 88].

Chapter 4

Feature Selection Using Support Vector Machine

4.1 Significance and Difficulty of Feature Selection in Chemical Data Processing

Feature selection is one of the most important steps in chemical data processing. In many cases, the accuracy of feature selection is the decisive factor for the successful solution of practical problems by chemical data processing.

Most of the chemical data processing problems are dealing with very complicated systems. It is usually impossible to solve these problems by some single equation or single influencing factor. It is also impossible to solve these problems exactly by some equations based on first principle. In many cases it is even impossible to assure how many features can form a complete feature set that can exactly describe the object of investigation, because the detailed mechanism of the process is not completely understood yet. So the only feasible strategy that we can adopt is to include all possibly relevant features to form a large initial data set, and then delete a part of them by feature selection.

The purpose of feature selection in chemical data processing is twofold: firstly it can help us to find the most influential features for the target function that we are interested in; secondly it can help us to delete the irrelevant, redundant features or the noisy features, in order to improve the quality of data set suitable for data processing.

A feature selection process mainly consists of two parts: *feature subset generation* and *feature subset evaluation*. Feature subset generation is a searching procedure. It generates subsets of features for evaluation. If the number of features of the original data set is relatively large and a large part of the features have to be eliminated, the number of subsets is so large that the subset evaluation work shall be impracticable. In the traditional techniques of chemical data processing, the features are removed one after another without recurring steps until the end of deletion. By this way, the feature subsets generated include only a small fraction of all the feature subsets possible, and the selected feature subset is only the best one among a part of feature subsets, not the best of *all* subsets. *This is a serious disadvantage of these simple feature subset generation methods*. In order to avoid this disadvantage, more advanced feature subset generation methods have been proposed. These methods can be classified into several types:

(1) Exhaustive search: the exhaustive search can get the best subset but it is only suitable for small data sets since it is computer time-consuming.

(2) Heuristic search: It is suitable for data sets of middle size.

(3) Non-deterministic search: such as search with genetic algorithm and others. It can be used for large data sets.

The second part of feature selection is *feature evaluation*, i.e., to determine which feature should be eliminated or be chosen in feature selection. In traditional technique of chemical data processing, the criterion of the importance of a feature is its contribution to the accuracy of data fitting of training set. According to statistical learning theory, best fitting accuracy of training set cannot assure the best generalization ability of the mathematical model obtained after feature selection. In recent years, it has been widely recognized that the generalization ability has to be measured by cross validation methods. So a series of new feature selection methods has been proposed based on the new criterion by accounting *the contribution for generalization ability tested by cross validation methods*. This new criterion of feature evaluation can give mathematical model with better generalization ability [84].

In this chapter, we will recommend two new kinds of feature selection methods for chemical data processing. One is *SVM-BFS method*, and another is *SVM-RFE method*. Both of them are based on *advanced feature subset generation techniques* and the *criterion of generalization ability contribution of features*.

Although the deletion of redundant features can improve the generalization ability of resultant mathematical model, in many cases, the information of the deleted features can be recovered by using these deleted features as a part of output of learning machine in mathematical modeling. This novel method is called *multitask learning*. This method will also be briefly described in this chapter.

4.2   SVM-BFS — Application of Wrapper Method and Floating Search Method

4.2.1   *Wrapper methods [79]*

Wrapper methods utilize the prediction ability of some learning machine (SVM in this book) to evaluate the feature subset. Compared with other methods, wrapper method can assure to get a feature subset with higher accuracy by using the specified learning machine (SVM here). The principle of wrapper methods is to minimize

$$W = R_{err}(S_w) \tag{4.1}$$

where S_w is the feature subset candidate, and R_{err} is defined by

$$R_{err} = \frac{1}{\ell}\sum_{j=1}^{\ell}(y_j{}' \neq y_j)^2 \tag{4.2}$$

where ℓ is the number of training samples, y' is the SVM-predicted value. The feature subset with smallest W will be selected as the best feature subset.

Wrapper methods, as a type of subset evaluation methods, can be used in coordination with different methods for feature subset generation. In this chapter, however, the coordination of wrapper methods with backward floating search (BFS) method will be emphasized, since the floating search method is one of the best heuristic methods for feature subset generation. The feature subset obtained by this method is nearly optimal [73]. And the backward floating search method is especially suitable for the data set with features not completely independent with each other. As we know, many data sets in chemical data processing belong to this category.

4.2.2 *Floating search method*

Floating search method proposed by Pudil et al. [112] is used as feature subset generation procedure, and has been proved to be one of the best subset generation algorithms for moderately large or small data sets. Backward floating search (BFS) method needs more computer time than forward floating search method, but it can treat interactive features well, so we combine it with SVM based on wrapper method and name it SVM-BFS method.

BFS is based on sequential backward search (SBS) and sequential forward search (SFS), of which SBS eliminates the least important feature one by one, while SFS adds the most important feature one by one. The least important feature f_x means that for a subset S_w, there exists $R_{err}(S_w - f_x) < R_{err}(S_w - f_i)$, $\forall f_i \in S_w$, while the most important feature f_x means that for a subset S_w which satisfies $S_w \cup S_u = S$ and $S_w \cap S_u = \varnothing$, there exists $R_{err}(S_w + f_x) < R_{err}(S_w + f_i), \forall f_i \in S_u$ where $R_{err}(S_w \mp f_x)$ means S_w removes or adds f_x. $FiM(S_w, S_u)$ is a function used to find the most important feature for S_w from S_u. SVM-BFS can be divided into five steps as follows.

Step 0 (*initialization*): It uses SBS method to remove the least two important features and sets the value for the current number and the target number of features. Since we want to eliminate the features from the maximum to 1, we set the target number is 1.

Note, in the following steps, keep the feature number of current feature subsets two less than the feature number of the total feature set, otherwise go to step1.

Step 1 (*exclusion*): It uses the basic SBS method to remove the least important feature. In BFS this step is the main module.

Step 2 (*conditional inclusion*): It finds the most significant feature among the excluded features, if it is not the feature just eliminated, the basic SFS is used to add it. Else, go to step 4.

Step 3 (continuation of conditional inclusion): It continues to find the most significant feature among the excluded features with respect to the subset got in step 2. If R_{err} of the current subset is not lower than that of any other subset with the same number of features, then go to step1;

otherwise, go to step 3.

Table 4.1   The SVM-BFS approach.

Input: S, n

output: S_i, E_i $i = 0, ..., n-1$

Use SBS to remove two features and $n_c = n\text{-}2$, $n_i = 1$ /* Initialization*/

$\left[S_{n-n_c}, E_{n-n_c}, f_c \right]$ = SBS (S_{n-n_c-1}); n_c - -; /*Exclusion*/

f_b = FiM $(S_{n-n_c-1}, S - S_{n-n_c-1})$; /*Conditional inclusion*/

if$(f_c \mathrel{!=} f_b)\{$

  S_b = SFS (S_{n-n_c-1}, f_b); n_c ++;

  if$(n - n_c == 2)$

    $S_{n-n_c-1} = S_b$; $E_{n-n_c-1} = R_{err}\ (S_{n-n_c-1})$;

  else while (1) {/*Continuation of conditional inclusion*/

      f_b = FiM $(S_b, S - S_b)$;

      S_t = SFS (S_b, f_b);

      if$(R_{err}\ (S_t) > E_{n-n_c-2})\{$

        $S_{n-n_c-1} = S_b$; $E_{n-n_c-1} = R_{err}\ (S_{n-n_c-1})$;

       break;

      }

      else {

        $S_b = S_t$; n_c ++

        if $(n - n_c == 2)$

          $S_{n-n_c-1} = S_b$; $E_{n-n_c-1} = R_{err}\ (S_{n-n_c-1})$;

         break;

        }

      }

     }

  }

  if$(n_c > n_t)$

    goto Exclusion;

  else

    display S_i, E_i $i = 0, ..., n-1$ /*display*/

Step 4 (*display*): If the feature number of current subset is larger than the target number, go to step 1; otherwise, display the optimal feature subset and the merit of subset with different number of features. In summary, the pseudo code of SVM-BFS is shown in Table 4.1 and the notations are listed in Table 4.2.

Table 4.2   Notation in the pseudo code of SVM-BFS.

Notation	Meaning
$[S_u, E_u, f_u] = \text{SBS}(S_w)$	to eliminate the least important feature f_u from S_w and get the subset S_u as well as the $E_u = R_{err}(S_u)$
$S_u = \text{SFS}(S_w, f_b)$	to add the most important feature f_b to S_w and get the subset S_u
E	the merit of the generated feature subsets
f_b	the best feature not in current feature subset
f_c	the feature being removed currently
n	the feature number of total feature set
n_c	the feature number of current subset
n_t	the feature number of target feature subset
S	the training set
S_b	R_{err} got with the subset is lower than that with the current subset
S_t	a temporary subset

4.3   SVM-RFE: Application of Optimal Brain Damage and Recursive Feature Elimination

Nowadays, as the rapid development of data collection techniques, more and more data are collected, and wrapper methods cannot meet the need of rapid data processing. Then, a number of input pruning methods using SVM have been developed. Guyon and his co-workers proposed to use optimal brain damage as the feature evaluation method and combined it with recursive feature elimination to perform gene selection [64].

4.3.1   *Optimal brain damage method*

In optimal brain damage (OBD) [43] a criterion, S_i, is defined as a variable describing the importance of the $i\text{-}th$ feature

$$S_i = \frac{1}{2}\frac{\partial^2 L}{\partial (w^i)^2}(Dw^i)^2 \tag{4.3}$$

where L is the object function introduced in Chapter 2, and w^i is the weight of the $i\text{-}th$ input feature, a larger S_i means the $i\text{-}th$ feature is more important. The criterion gives satisfactory results in neural networks [40], and has also been used in linear support vector machine and obtained excellent performance in gene selection. Guyon also gave the nonlinear form with SVM. With the non-linear SVM, OBD can be written as:

$$S_i = \frac{1}{2}\boldsymbol{\alpha}^T K(\mathbf{x}_k,\mathbf{x}_h)\boldsymbol{\alpha} - \frac{1}{2}\boldsymbol{\alpha}^T K(\mathbf{x}_k^{-i},\mathbf{x}_h^{-i})\boldsymbol{\alpha} \tag{4.4}$$

where the notation $-i$ in $K(\mathbf{x}_k^{-i},\mathbf{x}_h^{-i})$ means that component i has been removed. The feature corresponding to the smallest S_i should be removed.

In his paper, Rakotomamonjy claimed that OBD was a rather effective criterion for feature selection as compared with some other feature selection criteria[113].

4.3.2 *Recursive feature elimination method*

Recursive feature elimination (RFE) is a heuristic method to search feature subset, which can remove the irrelevant features efficiently, and is adequate to be used in the data sets whose relevant features are relatively small. Guyon combined RFE with OBD to perform gene selection and proposed SVM-RFE algorithm.

SVM-RFE Algorithm

Remaining feature subset $\mathbf{u} = [1, 2, \ldots, N]$ and feature eliminated list $\mathbf{r} = [\ ]$ are initialized firstly. Then, training samples $\mathbf{x}_0 = [\mathbf{x}^1, \ldots, \mathbf{x}^i, \ldots, \mathbf{x}^N]^T$ and the target value $\mathbf{y}$ are input into SVM-RFE.

Step 1: Restrict training samples to good feature indices $\mathbf{x} = \mathbf{x}(\mathbf{u})$, and in the first iteration, $\mathbf{x} = \mathbf{x}_0$.

Step 2: Train SVM to get $\boldsymbol{\alpha}$ = SVM-train$(\mathbf{x}, \mathbf{y})$, and compute the selection criteria c_i = S$(\boldsymbol{\alpha}, \mathbf{x})$ for all i , S means the evaluation method in section 4.3.1.

Step 3: Find the feature with the smallest selection criterion h = arg min$(\mathbf{c})$.

Step 4: Update feature eliminated list $\mathbf{r}$ = $[\mathbf{u}(h), \mathbf{r}]$ and eliminate the feature with the smallest selection criterion $\mathbf{u}$ = $\mathbf{u}(1 : h\text{-}1, h + 1 : \text{length}(\mathbf{u}))$. If $\mathbf{u} \neq []$ go to step 1.

Step 5: Output feature eliminated list $\mathbf{r}$.

4.4 Multitask Learning

After feature selection, there are usually many features not selected, and the information hidden in these discarded features shall be lost. In recent years, a new concept named *multitask learning* (MTL) has been proposed to *reuse the redundant features* [18; 19]. The basic idea of multitask learning is to use the selected features as the input feature set, and combine the target values with some of the discarded features as the target output. It has proved that MTL can improve the accuracy of prediction of the base learning method.

Table 4.3 The terms used in multitask learning.

Term	Explanation.
Main Task	The output target values to be learned.
Selected Inputs	The features selected as inputs.
Extra Features	The features selected from the discarded features.
Extra Inputs	The Extra Features when used as inputs.
Extra Outputs	The same Extra Features when used as outputs
STD	Standard PLS using Selected Inputs as inputs and only Main Task as outputs.
STD+IN	PLS using Extra Features as Extra Inputs to learn Main Task.
STD+OUT	PLS using Extra Features as Extra Outputs in paralleled with Main Task and using Selected Inputs as inputs.

In this chapter, an example of MTL applied to PLS method will be demonstrated. The terms and procedure used is shown in Table 4.3 and

Table 4.4 respectively.

Table 4.4   The procedure for multitask learning using PLS.

Step	Procedure
1	Select the Selected Inputs and Extra Features using the feature selection method in the above sections
2	Train an STD model using Selected Inputs as inputs and only Main Task as outputs
3	Train an STD+IN model using Selected Inputs plus Extra Features as inputs and only Main Task as outputs
4	Train an STD+OUT model using Selected Inputs as inputs and Main Task plus Extra Features as out puts

4.5   Computer Experiments: Feature Selection of Artificially Generated Data Set

In order to demonstrate the details of applications of SVM-BFS, SVM-RFE and MTL methods, it is instructive to use these methods to treat three artificially generated data sets, or toy data sets.

4.5.1   *Toy data sets*

To demonstrate the ability of the feature selection methods, we construct three data sets. For each one, 200 samples and 9 features are created, in which , $x^i (i = 1, 2, 3, 4, 5)$ is uniformly distributed random numbers chosen from a uniform distribution on the interval $[0.0, 1.0]$ generated randomly.

(1) Linear data set: The target can be described as:

$$target = x^1 + x^2 + x^3 + x^4 \tag{4.5}$$

In this data set, the other features are generated as follows, $x^6 = \sin(x^1), x^7 = \log x^2, x^8 = \tan(x^3), x^9 = (x^4)^2$ all features are scaled to the interval of $[-1, 1]$ using an affine transformation.

(2) Quadratic data set: The target can be described as:

$$target = x^1 + x^2 + x^3 x^4 \tag{4.6}$$

In   this   data   set,   the   other   features   are   generated   as   follows,

$x^6 = \sin(x^1), x^7 = \log x^2, x^8 = \tan(x^3), x^9 = (x^4)^2$ all features are scaled to the interval of $[-1,1]$ using an affine transformation.

(3) Nonlinear data set: The target can be described as:

$$\text{target}=x^1 + x^2 + \sqrt{x^3 + x^4} \qquad (4.7)$$

In this data set, the other features are generated as follows, $x^6 = \sin(x^1), x^7 = \log x^2, x^8 = \tan(x^3), x^9 = (x^4)^2$ all features are scaled to the interval of $[-1,1]$ using an affine transformation.

By this way, the nine features defined here can be categorized into three kinds: x^1, x^2, x^3, x^4 are the original features, x^6, x^7, x^8, x^9 are the functions of the four original features, and can be called as functional features, while the feature x^5 is irrelevant to the target values, and therefore can be deleted in all cases. It should be noted that among nine features there are four mutually independent features.

A correct feature selection should satisfy the following three requirements: (1) feature x^5 must be deleted as a noisy feature; (2) the finally selected feature subset must include x^1, x^2, x^3, x^4 or some functions of all these features, i.e., the completeness of the feature set must be assured; (3) the number of features of the finally selected subset should be equal to the number of independent features, i.e., it should be equal to 4.

4.5.2 *SVM-BFS method*

Using the SVM-BFS method, we select the features for the three artificial problems. The selected subset and its corresponding errors of each step are listed in Table 4.5. The SVM used in the feature selection of these three data sets are linear SVM, nonlinear SVM with radial basis function (RBF) kernel $(\sigma = 1.54)$ and SVM with RBF kernel $(\sigma = 1.22)$ respectively.

In Table 4.5, '1' in subset string means the corresponding feature is chosen and '0' means the feature is discarded. From this table, we can see that 1) at the fifth step of SVM-BFS five out of nine features have been deleted, they are x^5 and the functional features in the linear data set and the quadratic data set, while the functional features are chosen in the nonlinear data set, but the feature subset is complete because it also contains 4 independent features. So, it can be concluded that SVM-BFS can select the relevant feature subset correctly.

Table 4.5   The selection process by the SVM-BFS method on the three data sets.

Step	Linear		Quadratic		Nonlinear	
	Subset	Error*	Subset	Error*	Subset	Error*
1	111111101	0.00034	111101111	0.129	111101111	0.071
2	111101101	0.00034	111101110	0.105	101101111	0.055
3	111110100	0.00030	111101100	0.086	101101110	0.043
4	111110000	0.00014	111100100	0.068	100101110	0.033
5	111100000	2.8e-5	111100000	0.060	100100110	0.024

*Here the error denotes the averaged absolute error in LOO cross-validation test.

2) For the quadratic and nonlinear data sets, SVM-BFS can detect the irrelevant feature at the first step, while it cannot detect the irrelevant one for the linear data set, we think the reason is that the nonlinear redundant features can cause larger error than the random feature using the linear SVM.

4.5.3   *SVM-RFE method*

Using the spider package [137], we demonstrate the feature selection process on the three artificial data sets using SVM-RFE.

(1) Linear data set

Table 4.6   The S_i values in selection process by SVM-RFE with linear kernel on the linear data set.

Step	S_1	S_2	S_3	S_4	S_5	S_6	S_7	S_8	S_9	E*
1	0.0778	0.2318	0.1125	0.1584	0	0.0460	0.0003	0.0297	0.0082	5
2	0.0830	0.2270	0.1159	0.1581	-	0.0421	0	0.0267	0.0080	7
3	0.0738	0.2321	0.1087	0.1435	-	0.0308	-	0.0171	0	9
4	0.0224	0.1825	0.0442	0.1899	-	0.0117	-	0	-	8
5	0.0072	0.1629	0.1722	0.1656	-	0	-	-	-	6

*E denotes the label of feature deleted in this step.

For the linear data set, we use SVM with the linear kernel, the process can be seen in Table 4.6, the remaining features ranking by their weights is 3, 4, 2, 1, we can see that the irrelevant feature is eliminated first, then the four functional features are eliminated. So, the selection succeeds.

(2) Quadratic data set

For the quadratic data set, we first use SVM with the Polynomial kernel ($n = 2$), the process can be seen in Table 4.7, the remaining features ranking by their weights is 6, 2, 4, 3, we can see that the irrelevant feature is eliminated firstly, then one original and three functional features are eliminated. The selected subset is complete, so, the selection succeeds.

Table 4.7   The S_i values in selection process using SVM-RFE with Polynomial kernel on the quadratic data set.

Step	S_1	S_2	S_3	S_4	S_5	S_6	S_7	S_8	S_9	E^*
1	0.0199	0.0746	0.0146	0.0171	0	0.0202	0.0389	0.0095	0.0095	5
2	0.0105	0.0665	0.0060	0.0087	-	0.0107	0.0300	0.0002	0	9
3	0.0107	0.0667	0.0068	0.0279	-	0.0110	0.0305	0	-	8
4	0	0.0556	0.0249	0.0282	-	0.0005	0.0219	-	-	1
5	-	0.0328	0.0001	0.0035	-	0.0345	0	-	-	7

* E denotes the label of feature deleted in this step.

Table 4.8   The S_i values in selection process using SVM-RFE with RBF kernel on the quadratic data set.

Step	S_1	S_2	S_3	S_4	S_5	S_6	S_7	S_8	S_9	E^*
1	1.5864	3.9032	1.3252	1.3036	0	1.4438	1.4519	1.4169	1.2649	5
2	0.3526	2.7768	0	0.2631	-	0.2122	0.2078	0.1686	0.0571	3
3	0.1480	2.4052	-	0.3356	-	0	0.1007	0.3958	0.1233	6
4	0.8753	2.0144	-	0.2034	-	-	0	0.1223	0.0636	7
5	0.8731	2.1572	-	0.1659	-	-	-	0.0532	0	9

*E denotes the label of feature deleted in this step.

For the quadratic data set, we also use SVM with the RBF kernel ($\sigma = 1.54$) , the process can be seen in Table 4.8, the remaining features ranking by their weights is 4, 2, 8, 1, we can see that the irrelevant feature is eliminated firstly, then one original and three functional features are eliminated. The selected subset is complete, so, the selection succeeds. From Table 4.7 and 4.8, we find that SVM using different kernels may obtain different feature subset, but the feature subset is complete, so they are both satisfactory.

(3) Nonlinear data set

For the nonlinear data set, we use SVM with the RBF kernel $(\sigma = 1.22)$, the process can be seen in Table 4.9, the remaining features ranking by their weights is 8, 4, 7, 6, we can see that the irrelevant feature is eliminated first, then one original one and three functional features are eliminated, the subset is complete. So, the selection succeeds.

From the above experiments, we can see that the feature selection using support vector machine can efficiently eliminate the irrelevant feature and the redundant features. In the case of nonlinear data set, some features eliminated are not the functional but the original ones. Since the functional features can be also used to form a complete independent feature subset. It can be concluded that the result is also satisfactory.

Table 4.9　　The S_i values in selection process using SVM-RFE with RBF kernel on the nonlinear data set.

Step	S_1	S_2	S_3	S_4	S_5	S_6	S_7	S_8	S_9	E^*
1	0.4099	0.5811	0.4525	0.4638	0	0.4266	0.5365	0.4707	0.3416	5
2	0.0347	0.2008	0.0443	0.1320	-	0.0489	0.1486	0.0736	0	9
3	0	0.0805	0.0487	0.2171	-	0.0112	0.1311	0.0583	-	1
4	-	0.0080	0	0.0645	-	0.0507	0.0844	0.0306	-	3
5	-	0	-	0.2469	-	0.0641	0.1553	0.5119	-	2

*E denotes the label of features deleted in this step.

4.5.4　*Multitask learning*

As mentioned above, three artificially generated data sets have been preprocessed by feature selection. Based on this selection, we have performed multitask learning using the partial least square method with LOO cross-validation on these three data sets. The values of errors computed for different inputs and outputs are listed in Table 4.10.

Table 4.10   Errors of multitask learning using PLS on the three data sets.

Input Subset	Output subset	Linear	Quadratic	Nonlinear
111111111	Target	0.1723	0.1878	0.1191
111101111	Target	0.1710	0.1855	0.1191
000001111	Target	0.2699	0.2858	0.1597
111100000	Target	0.0790	0.1188	0.0918
111100000	Target+000000111*	$<10^{-4}$	0.1125	0.0848

*Target+000000111 means the output target vector contains the original target vector and the last three features of the original feature set, but the error is computed only on the original target vector.

From Table 4.10, it can be seen that the multitask learning improves the accuracy of prediction in this case.

Chapter 5

Principle of Atomic or Molecular Parameter-Data Processing Method

5.1 Two Different Strategies for Structure-Property Relationship Investigation

One of the chief tasks of chemistry is to study the relationships between the atomic or molecular structure and the macroscopic physico-chemical properties of various substances. Up to now, there are two parallel strategies for the solution of this problem. The first one is starting from the first principle to make deduction, namely, to start from quantum mechanics and statistical mechanics, or using molecular dynamics and molecular mechanics of computer simulation based on inter-atomic or inter-molecular potential, to calculate the macroscopic properties of substances. Although these methods are theoretically strict, but up to now these methods cannot be used to calculate the properties directly for most of chemical substances yet. In order to solve most of the practical problems of chemistry, we have to use the second strategy: semi-empirical methods. In these methods, based on the conclusions or concepts of the first principle, including some computational results of quantum chemical calculation, a series of atomic parameters or molecular descriptors have been designed to describe the structure of substances, and then to find the relationships between the experimental data and these atomic parameters or molecular descriptors by data processing. These methods are not very theoretically strict, since it is usually very difficult to know if a set of atomic parameters or molecular descriptors used is complete enough for the description of a macroscopic property exactly. But, anyway, this is a practical and useful strategy.

With the development of computer science and technology, especially the achievement of technique of machine learning, the second strategy has become much more powerful for problem-solving. Because most of the problems of chemical phenomena are multivariate problems, the coordination of atomic parameter and machine learning techniques can provide widely useful, very powerful methods for the computerized modeling and prediction of the physico-chemical properties of substances. For convenience, these methods are called "atomic parameter-data processing methods" or "atomic parameter-pattern recognition methods".

For organic compounds with complicated molecular configurations, the use of atomic parameters is not enough for the description of the structure of molecules. Organic compounds with the same kind and same number of atoms may be different substances. The first reason is the existence of constitutional isomers (for example, the isomers of hydrocarbon molecules usually have different properties. This is very important for petrochemical industry), and the second reason is the existence of stereo-isomerism (for example, most of biologically active molecules are chiral. It means that the enantiomers of them show marked differences in their properties. For example, their smell or taste, and even solubilities in chiral solvents may be completely different. This is very important in biochemistry and medical chemistry). The third reason is that it is usually necessary to describe the conformation of molecules, since the change of conformation may be important for some biochemical reactions. For example, the conformations of some proteins are known to be changed in the reaction with some receptors. If we want to have some molecular descriptors for QSPR/QSAR computation, the most convenient way is to design a series of numbers or parameters used as the independent variables for regression work. In order to meet these requirements, some topological indices are designed to differentiate the structure of constitutional isomers, i.e., the isomerism resulting from differences in vicinity relationships between the atoms in molecules. Although such kind of topological indices have good ability to describe many properties of molecules, they cannot differentiate enantiomers or molecules with different conformations. For this reason, a series of

3-dimentional molecular descriptors have been proposed. The 3-dimensional molecular design is very useful for drug design work.

According to the types of chemical bonds of substances, chemical substances can be roughly classified into three categories: substances with ionic bond, covalent bond or metallic bond. Substances with different chemical bond should be described by different sets of atomic parameters or molecular descriptors. In this chapter, we firstly describe various commonly used atomic parameters and molecular descriptors, and then discuss the suitable parameter sets for the substances with different chemical bonds.

5.2   Number of Valence Electrons of Atoms

It is well-known that most of chemical phenomena are directly related to the valence electrons of atoms, and the electrons in inner shells of atoms usually play a minor or indirect role in chemical phenomena, so it is easy to understand that the number of valence electrons should be one of the most important atomic parameters for the description of the physico-chemical properties of chemical substances. We will use Z to denote this atomic parameter in this book.

The atoms of many elements have only unique number of Z, because their potential level differences between the valence electrons and inner shell electrons are very large, so the inner shell electrons are normally impossible to take part in common chemical phenomena. But there are still other elements for which the potential levels between the valence electrons and some inner electrons are relatively close, so that these elements can have different valence numbers. For the compounds with definite valence numbers, we have to select this atomic parameter according to their valency. A more complicate problem happens for substances with metallic bond, because there is no definite valence number for these substances. In some cases, it seems reasonable to use the lower number of Z. For example, we can use $Z=2$ for lead or tin, and use $Z=3$ for bismuth, etc. A more complicated problem happens for the intermetallic compounds of transition metals, rare earth metals and actinides, since the d electrons (sometimes even the f electrons) and vacant d orbitals of the next outermost shells also take part in the metallic bond formation. In these cases, we usually use the number of d electrons as an additional atomic parameter in computation.

For ionic compounds, the ionic charge number is a very important atomic parameter. It is usually related or equal to the number of valence electrons. In these cases, we usually denote the charge number of cations or anions by z_+ or z_- respectively, instead of Z for computation work in data processing.

5.3   Ionization Potential of Atoms

When an atom is in isolated state, the energy levels of its electrons are the characteristic values for this atom. When this atom combines with other atoms to form molecules or condensed states, the electronic energy levels will be changed. And that of valence electrons changes more evidently. For this reason, we cannot use the data of ionization potential directly for the computation of the physico-chemical properties of chemical substances. But the values of the ionization potentials of isolated atoms should be eventually correlated to the energy levels in molecules or condensed states in some complicated manner. So the values of ionization potential of isolated atoms of elements should also be a useful atomic parameter in atomic parameter pattern recognition method.

The values of the ionization potential of isolated atoms are listed in Table 5.1. In our computation, it is usually denoted by I.

Table 5.1   Ionization potential of atoms (ev).

	I	II	III	IV	V	VI	VII	VIII
H	13.595							
He	24.581	54.403						
Li	5.390	75.619						
Be	9.320	18.206	153.850					
B	8.296	25.149	37.920	259.298				

C	11.256	24.376	47.871	64.476	391.986			
N	14.53	29.593	47.426	77.450	97.863	551.925		
O	13.614	35.108	54.886	77.394	113.873	138.080	739.114	
F	17.418	34.98	62.646	87.14	114.214	157.117	185.139	953.6
Na	5.138	47.29						
Mg	7.644	15.031	80.12					
Al	5.984	18.823	28.44	119.96				
Si	8.149	16.34	33.46	45.13	166.73			
P	10.484	19.72	30.156	51.354	65.007	220.414		
S	10.357	23.4	35.0	47.29	72.5	88.029	280.99	
Cl	13.01	23.80	33.9	53.5	67.80	96.79	114.27	348.3
K	4.339	31.81						
Ca	6.111	11.868	51.21					
Sc	6.54	12.80	24.75	73.9				
Ti	6.82	13.57	27.47	43.24	99.8			
V	6.74	14.65	29.31	48	65	129		
Cr	6.764	16.49	30.95	50	73	91	161	
Mn	7.432	15.636	33.69	-	76	-	119	196
Fe	7.87	16.18	30.643	-				
Co	7.86	17.05	33.49					
Ni	7.633	18.15	35.16					
Cu	7.724	20.29	36.83					
Zn	9.391	17.96	39.70					
Ga	6.00	20.51	30.70	64.2				
Ge	7.88	15.93	34.21	45.7	93.4			
As	9.81	18.63	28.34	50.1	62.6	127.5		
Se	9.75	21.5	32	43	68	82	155	
Br	11.84	21.6	35.9	47.3	59.7	88.6	103	193
Rb	4.176	27.5						

Sr	5.692	11.027	-	57		
Nb	6.88	14.32	25.04	38.3	50	103
Mo	7.10	16.15	27.13	46.4	61.2	68	126
Ru	7.364	16.76	28.46			
Rh	7.46	18.07	31.05			
Pd	8.33	19.42	32.92			
Ag	7.574	21.48				
Cd	8.991	16.904	37.47			
In	5.785	18.86	28.03	54.4		
Sn	7.342	14.628	30.49	40.72	72.3	
Sb	8.639	16.5	25.3	44.1	56	108
Te	9.01	18.6	31	38	60	72	137
Cs	3.893	25.1				
Ba	5.210	10.001				
La	5.61	11.43	19.17			
Au	9.22	20.5				
Hg	10.43	18.751	34.2			
Tl	6.106	20.42	29.8	50.7		
Pb	7.415	15.028	31.93	42.31	68.8	
Bi	7.287	16.68	25.56	45.3	56.0	88.3

5.4 Atomic Radii and Ionic Radii

Since electrons, as a kind of Fermi particles, obey Pauli exclusion principle, the overlap of the orbitals of outer electrons when two atoms approach each other induces a strong short range repulsion. For this reason, we can consider an atom or monoatomic ion as a spherical globe having definite radius in some kind of environments. This is of course only an approximation, since the inter-atomic charge transfer surely changes the size of atoms significantly. But we cannot consider this value as a variable in atomic parameter method; otherwise we cannot make our computation. Fortunately in our data processing work, the meaning of a parameter is different from the meaning of a constant in the method based on first principles. In data processing methods, any number relevant to the target of research can be used as one of the parameters in data processing to reveal the regularities. For example, we can use them as independent variables in regression method, or as features in pattern recognition method. In practice, it is found that atomic radius or ionic radius is a very useful atomic parameter. In this book, it is denoted by R.

Since the atomic sizes are different in the substances with different kinds of chemical bond. It is necessary to assign the atoms of elements three kinds of system of atomic radius. In ionic systems, we call the size of ions as ionic radius. In covalent compounds, system of covalent radius is used. And the atomic radius in metallic systems is called metallic radius of elements. In his classical work, Pauling has assigned the values of these three kinds of radius. Pauling's values of atomic or ionic radii have been widely used till now. But there are still other systems proposed by other authors in later years.

One of the reasons to use other systems is that the systems proposed by Pauling were not complete. The data of many elements such as rare earth elements and actinide elements were absent. The description of the chief systems of atomic or ionic radius will be given as follows:

(1) Chief systems for ionic radius

Table 5.2   Ionic radius of elements (Pauling scale).

				H^-	Li^+	Be^{2+}	B^{3+}	C^{4+}	N^{5+}	O^{6+}	F^{7+}					
				2.08	0.60	0.31	0.20	0.15	0.11	0.09	0.07					
C^{4-}	N^{3-}	O^{2-}	F^-	Na^+	Mg^{2-}	Al^{3+}	Si^{4+}	P^{3+}	S^{6+}	Cl^{7+}						
2.60	1.71	1.40	1.36	0.95	0.65	0.50	0.41	0.34	0.29	0.26						
Si^{4-}	P^{3-}	S^{2-}	Cl^-	K^+	Ca^{2+}	Sc^{3+}	Ge^{4+}	As^{5+}	Se^{6+}	Br^{7+}	Ti^{4+}	V^{5+}	Cr^{6+}	Cu^+	Zn^{2+}	Ga^{3+}
2.71	2.12	1.84	1.81	1.33	0.99	0.81	0.53	0.47	0.42	0.39	0.68	0.59	0.52	0.96	0.74	0.62
Ge^{4-}	As^{3-}	Se^{2-}	Br^-	Rb^+	Sr^{2+}	Y^{3+}	Sn^{4+}	Sb^{5+}	Te^{6+}	I^{7+}	Zr^{4+}	Nb^{5+}	Mo^{6-}	Ag^+	Cd^{2+}	In^{3+}
2.72	2.22	1.98	1.95	1.48	1.13	0.93	0.71	0.62	0.56	0.50	0.80	0.70	0.62	1.26	0.97	0.81
Sn^{4-}	Sb^{3-}	Te^{2-}	I^-	Cs^+	Ba^{2+}	La^{3+}	Pb^{4+}	Bi^{5+}						Au^+	Hg^{2+}	Tl^{3+}
2.94	2.45	2.21	2.16	1.69	1.35	1.15	0.84	0.74						1.37	1.10	0.95

*The unit of atomic or ionic radius in this chapter is Å.

Table 5.3 Ionic radius proposed by Sandersen, Bokii,Ahrens and Stockar*.

	1a	2a	3a	4a	5a	6a	7a	8a			1b	2b	3b	4b	5b	6b
II	Li	Be											B	C	N	
	0.68	0.47											0.41	0.34	0.31	
	0.68	0.34											(0.20)	0.20	0.15	
	0.68	0.35											0.23	0.16	0.13	
	0.50	0.32											0.21	0.15	0.11	
III	Na	Mg											Al	Si	P	S
	0.98	0.74											0.62	0.53	0.48	0.43
	0.98	0.74											0.57	0.39	0.35	0.29
	0.97	0.66											0.51	0.42	0.35	0.30
	0.86	0.65											0.51	0.41	0.34	0.29
IV	K	Ca	Sc	Ti	V	Cr	Mn	Fe +2 +3	Co +2+3	Ni +2 +3	Cu	Zn	Ga	Ge	As	Se
	1.34	1.00	0.79	0.71	0.62	0.59	0.56	0.81,0.73	0.80, 0.72	0.80, 0.72	1.03	0.86	0.75	0.68	0.66	0.60
	1.33	1.04	0.83	0.64	0.40	0.35	0.46	0.80, 0.67	0.78, 0.64	0.74, -	0.98	0.83	0.62	0.44	0.47	0.35
	1.33	0.99	0.81	0.68	0.59	0.52	0.46	0.74, 0.64	0.72, 0.63	0.69, -	0.96	0.74	0.62	0.53	0.46	0.42
	1.22	0.99	0.81	0.68	0.59	0.52	0.46	0.84, 0.72	0.82, 0.70	0.79, -	0.90	0.74	0.62	0.54	0.48	0.43
V	Rb	Sr	Y	Zr	Nb	Mo	Tc	Ru +2 +4	Rh +3+4	Pd +2 +3	Ag	Cd	In	Sn	Sb	Te
	1.46	1.09	0.88	0.77	0.68	0.65	-	0.86, 0.71	0.78, 0.72	0.88, 0.73	1.14	0.96	0.85	0.77	0.73	0.68
	1.49	1.20	0.97	0.82	0.66	0.65		-, 0.62	0.75, 0.65	-, 0.67	1.13	0.99	0.92	0.67	0.62	0.56
	1.47	1.12	0.92	0.79	0.69	0.62		-, 0.67	0.68, -	0.80, 0.65	1.26	0.97	0.81	0.71	0.62	0.56
	1.40	1.15	0.95	0.80	0.70	0.61		-, 0.76	0.87, -	1.01, 0.74	1.18	0.97	0.81	0.70	0.62	0.55
VI	Cs	Ba	La	Hf	Ta	W	Re	Os +2 +4	Ir +2 +4	Pt +2 +4	Au	Hg	Tl	Pb	Bi	-
	1.66	1.23	0.95	0.82	0.73	0.70	0.67	0.91, 0 .76	0.91, 0.76	0.92, 0.78	1.16	1.00	0.90	0.78	0.78	
	1.65	1.38	1.04	0.82	0.66	0.65	-	-, 0.65	-, 0.65	-, 0.64	(1.37)	1.12	1.05	0.76	0.74	
	1.67	1.34	1.14	0.78	0.68	0.62	0.57	-, 0.69	-, 0.68	0.80, 0.65	1.37	1.10	0.95	0.84	0.74	
	1.61	1.37	1.16	0.91	0.80	0.72	-	-, 0.88	-, 0.87	1.12 ,0.87	1.29	1.10	0.95	0.84	0.74	

Th	Pa	U	Np	Pu	Am
+4	+4 +5	+3 +4 +5 +6	+3 +4 +5 +6	+3 +4 +5 +6	+3 +4 +5 +6
0.90	0.89,0.84	0.96,0.89,0.83,0.79	0.96,0.88,0.83,0.78	0.95,0.88,0.82,0.78	0.95,0.88,0.82,0.77
0.99	0.96,0.90	1.03,0.93,0.87,0.83	1.01,0.92,0.88,0.82	1.00,0.90,0.87,0.81	0.99,0.89,0.86,0.80

*The values proposed by Sandersen, Bokii, Ahrens and Stockar are listed in the first, second, third and fourth rows respectively. The second row for actinides is given by Zachariasen.

Compared with some other atomic parameters, the values of ionic radii have more definite physical meaning. Although the ionic radius changes with coordination number, the change is not very large.

Table 5.2 lists the values of ionic radii proposed by Pauling [107]. It is a classical scale for ionic radius. In early years, geochemist Goldschmidt also proposed a system for ionic radii. Since the values proposed by Goldschmidt were very similar with those proposed by Pauling, it is not necessary to list them in this book.

Some relatively influential systems of ionic radius of elements are listed in Table 5.3 and Table 5.4 [118]. The ionic radius systems proposed by Quill are more complete for more elements.

Table 5.4   Ionic radius proposed by Quill.

Ion	Radius	Ion	Radius	Ion	Radius
Li^+	0.71	V^{2+}	(0.82)	Cd^+	(1.30)
Na^+	0.98	V^{3+}	0.69	Cd^{2+}	0.96
K^+	1.33	Nb^{2+}	(0.85)	Hg^+	1.50
Rb^+	1.47	Nb^{3+}	(0.78)	Hg^{2+}	1.10
Cs^+	1.74	Ta^{2+}	(0.88)	Ga^+	(1.08)
Be^{2+}	0.38	Cr^{2+}	(0.80)	Ga^{2+}	(0.76)
Mg^{2+}	0.66	Cr^{3+}	0.62	Ga^{3+}	0.60
Ca^{2+}	0.99	Mo^{2+}	(0.83)	In^+	1.40
Sr^{2+}	1.15	W^{2+}	(0.87)	In^{2+}	(1.02)
Ba^{2+}	1.37	Mn^{2+}	0.78	In^{3+}	0.81
Ra^{2+}	1.50	Mn^{3+}	0.66	Tl^+	1.59
B^{3+}	0.25	Re^{2+}	(0.86)	Tl^{2+}	(1.19)

Ion	Value	Ion	Value	Ion	Value
Al^{3+}	0.52	Re^{3+}	(0.78)	Tl^{3+}	0.95
Sc^{3+}	0.81	Fe^{2+}	0.76	Ge^{2+}	(0.84)
Y^{3+}	0.96	Fe^{3+}	0.64	Sn^{2+}	1.10
La^{3+}	1.16	Co^{2+}	0.74	Sn^{4+}	0.71
Ce^{2+}	(1.20)	Co^{3+}	0.63	Pb^{2+}	1.27
Ce^{3+}	1.14	Ni^{2+}	0.73	P^{3+}	(0.55)
Ce^{4+}	1.01	Ru^{2+}	(0.81)	P^{5+}	0.36
Pr^{3+}	1.12	Ru^{3+}	(0.72)	As^{3+}	0.69
Pr^{4+}	0.99	Rh^{+}	(1.07)	As^{5+}	0.46
Nd^{3+}	1.10	Rh^{2+}	(0.80)	Sb^{3+}	0.92
Sm^{2+}	(1.16)	Rh^{3+}	0.72	Sb^{5+}	0.63
Sm^{3+}	1.07	Pd^{2+}	(0.85)	Bi^{+}	1.70
Eu^{2+}	(1.14)	Pd^{3+}	(0.74)	Bi^{2+}	(1.32)
Eu^{3+}	1.05	Os^{2+}	(0.88)	Bi^{3+}	1.08
Gd^{3+}	1.03	Os^{3+}	(0.78)	Bi^{4+}	(0.88)
Tb^{3+}	1.02	Ir^{+}	(1.16)	Bi^{5+}	0.75
Tb^{4+}	0.91	Ir^{2+}	(0.92)	H^{-}	2.12
Dy^{3+}	1.00	Ir^{3+}	(0.80)	F^{-}	1.34
Ho^{3+}	0.99	Pt^{+}	(1.24)	Cl^{-}	1.80
Er^{3+}	0.98	Pt^{2+}	(0.96)	Br^{-}	1.90
Tm^{3+}	0.96	Pt^{3+}	(0.83)	I^{-}	2.23
Yb^{2+}	(1.06)	Cu^{+}	0.93	O^{2-}	1.35
Yb^{3+}	0.95	Cu^{2+}	(0.72)	S^{2-}	1.84
Lu^{3+}	0.93	Ag^{+}	1.21		
Ac^{3+}	1.27	Ag^{2+}	(0.91)		
Ti^{2+}	(0.85)	Au^{+}	1.37		
Ti^{3+}	0.64	Au^{2+}	(1.02)		
Zr^{2+}	(0.9)	Au^{3+}	(0.87)		
Zr^{3+}	(0.82)	Zn^{2+}	0.72		

In more recent years, Shannon and Preweitt have proposed two series of ionic radius: the "effective ionic radii" based on the assumption of oxide anion with coordination number 6 is 1.40 Å, and the "crystal radii" based on the assumption of fluoride anion with coordination number 6 is equal to 1.19Å. They are more accurate because of using more accurate crystallographic data. The values of effective ionic radii are listed in Table 5.5 [124; 125].

Table 5.5   Ionic radius proposed by Shannon and Preweitt (Å).

Ion	N	Radius	Ion	N	Radius	Ion	N	Radius	Ion	N	Radius
Li^+	4	0.59	Cr^{4+}	6	0.55	Rh^{3+}	6	0.665	Yb^{2+}	6	1.02
Li^+	6	0.76	Mn^{2+}	4	0.66	Rh^{4+}	6	0.60	Yb^{3+}	6	0.868
Be^{2+}	4	0.27	Mn^{2+}	6	0.830	Pd^{2+}	6	0.86	Lu^{3+}	6	0.861
Be^{2+}	6	0.45	Mn^{3+}	6	0.645	Pd^{4+}	6	0.615	Hf^{4+}	4	0.58
B^{3+}	4	0.11	Mn^{4+}	6	0.530	Ag^+	4	1.00	Hf^{4+}	6	0.71
B^{3+}	6	0.27	Fe^{2+}	4	0.63	Ag^+	6	1.15	Ta^{3+}	6	0.72
C^{4+}	4	0.15	Fe^{2+}	6	0.78	Ag^{2+}	6	0.94	Ta^{4+}	6	0.68
C^{4+}	6	0.16	Fe^{3+}	4	0.49	Ag^{3+}	6	0.75	Ta^{5+}	6	0.64
N^{-3}	4	1.46	Fe^{3+}	6	0.645	Cd^{2+}	4	0.78	W^{4+}	6	0.66
O^{2-}	4	1.38	Co^{2+}	4	0.58	Cd^{2+}	6	0.95	Re^{4+}	6	0.63
O^{2-}	6	1.40	Co^{2+}	6	0.745	In^{3+}	4	0.62	Os^{4+}	6	0.630
F^-	4	1.31	Co^{3+}	6	0.61	In^{3+}	6	0.800	Ir^{3+}	6	0.68
F^-	6	1.33	Ni^{2+}	4	0.55	Sn^{4+}	4	0.55	Pt^{2+}	6	0.80
Na^+	4	0.99	Ni^{2+}	6	0.69	Sn^{4+}	6	0.690	Pt^{4+}	6	0.625
Na^+	6	1.02	Ni^{3+}	6	0.60	Sb^{3+}	6	0.76	Au^+	6	1.37
Na^+	8	1.18	Cu^+	4	0.60	Sb^{5+}	6	0.60	Au^{3+}	4	0.68
Na^+	12	1.39	Cu^+	6	0.77	Te^{2-}	6	2.21	Au^{3+}	6	0.85
Mg^{2-}	4	0.57	Cu^{2+}	4	0.57	I^-	6	2.20	Hg^{2+}	4	0.96
Mg^{2-}	6	0.72	Cu^{2+}	6	0.73	Cs^+	6	1.67	Hg^{2+}	6	1.02
Al^{3+}	4	0.39	Zn^{2+}	4	0.60	Cs^+	8	1.74	Hg^{2+}	8	1.14
Al^{3+}	6	0.535	Zn^{2+}	6	0.74	Cs^+	12	1.88	Tl^+	6	1.50

Ion	N	Radius	Ion	N	Radius	Ion	N	Radius	Ion	N	Radius
Si^{4+}	4	0.26	Ga^{3+}	4	0.47	Ba^{2+}	6	1.35	Tl^{+}	12	1.70
Si^{4+}	6	0.400	Ga^{3+}	6	0.620	Ba^{2+}	8	1.42	Tl^{3+}	4	0.75
P^{3+}	6	0.44	Ge^{4+}	4	0.390	Ba^{2+}	12	1.61	Tl^{3+}	6	0.885
P^{5+}	4	0.17	Ge^{4+}	6	0.530	La^{3+}	6	1.032	Pb^{2+}	6	1.19
P^{5+}	6	0.38	As^{3+}	6	0.58	La^{3+}	8	1.160	Pb^{2+}	8	1.29
S^{-2}	6	1.84	As^{5+}	4	0.46	La^{3+}	12	1.36	Pb^{2+}	12	1.49
Cl^{-}	6	1.81	Se^{2-}	6	1.98	Ce^{3+}	6	1.01	Pb^{4+}	6	0.775
K^{+}	4	1.37	Br^{-}	6	1.96	Ce^{4+}	6	0.87	Bi^{3+}	6	1.03
K^{+}	6	1.38	Rb^{+}	6	1.52	Pr^{3+}	6	0.99	Bi^{5+}	6	0.76
K^{+}	8	1.51	Rb^{+}	8	1.61	Pr^{4+}	6	0.85	Po^{4+}	6	0.94
K^{+}	12	1.64	Rb^{+}	12	1.72	Nd^{3+}	6	0.983	At^{7+}	6	0.62
Ca^{2+}	6	1.00	Sr^{2+}	6	1.18	Nd^{3+}	8	1.109	Th^{4+}	6	0.94
Ca^{2+}	8	1.12	Sr^{2+}	8	1.26	Nd^{3+}	12	1.27	Th^{4+}	8	1.05
Ca^{2+}	12	1.34	Sr^{2+}	12	1.44	Pm^{2+}	6	0.97	Th^{4+}	12	1.21
Sc^{3+}	6	0.745	Y^{3+}	6	0.900	Sm^{3+}	6	0.958	Pa^{4+}	6	0.90
Sc^{3+}	8	0.870	Y^{3+}	8	1.019	Eu^{2+}	6	1.17	U^{3+}	6	1.025
Ti^{2+}	6	0.86	Zr^{4+}	4	0.59	Eu^{2+}	8	1.25	U^{4+}	6	0.89
Ti^{3+}	6	0.670	Zr^{4+}	6	0.72	Eu^{3+}	6	0.947	U^{4+}	12	1.17
Ti^{4+}	4	0.42	Nb^{3+}	6	0.72	Eu^{3+}	8	1.066	Np^{4+}	6	0.87
Ti^{4+}	6	0.605	Nb^{4+}	6	0.68	Gd^{3+}	6	0.938	Pu^{4+}	6	0.86
V^{2+}	6	0.79	Mo^{3+}	6	0.69	Tb^{3+}	6	0.923	Am^{3+}	6	0.975
V^{3+}	6	0.640	Mo^{4+}	6	0.650	Dy^{3+}	6	0.912	Am^{4+}	6	0.85
V^{4+}	6	0.58	Tc^{4+}	6	0.645	Ho^{3+}	6	0.901	Cm^{3+}	6	0.97
Cr^{2+}	6	0.80	Ru^{3+}	6	0.68	Er^{3+}	6	0.890	Cm^{4+}	6	0.85
Cr^{3+}	6	0.615	Ru^{4+}	6	0.62	Tm^{3+}	6	0.880	Bk^{4+}	6	0.83
									Cf^{4+}	6	0.821

[*]N denotes the coordination number of ions.

Although there are some little differences between the values of ionic radii proposed by different authors, it is not very important in chemical data processing work so far as the calculation of the detailed crystal structure is not concerned.

For most of typical ionic compounds, the values of ionic radius are relatively insensitive to their environments, so that it can be used for the investigation of geometry of crystalline solids or ionic liquids. Owing to this reason, the difference of ionic radii, or the radius ratio, R_1/R_2 is also very useful atomic parameter in QSPR studies.

It is generally recognized that partial covalency has shortening effect on the chemical bond. Therefore sometimes it is necessary to use the electronegativity difference as another atomic parameter, together with ionic radius in data processing.

(2) Systems of covalent radius

Table 5.6 lists the values of covalent radii proposed by Sanderson.

Table 5.6   Covalent radius of elements (Å).

Element	Covalent radius	Element	Covalent radius	Element	Covalent radius
H	0.37	Br	1.14	Tm	1.55
He	0.93	Kr	1.89	Yb	1.70
Li	1.32	Rb	2.15	Lu	1.55
Be	0.96	Sr	1.91	Hf	1.44
B	0.84	Y	1.61	Ta	1.32
C	0.77	Zr	1.45	W	1.30
N	0.74	Nb	1.32	Re	1.25
O	0.74	Mo	1.32	Os	1.27
F	0.72	Tc	1.20	Ir	1.28
Ne	1.31	Ru	1.26	Pt	1.30
Na	1.58	Rh	1.27	Au	1.43
Mg	1.38	Pd	1.30	Hg	1.45
Al	1.26	Ag	1.42	Tl	1.50
Si	1.17	Cd	1.43	Pb	1.50
P	1.10	In	1.46	Bi	1.49
S	1.04	Sn	1.40	Rn	2.14
Cl	1.00	Sb	1.41	Fr	2.46
Ar	1.74	Te	1.37	Ra	2.35

K	2.02	I	1.33	Ac	1.79
Ca	1.74	Xe	2.09	Th	1.58
Sc	1.44	Cs	2.33	PaIV	1.64
Ti	1.32	Ba	1.97	PaV	1.52
V	1.21	La	1.68	UIV	1.62
Cr	1.19	Ce	1.62	UV	1.50
Mn	1.19	Pr	1.62	UVI	1.42
Fe	1.20	Nd	1.62	NpIV	1.60
Co	1.18	Pm	1.67	NpV	1.49
Ni	1.17	Sm	1.64	NpVI	1.41
Cu	1.25	Eu	1.62	PuIV	1.58
Zn	1.27	Gd	1.60	PuV	1.48
Ga	1.25	Tb	1.59	PuVI	1.40
Ge	1.22	Dy	1.58	AmIV	1.57
As	1.22	Ho	1.57	AmV	1.47
Se	1.17	Er	1.56	AmVI	1.39

Since the length of covalent bond changes with its partial ionicity, it is necessary to use electronegativity difference as another atomic parameter, together with covalent radii for the data processing work if the ionic nature of the chemical bond is obvious.

(3) Systems of metallic radius

Table 5.7 lists the values of metallic radii proposed by Pauling.

Table 5.7   Metallic radius of elements (Pauling scale).

Li	Be													
1.58	1.12													
Na	Mg	Al												
1.92	1.60	1.43												
K	Ca	Sc	Ti	V	Cr	Mn	Fe	Co	Ni	Cu	Zn	Ga	Ge	
2.38	1.97	1.66	1.47	1.35	1.29	1.37	1.26	1.25	1.25	1.28	1.37	1.53	1.39	
Rb	Sr	Y	Zr	Nb	Mo	Tc	Ru	Rh	Pd	Ag	Cd	In	Sn	Sb
2.53	2.15	1.82	1.60	1.47	1.40	1.35	1.34	1.34	1.37	1.44	1.52	1.67	1.58	1.61
Cs	Ba	La	Hf	Ta	W	Re	Os	Ir	Pt	Au	Hg	Tl	Pb	Bi
2.72	2.24	1.82	1.59	1.47	1.41	1.37	1.35	1.36	1.39	1.44	1.55	1.71	1.75	1.82

[*]The unit of atomic or ionic radius described in this chapter is Å.

A more complete table of metallic radii was given by Teatum and his coworkers. It is shown in Table 5.8.

Table 5.8　　Metallic radius of elements (Å).

Element	Valency	Metallic radius	Element	Valency	Metallic radius
H	-1	0.78	Sb	5	1.59
Li	1	1.562	Te	6	1.60
Be	2	1.128	Cs	1	2.731
B	3	0.98	Ba	2	2.243
C	4	0.916	La	3	1.877
N	5	0.88	α-Ce	3.6	1.715
O	6	0.89	γ-Ce	3.1	1.825
Na	1	1.911	Pr	3	1.828
Mg	2	1.602	Nd	3	1.821
Al	3	1.432	Pm	3	1.810
Si	4	1.319	Sm	2.9	1.802
P	-3	1.28	Eu	2.1	2.042
S	-2	1.27	Eu	3	1.799
K	1	2.376	Gd	3	1.802
Ca	2	1.974	Tb	3	1.782
Sc	3	1.641	Dy	3	1.773
Ti	4	1.462	Ho	3	1.766
V	5	1.346	Er	3	1.757
Cr	3	1.360	Tm	3	1.746
Cr	6	1.282	Yb	2	1.940
Mn	4	1.304	Yb	3	1.740
Mn	6	1.264	Lu	3	1.734
Fe	6	1.274	Hf	4	1.580
Co	6	1.252	Ta	5	1.467
Ni	6	1.246	W	6	1.408
Cu	1	1.278	Re	6	1.375
Zn	2	1.394	Os	6	1.353
Ga	3	1.411	Ir	6	1.357
Ge	4	1.369	Pt	6	1.387
As	5	1.39	Au	1	1.442

Se	6	1.40	Hg	2	1.573
Rb	1	2.546	Tl	3	1.716
Sr	2	2.151	Pb	4	1.750
Y	3	1.801	Bi	5	1.70
Zr	4	1.602	Po	6	1.76
Nb	5	1.468	Fr	1	2.80
Mo	6	1.400	Ra	2	2.26
Tc	6	1.360	Ac	3	1.878
Ru	6	1.339	Th	4	1.798
Rh	6	1.345	Pa	5	1.63
Pd	6	1.376	U	6	1.56
Ag	1	1.445	Np	6	1.555
Cd	2	1.568	α-Pu	5.0	1.58
In	3	1.663	δ-Pu	4.5	1.644
Sn	2	1.623	ε-Pu	4.8	1.591
Sn	4	1.545	Am	3	1.81

Strictly speaking, the values of atomic radius listed above can only be used to describe the size of atoms in the metallic state of pure elements, because the real atomic radius of metallic state are rather flexible to change significantly due to charge transfer between different kinds of atoms in intermetallic compounds or alloy systems. But even so these values are still very useful atomic parameters if we use them together with the difference of electronegativities for the describing of the physico-chemical phenomena related to intermetallic compounds or alloy systems.

(4) Thermochemical radius of nonspherical ions

Many ionic systems contain nonspherical ions. For example, anions SO_4^{2-}, CrO_4^{2-}, MoO_4^{2-}, WO_4^{2-} have tetrahedral structure, and anions CO_3^{2-}, NO_3^- have triangular structure. It is useful to define some concept of "effective radii" together with an effective parameter describing the degree of nonsphericity to describe these ions. Yashimirskii has proposed a system of so-called "thermochemical radius" for this purpose. He defined thermochemical ionic radius as follows: "thermochemical radius is a hypothetical radius of nonspherical ions, by this value a spherical ion will give the same lattice energy of the crystal lattice formed by this nonspherical ion." Roughly speaking, the thermochemical radius is equal to the radius of a hypothetical ion having the same volume of the nonspherical ion.

At the same time, Yashimirskii defined a nonsphericity parameter. It is the ratio of the thermochemical radius to the longest distance from the center to most remote border of the ion. For tetrahedral ions, this value is approximately equal to 0.8, and for the triangular ion it is nearly equal to 0.67.

Table 5.9 lists the thermochemical radii of ions proposed by Yashimirskii [143].

Table 5.9　　Thermochemical radius of ions (Å).

Ions	Thermochemical	Ions	Thermochemical
CN^-	1.82	HSO_4^-	2.06
CO_3^{2-}	1.85	SH^-	1.95
CNS^-	1.95	SO_4^{2-}	2.30
NO_2^-	1.55	CrO_4^{2-}	2.40
CNO^-	1.59	SeO_4^{2-}	2.43
IO_3^-	1.82	MoO_4^{2-}	2.54
NO_3^-	1.89	WO_4^{2-}	2.57
BrO_3^-	1.91	PO_4^{3-}	2.38
ClO_3^-	2.00	AsO_4^{3-}	2.48
ClO_4^-	2.36	SbO_4^{3-}	2.60
MnO_4^-	2.40	SiO_4^{4-}	2.40
OH^-	1.40	NH_4^+	1.43
HCO_3^-	1.63		

When the nonspherical ions are involved in the computation work of data processing, the nonsphericity parameter defined by Yashimirskii should be used together with the thermochemical radii in data processing.

5.5　Electronegativity

Electronegativity is another very useful atomic parameter [107; 21], although its clear physical meaning cannot be defined very strictly.

Roughly speaking, the electronegativity difference of two elements denotes the degree of charge transfer or electron transfer between two kinds of atoms in a chemical bond. Table 5.10 lists the electronegativity scale of elements proposed by Pauling.

In Pauling's original scale, the influence of valence was not considered. Basanov proposed a modified electronegativity scale in which some elements with different valency had different values of electronegativity. Table 5.11 lists the values of electronegativity of elements proposed by Basanov [10].

To explain the thermodynamic properties of metallic systems, Miedema proposed his electronegativity scale denoted by ϕ. The values of Miedema's electronegativity for metallic systems are listed in Table 5.12.

For organic molecules or polyatomic ions, the elctronegativity scale for atomic groups is sometimes useful. Table 5.13 lists the electronegativity of some atomic groups.

The parameter electronegativity is usually denoted by X.

Table 5.10   Pauling scale of electronegativity of elements.

Li	Be	B											C	N	O	F
1.0	1.5	2.0											2.5	3.0	3.5	4.0
Na	Mg	Al											Si	P	S	Cl
0.9	1.2	1.5											1.8	2.1	2.5	3.0
K	Ca	Sc	Ti	V	Cr	Mn	Fe	Co	Ni	Cu	Zn	Ga	Ge	As	Se	Br
0.8	1.0	1.3	1.5	1.6	1.6	1.5	1.8	1.8	1.8	1.9	1.6	1.6	1.8	2.0	2.4	2.8
Rb	Sr	Y	Zr	Nb	Mo	Tc	Ru	Rh	Pd	Ag	Cd	In	Sn	Sb	Te	I
0.8	1.0	1.2	1.4	1.6	1.8	1.9	2.2	2.2	2.2	1.9	1.7	1.7	1.8	1.9	2.1	2.5
Cs	Ba	La-Lu	Hf	Ta	W	Re	Os	Ir	Pt	Au	Hg	Tl	Pb	Bi	Po	At
0.7	0.9	1.1-1.2	1.3	1.5	1.7	1.9	2.2	2.2	2.2	2.4	1.9	1.8	1.8	1.9	2.0	2.2
Fr	Ra	Ac	Th	Pa	U	Np-No										
0.7	0.9	1.1	1.3	1.5	1.7	1.3										

Table 5.11   Basanov scale of electronegativity of elements.

																	H
																	2.15
Li	Be											B	C	N	O	F	
0.95	1.5											2.0	2.6	3	3.5	3.9	
Na	Mg											Al	Si	P	S	Cl	
0.9	1.2											1.5	1.9	2.1	2.6	3.1	
K	Ca	Sc	Ti	V	Cr	Mn	Fe	Co	Ni	Cu	Zn	Ga	Ge	As	Se	Br	
0.8	1.0	1.3	(2)1.1 (3)1.3 (4)1.6	(3)1.4 (4)1.7 (5)1.9	(2)1.4 (3)1.6 (4)2.2 (6)2.4	(2)1.4 (3)1.5 (4)2.1 (7)2.5	(2)1.7 (3)1.8	1.7	1.8	(1)1.8 (2)2.0	1.6	1.6	2.0	2.0	2.4	2.9	
Rb	Sr	Y	Zr	Nb	Mo	Tc	Ru	Rh	Pd	Ag	Cd	In	Sn	Sb	Te	I	
0.8	1.0	1.2	1.5	1.7	(4)1.6 (6)2.1	(5)1.9 (7)2.3	2.0	2.1	2.1	1.9	1.7	1.7	(2)1.7 (4)1.9	(3)1.8 (5)2.1	2.1	2.6	
Cs	Ba	La	Hf	Ta	W	Re	Os	Ir	Pt	Au	Hg	Tl	Pb	Bi	Po	At	
0.75	0.9	1.2	1.4	(3)1.3 (5)1.7	(4)1.6 (6)2.0	(5)1.8 (6)2.1 (7)2.2	2.1	2.1	2.2	2.3	1.8	(1)1.4 (3)1.9	(2)1.6 (4)1.8	1.8	2.0	2.2	
Fr	Ra																
0.7	0.9																
		Ce	Pr	Nd	Pm	Sm	Eu	Gd	Tb	Dy	Ho	Er	Tu	Yb	Lu		
		1.2	1.2	1.3	1.3	1.3	1.2	1.3	1.3	1.3	1.3	1.3	1.3	1.2	1.3		
		Th	Pa	U	Np	Pu	Am	Cm	Bk	Cf								
		(2)1.0 (4)1.4	(3)1.3 (5)1.7	(4)1.4 (6)1.9	(4)1.4 (6)1.9	1.3	1.3	1.3	1.3	1.3								

Table 5.12   Miedema's electronegativity scale of elements.

Li	Be											B	C	N
2.85	5.05											5.30	6.24	6.86
Na	Mg											Al	Si	P
2.70	3.45											4.20	4.70	5.55
K	Ca	Sc	Ti	V	Cr	Mn	Fe	Co	Ni	Cu	Zn	Ga	Ge	As
2.25	2.55	3.25	3.80	4.25	4.65	4.45	4.93	5.10	5.20	4.45	4.10	4.10	4.55	4.80
Rb	Sr	Y	Zr	Nb	Mo	Tc	Ru	Rh	Pd	Ag	Cd	In	Sn	Sb
2.10	2.40	3.20	3.45	4.05	4.65	5.30	5.40	5.40	5.45	4.35	4.05	3.90	4.15	4.40
Cs	Ba	La	Hf	Ta	W	Re	Os	Ir	Pt	Au	Hg	Tl	Pb	Bi
1.95	2.32	3.17	3.60	4.05	4.80	5.20	5.40	5.55	5.65	5.15	4.20	3.90	4.10	4.15
			Th	U										
			3.30	4.05										
Ce	Pr	Nd	Pm	Sm	Eu	Gd	Tb	Dy	Ho	Er	Tm	Yb	Lu	
3.18	3.19	3.19	3.19	3.20	3.20	3.20	3.21	3.21	3.22	3.22	3.22	3.22	3.22	

Table 5.13   Electronegativity of atomic groups.

Atomic group	Electronegativity	Atomic group	Electronegativity
SH	2.45	C_6H_5	2.70
NO_3	3.91	NH_2	2.99
SO_3	3.83	SCN	2.70
SO_4	3.83	CN	2.52
OCO	3.83	COOH	2.57
OH	3.51	CHO	2.61
CO	2.61		

5.6   Charge-Radius Ratio

Charge-radius ratio, denoted by Z/R_k, is the ratio of the number of valence electrons and the radius of cation after *all* valence electrons ionized.

Charge-radius ratio is a parameter describing the electrostatic field of the atomic core acting on the valence electrons surrounding it. The values of the charge-radius ratio of elements are listed in Table 5.14.

Table 5.14   Charge-Radius Ratio of elements.

1	2	3	4	5	6	7	8	9	10	11	12	13	14	15	16	17	18
H (16.0)																	
Li 1.67	Be 6.45											B 15.0	C 26.7	N 45.4	O 66.7	F 100	
Na 1.05	Mg 3.08											Al 6.00	Si 9.77	P 14.7	S 20.7	Cl 26.9	
K 0.75	Ca 2.02	Sc 3.70	Ti (4)5.90 (3)4.69	V (5)8.48 (3)4.62	Cr (6)11.5 (3)4.69 (2)2.56	Mn (7)15.2 (3)4.55 (2)2.56	Fe (3)4.68 (2)2.63	Co (2)2.70	Ni 2.74	Cu 1.04	Zn 2.70	Ga 4.84	Ge 7.55	As (5)10.6 (3)4.35	Se 14.3	Br 17.9	
Rb 0.68	Sr 1.77	Y 3.23	Zr 5.00	Nb 7.14	Mo 9.68	Tc —	Ru (4)5.96 (2)2.47	Rh (4)6.16 (3)4.17	Pd (4)5.41 (2)2.50	Ag 0.79	Cd 2.06	In (3)3.70 (1)0.71	Sn (4)5.64 (2)1.82	Sb (5)8.07 (3)3.26	Te 10.7	I 14.0	
Cs 0.59	Ba 1.48	La 2.61	Hf 4.94	Ta 7.04	W 8.7	Re 12.3	Os (4)5.80 (2)2.28	Ir (4)5.89 (2)2.20	Pt (4)6.15 (2)2.50	Au 0.73	Hg 1.82	Tl (3)3.16 (1)0.63	Pb (4)4.77 (2)1.57	Bi (5)6.76 (3)2.78	Po (8.8)	At —	
Fr —	Ra 1.33	Ac —	Th 4.04	Pa 5.55	U (6)7.23 (4)4.30 (2)2.91												

Ce	Pr	Nd	Pm	Sm	Eu	Gd	Tb	Dy	Ho	Er	Tu	Yb	Lu
(4)3.96 (3)2.63	2.68	2.73	2.78	(3)2.80 (2)1.73	(3)2.86 (2)1.75	2.91	2.97	3.00	3.03	3.06	3.13	3.16	3.23

Np	Pu	Am
(6)7.32 (4)4.35 (3)2.97	(6)7.41 (4)4.45 (3)3.00	(6)7.50 (4)4.49 (3)3.03

The charge-radius ratio is a parameter describing the metallic character of elements. It can be seen from Table 5.14 that the elements with charge-radius ratio less than 7.0 are metallic elements, and the elements with charge-radius ratio larger than 7.0 are non-metallic elements. It is interesting to see that the elements having charge-radius ratio slightly higher than 7.0 are semiconductors.

If we use the sum of charge-radius ratio, $\sum(Z/R_K)$ and the difference of electronegativity as coordinates, and plot the nature of chemical bond between the pair of elements on the map spanned by these coordinates, it can be found that the element-pairs forming different types (ionic bond, covalent bond and metallic bond) of chemical bond are distributed in different regions of this map (Fig. 5.1).

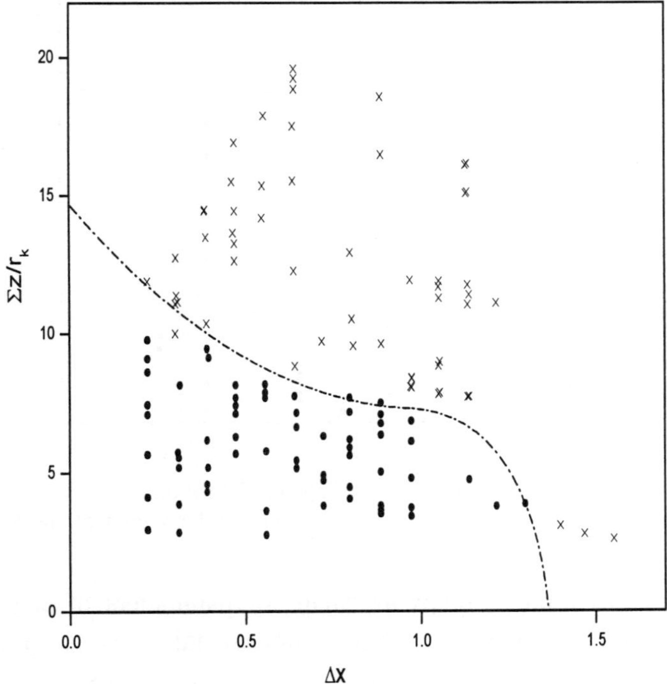

• compound with metallic bond   × compound with ionic or covalent bond
Fig. 5.1   Regularity of formation of ionic, covalent and metallic bond in inorganic compounds.

In our previous work, it has been found that to plot two-dimensional maps with some function of charge-radius ratio and that of electronegativity as coordinates is an effective way to find some rough regularities of chemical systems. For example, if we use the ratio of (Z/R_k) of two metallic elements as the coordinate, and use the difference of electronegativities of two metallic elements as abscissa, and plot the representative points of the binary metallic systems consisting of nontransition elements onto the map spanned by these two parameters, the intermetallic compound-forming systems and the systems without intermetallic compound formation can be plotted separately in different regions (Fig. 5.2) with clear-cut boundary between these regions.

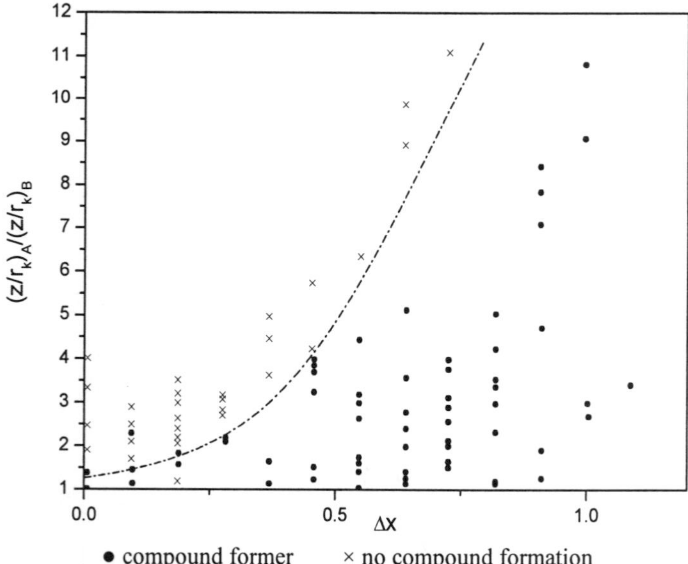

• compound former   × no compound formation

Fig. 5.2   Regularity of formation of intermetallic compounds between nontransition elements.

We called this kind of maps as "maps of parameters of chemical bond". It appears that such kind of two-dimensional maps can provide more powerful tools for structure-property investigation work.

Although it is obvious that to use a two-parameter map is more powerful than to use a single atomic parameter, it is easy to understand that only two parameters may be not enough for the description of more complicated chemical phenomena yet, and that the data processing in multi-dimensional space by pattern recognition methods (including SVM) with more than two atomic parameters should be a more powerful tool in chemical data processing and chemical knowledge discovery. We call this method *atomic parameter-pattern recognition method.*

5.7 Topological Parameters of Molecules and 3-D Molecular Descriptors

In order to investigate the structure-property relationships of the compounds with relatively complex molecules, it is not enough to use atomic parameters only, since the structure of molecules is also a key factor affecting the physico-chemical properties of a compound. This is particularly important for the description of the structure-property relationships of organic compounds.

5.7.1 *2-Dimensional molecular descriptors*

A group of very useful molecular structure descriptors is topological index. Since molecular structure can be usually described by molecular graphs and their matrices (see Fig. 5.3). Different kinds of invariants of molecular graphs can be used as the molecular parameters for the QSPR investigation. Since these invariants reflect the topological properties of molecules, these parameters are usually called topological indices.

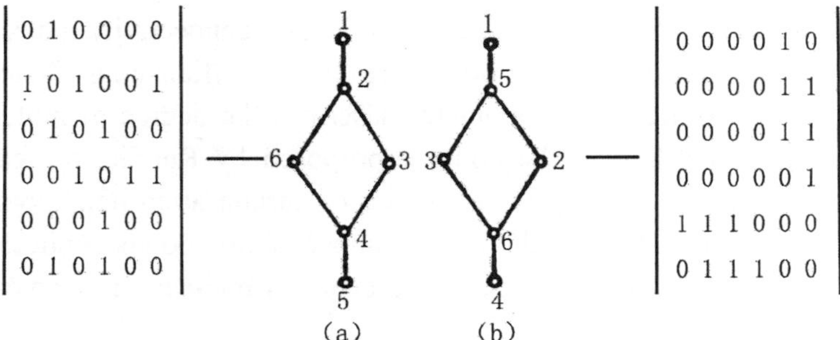

Fig. 5.3   Example of molecular graph and its matrix.

The first topological index was proposed by Wiener in 1947 [138]. Starting from the research work about the relationship between the boiling points and the topological structures of paraffin hydrocarbons, he suggested two topological parameters. One is w, called path number. It is defined as the sum of the distances between any two carbon atoms in the molecule in terms of carbon-carbon bonds. The other is p, called polarity number. It is defined as the number of pairs of carbon atoms which are separated by three carbon-carbon bonds. Later, it has been found that not only the boiling points of paraffin hydrocarbons, but also many physical and chemical properties of molecular compounds can be correlated with these two parameters. Roughly speaking, since the path number is calculated as the total distance between all carbon atoms, the smaller this total distance, the larger the compactness of the molecule. On the other hand, the physical meaning of polarity number (p) can be understood as follows: it is a semi-quantitative measure of intra-molecular attraction forces transmitted through the carbon chain. According to the theory of alternate polarities, the charges on the carbon atoms in a carbon atom chain change alternatively. So the oppositely charged neighbors of a carbon atom are its first, third, fifth neighbors, and so on. Since the number of carbon–carbon bond is the same for all isomers, and the influences of the fifth or more remote neighbors are too weak to be considered, the number of pairs of carbon atoms separated by three carbon-carbon bonds can be roughly considered as an influential parameter describing the properties of molecules.

Another important topological index was proposed by Randic in 1975 [114], this parameter is usually called molecular connectivity index. It is well-known that molecular branching is an influencing factor of molecular properties. This parameter describes the degree of molecular branching. According to the concept proposed by Randic, the carbon atoms can be classified by their number of carbon atom neighbors, the number of carbon atom neighbors or number of C-C bonds connected to a carbon atom i is denoted by p_i, so that every carbon-carbon bond can be described by a number

$$C_k = (p_i p_j)^{-1/2}$$

And the sum of C_k for all carbon-carbon bonds in a molecule is called the molecular connectivity index:

$$\chi = \sum C_k = \sum (p_i p_j)^{-1/2} \qquad (5.1)$$

After Randic proposed his topological parameter, Kier and Hall developed his concept and have built a series of molecular connectivity series [76]. The member of this series is usually denoted by $m\chi$, where m is the order of this member. For example, when the Randic's original χ is calculated, only a pair of carbon atoms on a C-C bond is considered, it means that only the sub-graphs consisting only one C-C bond of a molecular graph is considered each time. In the molecular connectivity series, it is denoted by 1χ. Similarly, if each time two connected C-C bonds are considered, we can have the member of second order:

$$^2\chi = \sum (p_i p_j p_k)^{-1/2} \qquad (5.2)$$

Molecular connectivity series is widely used in QSPR research work.

In order to extend the molecular connectivity indices to investigate unsaturated compounds, Kier suggested using quantity δ instead of p:

$$\delta_i = Z_i^{\text{V}} - h_i \qquad (5.3)$$

where Z_i^{V} is the number of the valence electrons of atom i, and h_i denotes the number of hydrogen atoms connected to carbon atom i.

In order to investigate the hetero-atom containing molecules, empirical δ values of some atoms and atomic groups have been proposed.

Balaban proposed another connectivity index based on the distance matrix of molecular graph. He defined the sum of a column or a row of distance matrix as the distance sum of an atom.

Although the topological indices have been widely used in QSPR/QSAR research, one of the most frequent criticisms is the absence of a clear physico-chemical interpretation. So it is desirable to find the physical meaning of the regularities found by topological indices. Some of work in this direction has been done. Galvez has proved that the molecular connectivity index $^0\chi$ is intimately related to the molecular volume, and $^1\chi$ is intimately related to the molecular surface of alkanes [55].

5.7.2   *3-Dimensional molecular descriptors*

Although the 2-dimensional molecular descriptors are very useful due to their simplicity (this is beneficial for computer time saving) and effective in many cases, they can not differentiate the enantiomers and the differences of conformation. Since the difference of enantiomers and conformations are very important in biochemistry and medical chemistry, it is necessary to design more sophiscated molecular descriptors. And it is also a tendency that these molecular descriptors should be combined with quantum chemical calculation and molecular mechanics calculation [6].

The 3-dimensional Wiener number (3-W) was introduced by Trinajstic and his coworkers, and was based on topographic (geometric) distance matrix. Thus, one can encode information on geometric diastereomers and conformers. Compared with the 2-dimensional Wiener topological index based on topological distances D_{ij}:

$$(2\text{-}W) = (1/2)\sum\sum(2D_{ij}) \tag{5.4}$$

the 3-dimensional Wiener index can be expressed as follows:

$$(3\text{-}W) = (1/2)\sum\sum(3D_{ij}) \tag{5.5}$$

The 3-D Wiener index can include or exclude the hydrogen atoms in molecules. The advantage of 3-W over 2-W is obvious. For example, the 2-W indices for all heptanes present two pairs of degenerate values: 3-ethyl-heptane and 2, 4-dimethylpentane have two 2-W equal to 48; 2, 2- and 2, 3-dimethylpentene have 2-W equal to 46. Neither the hydrogen-exclusive nor hydrogen-inclusive 3-W values are degenerate for any conformer of heptanes. Actually, so far no degeneracy was found for 3-W in the case of staggered hydrogen-inclusive alkanes.

The 3-W index is useful for QSPR/QSAR. Normal boiling points for the first 54 alkanes are correlated by following equation with r = 0.998, s = 5.8 C:

$$T_b = 395(3\text{-}W)^{0.0986} - 682 \tag{5.6}$$

Satisfactory correlations are also obtained for the enthalpies of first 21 alkanes (excluding methane) both in terms of 2-W and 3-W numbers:

$$3\text{-W} = 0.0849(2\text{-W}) - 13 \quad r = 0.997, s = 3.66\ ^\circ C \tag{5.7}$$

Molecular shape has a fundamental influence on both the static and dynamic properties of molecules. For example, the shapes of nuclear and electronic distributions determine the molecular dipole moments as well as the likely sites of approach by a nucleophilic reagent. The evolution of concepts and models used by chemists for the description of the molecular shape closely mirrors the advances made in our understanding of molecular behavior. Whereas most of the early models focused on the nuclear arrangements, the more advanced recent approaches have placed increasingly more emphasis on the electronic distribution. The nuclear position descriptors, however, is still very important, since the nuclei in molecules are more "particle-like" than electrons and therefore much easier to be described. For the description of the shape of complex molecules, the concept of similarity appears to be very useful.

Similarity has played an important role in the discussion of structure-property relationships, particularly in discussions of biological activity of chemicals. The early recognition of the role of molecular similarity in structure-activity studies may be traced to Email Fischer's "lock and key" model for the interaction of proteins and drugs. Much of the works in QSAR studies rests on the paradigm that similar structures have similar properties. In recent years, attempts have been made to express the degree of mutual similarity quantitatively. One of the ways to describe similarity is to express it by a series of 2-dimensional and 3-dimensional molecular descriptors as follows:

$$\text{Similarity} = \left[(M_1\text{-}M_1')^2 + (M_2\text{-}M_2')^2 + (M_3\text{-}M_3')^2 + ... \right]^{1/2} \tag{5.8}$$

where M and M' are some molecular descriptors of two molecules respectively.

5.7.3 *Quantum chemical parameters used in QSPR/QSAR computation*

In data processing by SVR or other algorithms, following quantum chemical parameters are usually used in computation works: HOMO, LUMO, net atomic charges of different atoms, dihedral angles of molecules optimized by molecular mechanics calculations. These parameters will be described in detail in section 9.3 of chapter 9.

5.8   Atomic Parameters for Ionic Systems

Ionic solids and molten salts are typical ionic systems. Electrolyte solutions are the mixtures of ions and molecules, but may be considered as ionic systems as an approximation if the solvent is considered as a continuous medium [109].

The cohesive energy of ionic systems includes three terms: electrostatic potential energy, inter-ionic short range repulsion energy and van der Waals potential energy. The electrostatic potential energy includes the attractive potential energy between ions of different sign of charges and the repulsive potential energy between ions of the same sign of charges. Electrostatic attraction between cations and anions makes them come together. But the short range repulsion potential increases very quickly when the inter-ionic distance approaches to the sum of cationic and anionic radii. The van der Waals potential is a relatively weak and short range action. Therefore the physico-chemical properties are usually chiefly determined by the electrostatic potential energy and the short range repulsion energy. Since the potential energy-distance curves of the short range repulsion between ions are rather steep, as a rough approximation, the ions can be considered as *hard spheres and electrostatic force acts between these spheres*. So we can have the "*electrostatic-hard sphere model*" of ionic systems. Based on this model, Reiss has proposed the *theory of corresponding state of ionic systems* [116]. According to this theory, the configuration integral of an ionic system can be expressed by some function of a series of dimensionless numbers[*]:

$$I = f\left[(kT)/(Z_{+}Z_{-}/\sum R),(R_{+}/R_{-}),...\right] \qquad (5.9)$$

---

[*]In Reiss's original paper, the geometrical dimensionless number (R_1/R_2) was not listed as an independent dimensionless number. But in our work we have found that it is a reasonable and very effective dimensionless number in ionic system studies [25].

So many thermodynamic properties should be some unknown functions of the dimensionless numbers listed in the above-mentioned equation. Since this is a multivariate problem, SVM or other statistical algorithms can be used to find the empirical relationships between the thermodynamic properties and these dimensionless numbers.

The crystal type of ionic compound is chiefly determined by the ionic radii of constituent elements. For compounds of definite valence type, the regularities of the crystal types of ionic compounds can be roughly expressed by so-called structure maps [102], as shown by an example in Fig. 5.4.

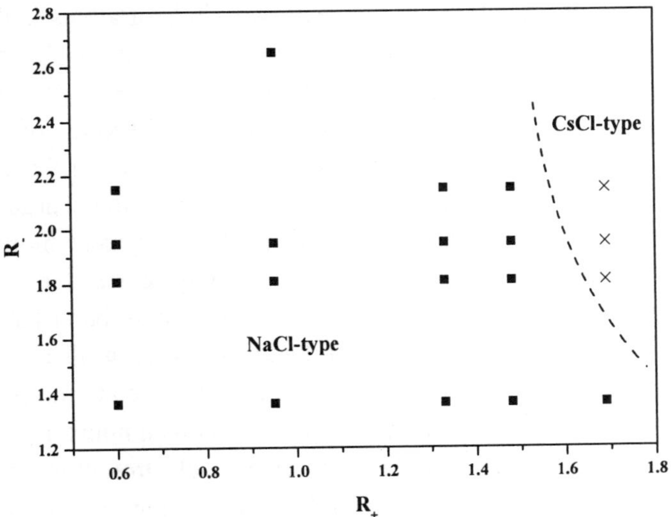

Fig. 5.4   An example of structure maps.

5.9   Atomic Parameters for Covalent Compounds

Most of covalent compounds are molecular compounds. The properties of these compounds are chiefly determined by their molecular structure. It is usually to use many quantum chemical parameters such as HOMO, LUMO, atomic net charges and the topological indices or 3-dimensional molecular descriptors as the useful parameters for the investigation of these compounds.

Some inorganic covalent compounds are crystalline solids built by covalent bonds. Some of these types of compounds are important functional materials, such as compound semiconductors. Atomic parameters such as covalent radius and electronegativity are useful atomic parameters for the investigation of these materials.

5.10   Atomic Parameters for Metallic Systems

The cohesive energy of intermetallic compounds and metallic systems is mainly affected by three factors [108]: (1) geometric factor: Metallic systems tend to become more stable by the closest packing of atoms. Since the ratio of metallic radii between two kinds of atoms is the chief factor which affect the closest packing, this factor can be described by the ratio of the metallic radii of different elements (R_1/R_2); (2) charge transfer factor: The charge transfer between different kinds of atoms is due to the different energy levels of valence electrons and the electrons move from high energy level to the lower one. Hence the charge transfer makes the decrease of total energy, and makes the system more stable. The extent of charge transfer can be roughly correlated with the electronegativity difference between atoms; (3) energy band factor: The change of the structure of the energy band of a system also makes the total energy changed. It is difficult to find an effective atomic parameter to correlate this effect, but it is related to the averaged number of valence electrons. We can use the ratio between valence electron number and the number of atoms in a system as an indirectly relevant parameter in data processing work.

A semi-empirical model proposed by Miedema, called "cellular model" [99] of alloy phases, has been proved as an effective mathematical model for the estimation of the thermodynamic properties of metallic systems, including the thermodynamic properties of intermetallic compounds and liquid alloys. According to this model, the difference of electronegativities, $\Delta\phi$, denoting the charge transfer between the atoms of different elements, is the driving force for the formation of intermetallic compounds; while another parameter, the difference of parameter $n_{ws}^{1/3}$ of different elements (Table 5.15), is the "resistance" of intermetallic compound formation. $n_{ws}^{1/3}$ represents the valence electron density of the Wagner-Seitz cell in the metallic phases. Different metallic elements have different values of $n_{ws}^{1/3}$. If an intermetallic compound is formed between metallic elements A and B. according to the cellular model proposed by Miedema, the interface between the Wagner-Seitz cells of atom A and atom B shall have a gradient of valence electron density. Miedems assumed that such gradient makes the decrease of cohesive energy, so that $n_{ws}^{1/3}$ difference becomes the resistance of the formation of intermetallic compounds.

Table 5.15   Values of $n_{ws}^{1/3}$ of elements.

Li	Be												B	C	N
0.98	1.67												1.75	1.77	1.65
Na	Mg												Al	Si	P
0.82	1.17												1.39	1.50	1.65
K	Ca	Sc	Ti	V	Cr	Mn	Fe	Co	Ni	Cu	Zn		Ga	Ge	As
0.65	0.91	1.27	3.80	1.64	1.73	1.61	1.77	1.75	1.75	1.47	1.32		1.31	1.37	1.44
Rb	Sr	Y	Zr	Nb	Mo	Tc	Ru	Rh	Pd	Ag	Cd		In	Sn	Sb
0.60	0.84	1.21	3.45	1.64	1.77	1.81	1.83	1.76	1.67	1.36	1.24		1.17	1.24	1.26
Cs	Ba	La	Hf	Ta	W	Re	Os	Ir	Pt	Au	Hg		Tl	Pb	Bi
0.55	0.81	1.18	3.45	1.63	1.81	1.85	1.85	1.83	1.78	1.57	1.24		1.12	1.15	1.16
			Th	U											
			1.28	1.56											

Ce	Pr	Nd	Pm	Sm	Eu	Gd	Tb	Dy	Ho	Er	Tm	Yb	Lu
1.19	1.20	1.20	3.19	1.21	1.21	1.21	1.22	1.22	1.22	1.23	1.23	1.23	1.23

Although the cellular model proposed by Miedema is rather successful for the semi-quantitative description of the thermodynamic property of intermetallic compounds and liquid alloys. The ignorance of geometrical factor makes this model to be somewhat inaccurate. In some of our work, inclusion of atomic radius ratio can give better results of computation by support vector regression.

Chapter 6

SVM Applied to

Phase Diagram Assessment and Prediction

6.1 Comprehensive Assessment and Computerized Prediction of Phase Diagrams

Phase diagrams describe the phase relations in equilibrated systems of chemical substances. The determination of phase diagrams is one of the commonest research topics in physical chemistry. After about more than one hundred years' work, a huge amount of experimental data has been accumulated. These experimental data have been recorded into many handbooks and databases. They are widely used in daily research work, plant design and other practical work in chemistry, chemical technology and materials science.

The number of published phase diagrams is already rather large. For example, up to now, more than ten thousand phase diagrams of binary and ternary alloy systems; more than ten thousand phase diagrams of oxide systems and more than 4000 phase diagrams of molten salt systems have been collected into relevant handbooks and databases. Although these experimentally measured phase diagrams are very useful, they are not sufficient for many practical applications, because there are still much more systems having no available phase diagram yet. For example, since the number of common inorganic salts is more than one hundred, the number of the binary and ternary molten salt systems should be near one million (the systems useful in many practical problems are multi-component systems. If the mixtures of four or more components have to be considered, the number of systems will be still more than this figure). Moreover, almost all published ternary phase diagrams are restricted within some cross sections of these phase diagrams, and the phase relation of other temperature or composition range is obscure. Generally speaking, there are still more than 99% of phase equilibrium problems having no enough experimental data for retrieval. Hence it is clear that "can we have some method for the computerized prediction of the unknown phase diagrams?" has been an emergent problem.

Thermodynamics is very useful for phase diagram prediction, provided that the relevant thermodynamic data are available. One of the deficiencies of thermodynamic method is that it is unable to predict unknown new phases and the properties of unknown new phases. In recent years, we have found that the atomic parameter-pattern recognition method is rather effective for the modeling and prediction of the chief characteristic quantities (such as: formation of intermediate compounds and their stoichiometries, crystal types, melting types, melting points, etc.) of phase diagrams. Since the data set for training in this method is usually not very large, support vector machine has been proved to be a useful tool for this work.

On the other hand, it is unfortunate that not all phase diagrams published in literatures are completely reliable. Some of them are in controversy among different authors because their experimental results are contradictory with each other. Some of them are not believable because their results are contradictory to phase rule or thermodynamic principle. And still others are also not believable because the conclusions of the authors are based on unreliable experimental data. For example, some authors used polythermal visualization method to determine phase diagrams, and concluded the existence of intermediate compound based on some small turning point of liquidus curve. Actually this is unreliable because the small "turning point" may be induced by small experimental error of the polythermal visualization method. In order to make the published materials more reliable, it is necessary to do systematic assessment work to eliminate or correct the unreliable data from the databases or handbooks. Up to now, the thermodynamic method for the assessment work of phase diagrams has given very plentiful results [83].

But it is still not enough if we use this method only. Thermodynamic method cannot make prediction of new phase formation because the thermodynamic function of unknown intermediate compound is not available.

Strictly speaking, the most reliable method of phase diagram assessment should consist of the following three steps:

(1) Thermodynamic assessment: If some phase diagrams are not consistent with phase rule or thermodynamic principle, we can assure that it is unreliable.

(2) Comparison of this phase diagram with similar systems: If the phase diagram is not contradictory to phase rule and thermodynamic principle, we can use atomic parameter-pattern recognition method to perform the second step of assessment. In this step, a series of other phase diagrams of the same valence type should be collected, and then we can use atomic parameter-pattern recognition method to try to find some regularities. If most of the phase diagrams of this valence type obey these regularities, but a few of them appear to be outliers, we can consider these few phase diagrams as suspicious phase diagrams. By this way, a few percent of suspicious phase diagrams can be found out from numerous phase diagrams.

(3) Experimental confirmation or negation of the suspicious phase diagrams found by atomic parameter-pattern recognition method: If some published phase diagrams have been found to be an outlier or suspicious, we should do experimental work to confirm or negate our suspicion.

In recent years, we have used this comprehensive strategy to make the computerized assessment of a series of phase diagrams. By this strategy, we have found that the coordination of atomic parameter method and various pattern recognition techniques (including SVM) is rather effective to find out the outliers or suspicious phase diagrams [30; 23; 26; 32].

6.2 Atomic Parameter-Pattern Recognition Method for Phase Diagram Prediction

From the logical point of view, the atomic parameter-pattern recognition method we used is a transduction method. The physico-chemical foundation of this method can be understood as follows: phase equilibrium, or phase diagram, is determined by the thermodynamic function of mixtures. And as a result of final analysis, the thermodynamic function of a mixture can be considered as that determined by the atom-atom interaction in the mixture. While the atom-atom interaction can be roughly described by some atomic parameters of the constituent atoms, such as atomic or ionic radius, ionic charges, number of valence electrons or d electrons in next outmost shell of atom and electronegativity of constituent elements, etc. So if some multi-dimensional space is spanned by these atomic parameters, and the representative points of the systems with known phase diagrams are plotted into this space as training points, then pattern recognition methods can be used to find the mathematical models describing the characteristics of phase diagrams. These mathematical models found should be useful for predicting or estimating the characteristics of unknown phase diagrams.

The above-mentioned method is especially suitable for the prediction of the formation, the structure and property of unknown new intermediate phases. So it is just a complementary tool of the thermodynamic method for the assessment and prediction of phase diagrams.

6.3 Prediction of Intermediate Compound Formation

The formation of intermediate compounds is one of the most important factors affecting the geometry of phase diagrams. Atomic parameter-pattern recognition method is an effective technique dealing with the assessment or prediction problems about the formability of intermediate compounds.

6.3.1 *Regularities of formability of intermediate compounds of ionic systems*

"Phase diagrams of ionic systems with similar ionic sizes, ionic charges, and electronegativities of components have similar geometry", this evidence has been already noticed in the middle of last century. The similarity between the phase diagram of KF-NiF_2 system and that of RbF-MgF_2 system due to the similarity of ionic radii was pointed out by Wagner and Balz in 1952 [14]. Based on this concept and pattern recognition technique, we have used atomic parameter and support vector machine to assess or predict the formation of the intermediate compounds for a series of ionic systems. Here two examples dealing with the intermediate compound formation in binary halide systems of AX-BX_2 type (X=F, Cl, Br, I) will be described to illustrate the applications of SVM in phase diagram assessment and prediction.

(1) The assessment of the phase diagram of $CsBr$-$CaBr_2$ system (Regularity of intermediate compound formation):

Table 6.1 lists the data of phase diagrams of the $MeBr$-$Me'Br_2$ type systems, where Me and Me' are monovalent and divalent metallic elements respectively. Here class "1" is the phase diagram having intermediate compounds, while class "2" is that having no intermediate compound formation.

Table 6.1   The formability of intermediate compounds of $MeBr$-$Me'Br_2$ systems.

System	Class[*]	R_+	R_{2+}	X_+	X_{2+}
$LiBr$-$BaBr_2$	2	0.60	1.35	0.95	0.9
$LiBr$-$CoBr_2$	2	0.60	0.69	0.95	1.7
$LiBr$-$SnBr_2$	2	0.60	1.10	0.95	1.8
$NaBr$-$MgBr_2$	2	0.95	0.65	0.9	1.2
$NaBr$-$CaBr_2$	2	0.95	0.99	0.9	1.0
$NaBr$-$SrBr_2$	2	0.95	1.13	0.9	1.0
$NaBr$-$BaBr_2$	2	0.95	1.35	0.9	0.9
$NaBr$-$CdBr_2$	2	0.95	0.97	0.9	1.7
$NaBr$-$CoBr_2$	2	0.95	0.69	0.9	1.7
$NaBr$-$MnBr_2$	2	0.95	0.80	0.9	1.4
$NaBr$-$SnBr_2$	2	0.95	1.10	0.9	1.8
$NaBr$-$PbBr_2$	2	0.95	1.21	0.9	1.6
KBr-$MgBr_2$	1	1.33	0.65	0.8	1.2

KBr-CaBr$_2$	1	1.33	0.99	0.8	1.0
KBr-SrBr$_2$	1	1.33	1.13	0.8	1.0
KBr-BaBr$_2$	1	1.33	1.35	0.8	0.9
KBr-CdBr$_2$	1	1.33	0.97	0.8	1.7
KBr-HgBr$_2$	1	1.33	1.10	0.8	1.9
KBr-FeBr$_2$	1	1.33	0.76	0.8	1.7
KBr-CoBr$_2$	1	1.33	0.69	0.8	1.7
KBr-MnBr$_2$	1	1.33	0.80	0.8	1.4
KBr-SnBr$_2$	1	1.33	1.10	0.8	1.8
KBr-PbBr$_2$	1	1.33	1.21	0.8	1.6
RbBr-MgBr$_2$	1	1.48	0.65	0.8	1.2
RbBr-CaBr$_2$	1	1.48	0.99	0.8	1.0
RbBr-SrBr$_2$	1	1.48	1.13	0.8	1.0
RbBr-BaBr$_2$	1	1.48	1.35	0.8	0.9
RbBr-CdBr$_2$	1	1.48	0.97	0.8	1.7
RbBr-HgBr$_2$	1	1.48	1.10	0.8	1.9
RbBr-FeBr$_2$	1	1.48	0.76	0.8	1.7
RbBr-CoBr$_2$	1	1.48	0.69	0.8	1.7
RbBr-MnBr$_2$	1	1.48	0.80	0.8	1.4
RbBr-SnBr$_2$	1	1.48	1.10	0.8	1.8
CsBr-CaBr$_2$	2	1.69	0.99	0.75	1.0
CsBr-SrBr$_2$	1	1.69	1.13	0.75	1.0
CsBr-BaBr$_2$	1	1.69	1.35	0.75	0.9
CsBr-ZnBr$_2$	1	1.69	0.74	0.75	1.6
CsBr-CdBr$_2$	1	1.69	0.97	0.75	1.7
CsBr-HgBr$_2$	1	1.69	1.10	0.75	1.8
CsBr-CoBr$_2$	1	1.69	0.69	0.75	1.7
CsBr-MnBr$_2$	1	1.69	0.80	0.75	1.4
CsBr-SnBr$_2$	1	1.69	1.10	0.75	1.7
CsBr-PbBr$_2$	1	1.69	1.21	0.75	1.6
AgBr-CdBr$_2$	2	1.27	0.97	1.9	1.7
AgBr-HgBr$_2$	2	1.27	1.10	1.9	1.8
TlBr-CdBr$_2$	1	1.40	0.97	1.4	1.7
TlBr-HgBr$_2$	1	1.40	1.10	1.4	1.8
TlBr-MnBr$_2$	1	1.40	0.80	1.4	1.4

[*]Class "1" denotes the samples forming intermediate compound, class "2" denotes the samples without intermediate compound formation, R_+ and R_{2+} denote the ionic radii of monovalent and divalent cations respectively and X_+ and X_{2+} denote the electronegativities of monovalent and divalent metallic elements respectively.

The data listed in Table 6.1 are treated by support vector classification. It has been found that the published phase diagram of CsBr-CaBr$_2$ system cannot be classified correctly (The published phase diagram is a simple eutectic one, without intermediate compound formation [38], but the classification and leave-one-out (LOO) cross-validation always indicate the existence of intermediate compound). So it seems that the phase diagram published for this system may be wrong. Then we decide to do experimental work to confirm this suspicion.

The methods used are differential thermal analysis (DTA) and X-ray diffraction (including high temperature X-ray diffraction). The experimental result is that there is indeed a 1:1 congruently melting compound ($CsCaBr_3$) formed. The melting point of $CsCaBr_3$ is 821 ℃.

Based on the results of data processing of the phase diagrams of MeX-$Me'X_2$ (X=F, Cl, Br, I) systems, it has been found that larger monovalent cation Me^+, smaller divalent cation and smaller halide anion favor the intermediate compound formation. This fact can be explained as follows: According to *Pauling's fourth rule of the structure of ionic crystals, in a crystal containing different cations those with large valence and small coordination number tend not to share polyhedron with each other*. This rule implies that *cations with large electric charges tend to be as far apart from each other as possible in order to reduce their contribution to the repulsive Coulomb energy of the crystal*. In MeX-$Me'X_2$ systems, the electrostatic repulsive energy between Me'^{2+} cations can be reduced after the formation of the crystal lattice of intermediate compounds containing second kind of cation having smaller electric charge. Smaller Me^{2+} cation, larger Me^+ cation and smaller X^- anion will give rise to stronger tendency of the repulsive energy reduction, and therefore stronger tendency of intermediate compound formation.

Actually, the above-mentioned regularity is also applicable to the formability of intermediate compound in other valence types of ionic systems. Namely, the higher electric potential of highly charged cation, the lower electric potential of lowly charged cation, and the smaller the radius of anion, the stronger the tendency of intermediate compound formation of ionic systems.

(2) The assessment of phase diagram of CsF-CaF_2 system (relative stability of $KNiF_3$-type and K_2NiF_4-type structures):

In recent years, it has been found that there are many complex oxides and complex halides with perovskite-like structure having many valuable properties as functional materials. Among different perovskite-like structures, the layered structure formed by stacking of the layer of two-dimensional perovskite structure and two-dimensional rock salt structure is the most attractive for investigators, because many high temperature superconductors are of this structure. Since $KNiF_3$ and K_2NiF_4 are the simplest proto-type compounds with perovskite structure and layered perovskite-like structure respectively, the physico-chemical model and the regularity of the relative stability of $KNiF_3$-type compounds and K_2NiF_4-type compounds are interesting topics for materials scientists and chemists. Yokokawa and Rezniskii have tried to investigate this problem by assuming that the difference of the coordination number of monovalent ion is the chief factor affecting the relative stability of these two series of compounds. But they cannot find an effective criterion to explain the experimental facts of relevant fluoride or oxide systems. In one of our recent work, a crystal chemical model has been proposed and a mathematical model based on it has been proposed for solving this problem.

Figure 6.1 illustrates the typical lattice structures of $KNiF_3$ and K_2NiF_4, as the proto-types of the perovskite-type lattice and the simplest layered perovskite-like lattice structure.

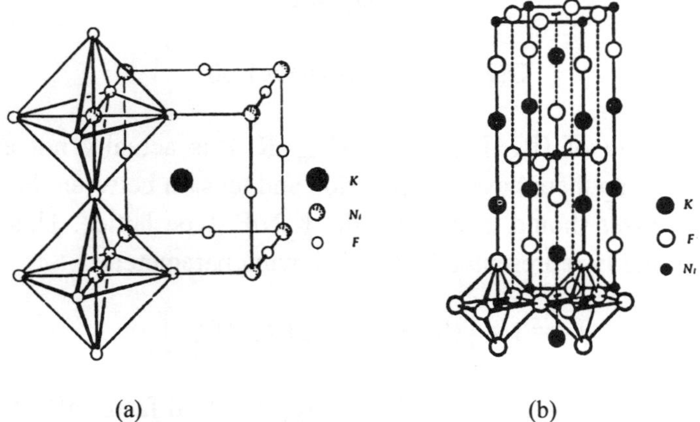

Fig. 6.1　The structure of lattices of $KNiF_3$ (a) and K_2NiF_4 (b).

From Fig. 6.1 it can be seen that there may be two chief factors affecting the relative stability of $KNiF_3$-type and K_2NiF_4-type structures: 1) *Coulombic repulsive energy reduction due to the longer distance between highly charged cations in K_2NiF_4 structure:* According to Pauling's fourth rule about the stability of complex ionic lattices, *cations with large electric charges tend to be as far apart from each other as possible to reduce their contribution to the Coulombic energy of the crystal.* It can be seen that the change from $KNiF_3$ structure to K_2NiF_4 structure the cationic repulsion between divalent small Ni^{2+} ions should be reduced and replaced by the weaker ionic repulsion between Ni^{2+} ion and large monovalent K^+ ion.

This energy change should be a factor to stabilize the K_2NiF_4 structure. This factor can be roughly represented by the parameter η:

$$\eta = \left[1/(R_{Ni}+R_F)\right] + \left[1/(R_K+R_F)\right] - \left[4/(R_{Ni}+R_K+2R_F)\right] \qquad (6.1)$$

where R_{Ni}, R_K and R_F are the ionic radii of nickel cation (or other bivalent cation), potassium cation (or other monovalent cation) and anion respectively. 2) *Internal strain induced by inter-layer matching between perovskite layer and rock salt layer*: Since the perovskite-type layer and rock salt-type have to stack together, the interionic distances D (K-F) and D(Ni-F) have to obey the following equation:

$$D(K\text{-}F)/ \sqrt{2}D(Ni\text{-}F) = 1.00 \qquad (6.2)$$

But the ratio of (R_K+R_F) to $\sqrt{2}$ $(R_{Ni}+R_F)$ is actually not exactly equal to unity, so there have compression and tension between these ions. This *misfit effect* should destabilize the K_2NiF_4-type lattice. This factor should be roughly represented by the following parameters:

$$1\text{-}t = 1 - \left[(R_K+R_F)/ \sqrt{2}(R_{Ni}+R_F)\right] \qquad (6.3)$$

Besides, it is reasonable that there may be a third factor affecting the relative stability. There may be some inter-ionic charge transfer effect due to the difference of the electronegativities between cationic elements.

The difference of electronegativites (Δx) should be a rough measure of this effect.

In order to correlate the above-mentioned parameters with the experimental facts about the relative stability of $KNiF_3$ type and K_2NiF_4 type compounds, a data file with the experimental data of 27 phase diagrams of fluoride systems having $KNiF_3$ type compounds has been built, and SVM computation is used to find the mathematical model for the coexistence of these two types of compounds. The phase diagrams with both types of compounds are defined as class "1", while the phase diagrams with $KNiF_3$ type compound only are defined as class "2". By support vector classification and LOO cross-validation method, it has been found that the predicted class of the phase diagram of $CsF\text{-}CaF_2$ system always disagrees with the class shown by the published diagram. This phase diagram determined by Bukhanova is shown in Fig. 6.2a. It includes only one intermediate compound ($CsCaF_3$). It is already known that $CsCaF_3$ is a perovskite-type compound. But the LOO prediction indicates that there should have another compound of K_2NiF_4 type. So it is interesting to do experimental work to see whether the prediction by SVM is correct or not.

After our experimental work with DTA and X-ray diffraction method, it is indeed proved the existence of an incongruent melting compound (Cs_2CaF_4). The phase diagram revised by us is shown in Fig. 6.2b [47].

It has been found that Cs_2CaF_4 is a compound with K_2NiF_4-type structure.

(3) The regularities of ternary compound formation in ternary systems:

The phase diagrams of ternary systems are very useful for metallurgists and materials scientists. It is unfortunate that the phase diagrams of many ternary systems are not determined yet, so it is desirable to use thermodynamic method to calculate ternary phase diagrams based on the data of known relevant binary systems. The thermodynamic method used here, however, cannot confirm whether there is some ternary intermediate compound formed or not. Since it is well known that many ternary systems are ternary new phase formers, it is impossible to make a complete computerized prediction of a ternary phase diagram without the consideration of the possibility of the existence of new ternary phases. So it is necessary to have some method to predict the unknown ternary compounds in ternary systems. The atomic parameter-pattern recognition method is just such an effective method for this purpose.

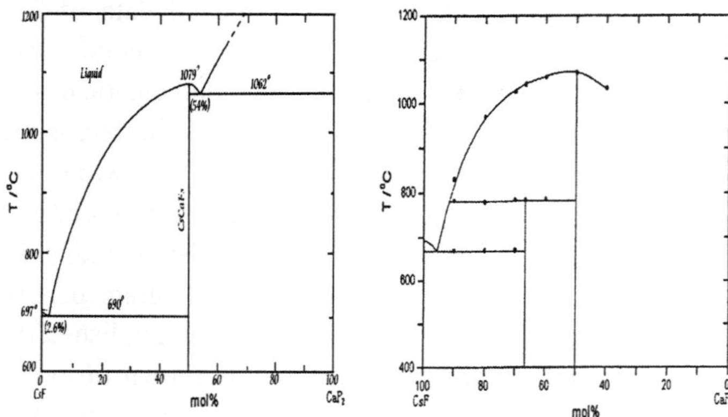

Fig. 6.2   Phase diagram of CsF-CaF$_2$ system.
a.   the phase diagram published by Bukhanova [123].
b.   the phase diagram determined in our laboratory, by Ding Yimin and Chi Liang.

As an example of this type of work, Table 6.2 lists the data of known phase diagrams of MeX$_n$-Me'X-Me''X system (here X = F, Cl, Br, I; Me is polyvalent metal; Me' and Me'' are monovalent metals). Class "1" denotes the systems with ternary intermediate compound formation, and class "2" denotes the systems without ternary intermediate compound formation. SVC method is used for the classification of systems of these two types.

The samples and their atomic parameters are presented in Table 6.2. The rate of correctness of classification of the samples listed in Table 6.2 is 100%. The rate of correctness of prediction by LOO cross-validation is more than 95%.

Table 6.2   Formation of ternary intermediate compounds of MeX$_n$-Me'X-Me''X systems.

System	Class	R_a^*	R_b^*	R_c^*	Z_a^*	X_a^*	X_x^*	R_x^*	R_x/R_a
Cr,Na,K|Cl	1	0.64	0.95	1.33	3	1.6	3.1	1.81	2.828
Cr,Na,Cc|Cl	1	0.64	0.95	1.69	3	1.6	3.1	1.81	2.828

Zr,Na,K|F	1	0.80	0.95	1.33	4	1.5	3.9	1.36	1.700
Bi,Li,Na|Cl	1	1.08	0.60	0.95	3	1.8	3.1	1.81	1.676
Be,Na,K|F	1	0.31	0.95	1.33	2	1.5	3.9	1.36	4.387
Be,Na,Rb|F	1	0.31	0.95	1.48	2	1.5	3.9	1.36	4.387
Cr,Na,Rb|Cl	1	0.64	0.95	1.48	3	1.6	3.1	1.81	2.828
Fe,Na,K|Cl	1	0.64	0.95	1.33	3	1.8	3.1	1.81	2.828
Th,Na,K|F	1	1.04	0.95	1.33	4	1.4	3.9	1.36	1.308
Y,Na,K|Cl	1	0.93	0.95	1.33	3	1.3	3.1	1.81	1.946
V,Li,K|F	1	0.69	0.60	1.33	3	1.4	3.9	1.36	1.971
V,Li,Cs|F	1	0.69	0.60	1.69	3	1.4	3.9	1.36	1.971
U,Na,Rb|F	1	0.93	0.95	1.48	4	1.4	3.9	1.36	1.462
U,Na,K|F	1	0.93	0..95	1.33	4	1.4	3.9	1.36	1.462
Be,Li,Rb|F	1	0.31	0.60	1.48	2	1.5	3.9	1.36	4.387
Be,Li,Na|F	1	0.31	0.60	0.95	2	1.5	3.9	1.36	4.387
Zr,Na,Rb|F	1	0.80	0.95	1.48	4	1.5	3.9	1.36	1.700
Al,Na,Cs|Cl	1	0.50	0.95	1.69	3	1.5	3.1	1.81	3.620
Al,Li,Rb|F	1	0.50	0.60	1.48	3	1.5	3.9	1.36	2.720
Al,Na,K|F	1	0.50	0.95	1.33	3	1.5	3.9	1.36	2.720
Al,Li,K|F	1	0.50	0.60	1.33	3	1.5	3.9	1.36	2.720
Pb,Na,K|Br	2	1.21	0.95	1.33	2	1.6	2.9	1.95	1.612
Pb,Na,Cs|Br	2	1.21	0.95	1.69	2	1.6	2.9	1.95	1.612
Pb,Na,K|I	2	1.21	0.95	1.33	2	1.6	2.6	2.15	1.777
Pb,Na,Cs|Br	2	1.21	0.95	1.69	2	1.6	2.9	1.95	1.612
Pb,K,Tl|I	2	1.21	1.33	1.44	2	1.6	2.6	2.15	1.777
Ba,Na,Cs|F	2	1.35	0.95	1.69	2	1.6	3.9	1.36	1.007
Pb,K,Tl|Br	2	1.21	1.33	1.44	2	1.6	2.9	1.95	1.612
Mn,Na,Cs|F	2	0.80	0.95	1.69	2	1.4	3.9	1.36	1.700
Pb,Na,Rb|Br	2	1.21	0.95	1.48	2	1.6	2.9	1.95	1.612
Mn,Na,K|Cl	2	0.80	0.95	1.33	2	1.4	3.1	1.81	2.263
Pb,K,Cs|Br	2	1.21	1.33	1.69	2	1.6	2.9	1.95	1.612
Mn,Na,K|F	2	0.80	0.95	1.33	2	1.4	3.9	1.36	1.700
Pb,Na,Tl|Br	2	1.21	0.95	1.44	2	1.6	2.9	1.95	1.612
Sc,Li,Cs|F	2	1.13	0.60	1.69	3	1.3	3.9	1.36	1.204
Mn,Li,Rb|F	2	0.80	0.60	1.48	2	1.4	3.9	1.36	1.700
Al,Li,Na|F	2	0.50	0.60	0.95	3	1.5	3.9	1.36	2.720
Al,Na,Cs|Br	2	0.50	0.95	1.69	3	1.5	2.9	1.95	3.900
Y,Li,Cs|F	2	0.93	0.60	1.69	3	1.2	3.9	1.36	1.462
Y,Na,K|F	2	0.93	0.95	1.33	3	1.2	3.9	1.36	1.462
Al,Li,K|Cl	2	0.50	0.60	1.33	3	1.5	3.1	1.81	3.620
Tl,Ag,K|Cl	2	0.95	1.27	1.33	3	1.8	3.1	1.81	1.905
Al,Na,Cs|Br	2	0.50	0.95	1.69	3	1.5	2.9	1.95	3.900
Th,Li,Na|F	2	1.04	0.60	0.95	4	1.4	3.9	1.36	1.308
Sr,Li,Cs|Cl	2	1.13	0.60	1.69	2	1.0	3.1	1.81	1.602
Sr,Li,K|F	2	1.13	0.60	1.33	2	1.0	3.9	1.36	1.204
Sr,Na,K|Br	2	1.13	0.95	1.33	2	1.0	2.9	1.95	1.726

Sr,Na,K|F	2	1.13	0.95	1.33	2	1.0	3.9	1.36	1.204
Th,K,Rb|F	2	1.04	1.33	1.48	4	1.4	3.9	1.36	1.308
Pb,Tl,Cs|Cl	2	1.21	1.44	1.69	2	1.6	3.1	1.81	1.496
Mn,Li,K|F	2	0.80	0.60	1.33	2	1.4	3.9	1.36	1.700
Mn,Li,Cs|F	2	0.80	0.60	1.69	2	1.4	3.9	1.36	1.700
Mn,K,Cs|F	2	0.80	1.33	1.69	2	1.4	3.9	1.36	1.700
Cd,K,Cs|Br	2	0.97	1.33	1.69	2	1.7	2.9	1.95	2.010
Ca,Na,K|F	2	0.99	0.95	1.33	2	1.0	3.9	1.36	1.374
Cd,K,Cs|Br	2	0.97	1.33	1.69	2	1.7	2.9	1.81	2.010
Cd,K,Tl|Br	2	0.97	1.33	1.44	2	1.7	2.9	1.95	2.010
Cd,K,Cs|I	2	0.97	1.33	1.69	2	1.7	2.6	2.15	2.216
Ca,Na,K|Cl	2	0.99	0.95	1.33	2	1.0	3.1	1.81	1.828
Ba,Li,K|F	2	1.35	0.60	1.33	2	0.9	3.9	1.36	1.007
Ba,Li,Rb|Br	2	1.35	0.60	1.48	2	0.9	2.8	1.95	1.444
Ca,Li,K|Cl	2	0.99	0.60	1.33	2	1.0	3.1	1.81	1.828
Ba,Li,Na|F	2	1.35	0.60	0.95	2	0.9	3.9	1.36	1.007
Ba,Li,Na|Cl	2	1.35	0.60	0.95	2	0.9	3.1	1.81	1.341
Cd,Na,Cs|Br	2	0.97	0.95	1.69	2	1.7	2.9	1.95	2.010
Cd,Na,Cs|I	2	0.97	0.95	1.69	2	1.7	2.6	2.15	2.216
Cd,Na,K|Br	2	0.97	0.95	1.33	2	1.7	2.9	1.95	2.010
La,Na,K|F	2	1.15	0.95	1.33	3	1.2	3.9	1.36	1.183
La,Li,Cs|F	2	1.15	0.60	1.69	3	1.2	3.9	1.36	1.183
Mg,Li,K|F	2	0.65	0.60	1.33	2	1.2	3.9	1.36	2.092
Mg,Na,K|F	2	0.65	0.95	1.33	2	1.2	3.9	1.36	2.092
Mg,Na,K|Cl	2	0.65	0.95	1.33	2	1.2	3.1	1.81	2.785
Al,Na,Cs|I	2	0.50	0.95	1.69	3	1.5	2.6	2.15	4.300
Ba,Li,K|Cl	2	1.35	0.60	1.33	2	0.9	3.1	1.81	1.341
Cd,Tl,Cs|Br	2	0.97	1.44	1.69	2	1.7	2.9	1.95	2.010
Cd,Na,K|Br	2	0.97	0.95	1.33	2	1.7	2.9	1.95	2.010
Cd,Na,K|I	2	0.97	0.95	1.33	2	1.7	2.6	2.15	2.216
Cd,Na,Tl|Br	2	0.97	0.95	1.44	2	1.7	2.9	1.95	2.010
Cd,Tl,Cs|Br	2	0.97	1.44	1.69	2	1.7	2.9	1.95	2.010
Ba,Na,Cs|Cl	2	1.35	0.95	1.69	2	0.9	3.1	1.81	1.341
Al,K,Cs|F	2	0.50	1.33	1.69	3	1.5	3.9	1.36	2.720

[*]R_a, R_b, R_c denote the ionic radii of Me^{n+}, Me'^{+} and Me''^{+} respectively; R_x denotes the ionic radius of X^-, X_a and X_x denote the electronegativity of Me and X respectively. Z_a denotes the charge number of Me^{n+} ion.

It has been found that the most influential factors for ternary compound formation are $Z_a/(R_a+R_x)$ and R_x/R_a in this type of systems. Larger $Z_a/(R_a+R_x)$ and larger R_x/R_a correspond to ternary intermediate compound formation. This fact can be explained as follows: According to Pauling's first rule of the crystal structure of complex ionic compounds: *A coordinated polyhedron of anions is formed about each cation, the cation-anion distance being determined by the radius sum and the ligancy of the cation by the radius ratio.* In the case of ternary intermediate compound formation problem, cation of Me^{n+} have the strongest electric field as compared with cations of Me'^{+} and Me''^{+}. So cation Me^{n+} preferentially combines with anion X^{-} to form polyhedra, or *complex anions.* $Z_a/(R_a+R_x)$ represents the strength of the electrostatic potential between X^{-} and Me^{n+}, hence it represents the stability of the polyhedron or *complex anion* formed, while R_x/R_a determines the ligancy of cation, or the *number of vertex* of the polyhedron. If both $Z_a/(R_a+R_x)$ and R_x/R_a are large, it means that there exist *stable anionic polyhedra with smaller number of vertex.* In other words, it means highly non-spherical complex anion can be formed in these systems.

It is well-known in crystal chemistry that non-spherical anions can provide different kinds of sites for cations in crystal lattice. For example, in the lattice of β-K_2SO_4, potassium ions are located in two different micro-environments: half potassium ions are coordinated to ten oxygen atoms, while the other half potassium ions are coordinated to nine oxygen atoms. It is easily understandable that the existence of two kinds of sites in crystal lattice favors the formation of ternary compounds since different micro-environments of different sites may be suitably occupied by different kinds of cations.

6.3.2 *Regularity of formability of intermediate compound in metallic systems*

The formability of intermetallic compounds can be investigated by SVM and the atomic parameters suitable for metallic systems, i.e., Midema's electronegativity (ϕ), metallic radius (R), number of valence electrons (Z) of free atom and parameter $n_{ws}^{1/3}$ and their functions. For example, Table 6.3 lists the data about the formability of ternary intermetallic compounds and related atomic parameters of known Mg-containing ternary alloy systems. By support vector classification with Gaussian kernel, the rate of correctness of classification is 100%, and the rate of correctness of prediction in LOO cross-validation is 94.9%.

Table 6.3   Formability of ternary intermetallic compounds in Mg-Me-Me' systems.

System	Class	$n_{ws}^{1/3}(1)^*$	$n_{ws}^{1/3}(2)^*$	$\phi(1)^*$	$\phi(2)^*$	$R(1)^*$	$R(2)^*$	$Z(1)^*$	$Z(2)^*$
Mg-Sr-Pb	1	0.84	1.15	2.40	4.10	2.151	1.750	2	4
Mg-Ca-Cd	1	0.91	1.24	2.55	4.05	1.974	1.568	2	2
Mg-Ca-Sb	1	0.91	1.26	2.55	4.40	1.974	1.590	2	5
Mg-Ca-Zn	1	0.91	1.22	2.55	4.10	1.974	1.394	2	2
Mg-Ba-Bi	1	0.81	1.16	2.32	4.15	2.243	1.700	2	3
Mg-Ba-Pb	1	0.81	1.15	2.32	4.10	2.243	1.750	2	4
Mg-Ba-Sb	1	0.81	1.26	2.32	4.40	2.243	1.590	2	5
Mg-Zn-Al	1	1.32	1.39	4.10	4.20	1.394	1.432	2	3
Mg-Ca-Sn	1	0.91	1.24	2.55	4.15	1.974	1.623	2	4
Mg-Na-Pb	1	0.82	1.15	2.70	4.10	1.911	1.750	1	4
Mg-Na-Sb	1	0.82	1.26	2.70	4.40	1.911	1.590	1	5
Mg-Li-Pb	1	0.98	1.15	2.85	4.10	1.562	1.750	1	4
Mg-Li-Sb	1	0.98	1.26	2.85	4.40	1.562	1.590	1	5
Mg-Li-Sn	1	0.98	1.14	2.85	4.15	1.562	1.623	1	4
Mg-Li-Tl	1	0.98	1.12	2.85	3.90	1.562	1.716	1	3
Mg-Li-Zn	1	0.98	1.32	2.85	4.10	1.562	1.394	1	2
Mg-Li-Hg	1	0.98	1.24	2.85	4.20	1.562	1.573	1	2
Mg-Li-Ga	1	0.98	1.31	2.85	4.10	1.562	1.411	1	3
Mg-K-Sb	1	0.65	1.26	2.25	4.40	2.376	1.590	1	5
Mg-Li-Cd	1	0.98	1.24	2.85	4.05	1.562	1.568	1	2
Mg-Li-Al	1	0.98	1.39	2.85	4.20	1.562	1.432	1	3
Mg-Na-Ga	1	0.82	1.31	2.70	4.10	1.911	1.411	1	3
Mg-K-Bi	1	0.65	1.16	2.25	4.15	2.376	1.700	1	3
Mg-In-Sb	2	1.17	1.26	3.90	4.40	1.663	1.590	3	5
Mg-In-Sn	2	1.17	1.24	3.90	4.15	1.663	1.623	3	4
Mg-Cd-Pb	2	1.24	1.15	4.05	4.10	1.568	1.750	2	4
Mg-Cd-Tl	2	1.24	1.12	4.05	3.90	1.568	1.716	2	3
Mg-Cd-Zn	2	1.24	1.32	4.05	4.10	1.568	1.394	2	2
Mg-Cd-Bi	2	1.31	1.16	4.10	4.15	1.411	1.700	3	3
Mg-Sn-Bi	2	1.24	1.16	4.15	4.15	1.623	1.700	4	3
Mg-Zn-Bi	2	1.32	1.16	4.10	4.15	1.394	1.700	2	3
Mg-Al-Pb	2	1.39	1.15	4.20	4.10	1.432	1.750	3	4
Mg-Ca-Al	2	0.91	1.39	2.55	4.20	1.974	1.432	2	3
Mg-Al-Sb	2	1.39	1.26	4.20	4.40	1.432	1.590	3	5
Mg-Al-Sn	2	1.39	1.24	4.20	4.15	1.432	1.623	3	4
Mg-Al-Bi	2	1.39	1.16	4.20	4.15	1.432	1.700	3	3
Mg-Sr-Al	2	0.84	1.39	2.40	4.20	2.151	1.432	2	3
Mg-Al-In	2	1.39	1.17	4.20	3.90	1.432	1.663	3	3

*Since all systems contain magnesium, the atomic parameters of magnesium are not used in computation. The $n_{ws}^{1/3}(1)$, $\phi(1)$, $R(1)$ and $Z(1)$ are the atomic parameters of Me and that with sign (2) are the atomic parameters of Me'.

6.3.3 *The regularities of the formation of hydrated salts in water-salt systems*

Hydrated salts are the intermediate compounds of water-salt systems. Some hydrated salts (such as $CaCl_2.6H_2O$, $Ca(NO_3)$ $6H_2O$) are good materials for heat storage, and many hydrated double salts(such as alum) are industrial products. The computerized prediction of the formation and stability of hydrated salts and hydrated double salts is crucial for the computerized prediction and assessment of the phase diagrams of water-salt systems.

Compared with non-aqueous salt systems, the regularities of formation of intermediate compounds in water-salt systems are more complicated. Since water molecules can make all ions or a part of ions hydrated. Water molecule can combine with anion or cation only, or combine with anion and cation simultaneously. Hydrogen bond often plays important roles in the formation of hydrated salts, especially for hydrated fluorides. Since most of above-mentioned interactions are related to the electric field strength of the cations and anions, it is still possible to use ionic charge, ionic radii, the nonsphericity parameter (for salt of oxy-acids, the number of oxygen atoms can be used as the nonsphericity parameter) and the electronegativity of elements as the atomic parameters to find some regularities by SVM or other pattern recognition methods. For example, for the salts of monobasic oxy-acids, icluding bromates, chlorates, iodates, nitrates, nitrites and perchlorates, the sample points of hydrate forming salts and those of salts without hydrate can be separated by a projection maps of PLS method (Fig. 6.3). By using SVM method, the following criterion for hydrate formation can be obtained:

$$9.58Z_+ -16.05R_+ +4.082X_+ +7.04 > 0 \qquad (6.4)$$

Here Z_+ and R_+ are the ionic charge and ionic radius respectively. And X_+ is the electronegativity of cationic elements. It implies that higher cationic charge number, smaller cationic radius (it means the stronger cationic potential) and larger electronegativity of atomic group of anion favor the hydrate formation.

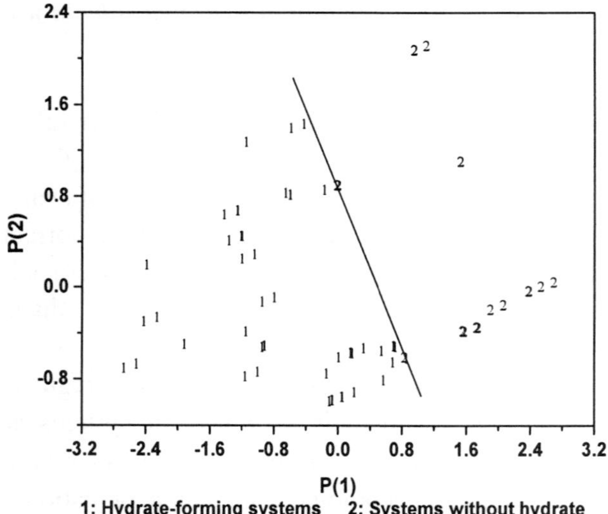

Fig. 6.3   Regularity of formation of hydrates.

6.4   Prediction of Formation of Extended Solid Solutions

6.4.1   *Principle of theory of solid solubility of inorganic systems*

The formation of extended solid solutions changes the geometry of phase diagrams significantly. Therefore, prediction of the formability of extended solid solutions is also an important step of the phase diagram prediction or assessment work.

For metallic systems, the formation of solid solutions is chiefly affected by size factor. As a rough criterion, the 15% rule proposed by Hume-Rothery is well-known. This rule states that the necessary condition of formation of extended solid solution between two metallic elements is that the relative difference of the atomic radii of these metals must be less than 15%. But size factor is not the only factor affecting the solid solubility between two metals. Electronegativity difference appears to be another important factor, since large electronegativity difference leads to intermetallic compound formation and so the solid solubility is depressed. Darken and Gurney proposed a two-dimensional plot to describe this regularity. Fig. 6.4 is an example of Darken-Gurney plot.

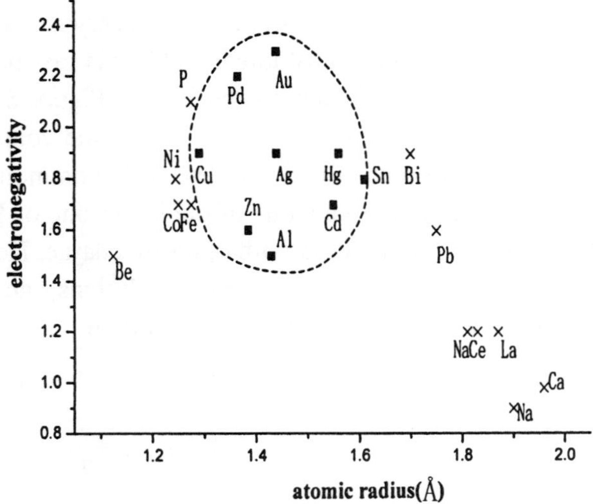

Fig. 6.4   An example of Darken-Gurney plot.

It can be seen that the elements forming extended solid solutions with silver distributed in a definite region in a map plotted by electronegativities versus atomic radii of atoms. Size factor also plays an important role in the formation of solid solutions between ionic solids. It is imaginable that a misfit effect and internal strain should occur when certain ion in a crystal lattice is substituted by another ion with different size. And the mechanism to relax the internal strain should be an influential factor affecting the energy of solution and the mutual solubility. According to the dynamic relaxation model proposed by Basanov [11], the deformation of dynamic fissures formed by the thermal motion of ions plays an important role in the relaxation process of the internal strain induced by ion substitution. For simple ionic solids (such as alkali halides with rock salt structure), it is reasonable to assume that the internal strain induced by ion substitution is proportional to the difference of ionic radii, and inversely proportional to the size of unit cell or the sum of cation-anion distance. Therefore a parameter δ can be proposed to denote the degree of internal strain induced by the substitution of ions with different sizes:

$$\delta = (R - R')/(2R'' + R + R') \qquad (6.5)$$

here R is ionic radius of the ion substituted into the crystal lattice, and R' is ionic radius of the ion to be substituted, and R'' is the ionic radius of the ion with opposite charge in the crystal lattice. Figure 6.5 illustrates the influence of δ and the radius ratio R_{Large}/R'' to the continuous solid solution formability between alkali halides with common cations or anions. Here R_{Large} is equal to the value of the larger one of R or R', and R'' is the radius of the common ion with opposite charge. Here class "1" denotes the systems with continuous solid solutions, and class "2" denotes the systems without continuous solid solution.

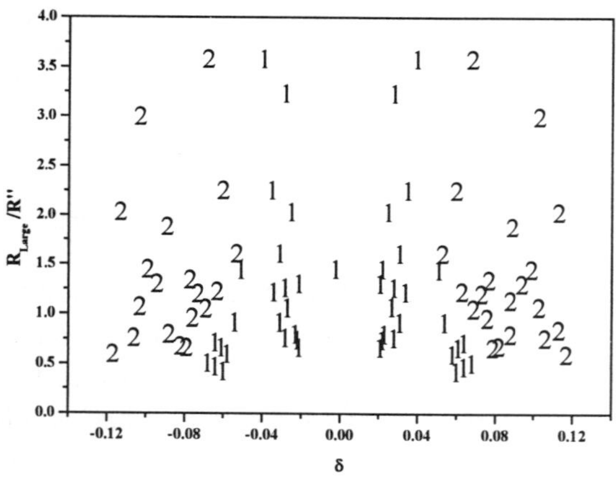

Fig. 6.5   Solid solubility between alkali halides.
1. continuous solid solution former   2. no continuous solution.

For some typical ionic systems, such as binary systems of alkali halides or the binary systems of alkali metal salts of dibasic oxy-acids, the solid solubility is chiefly dependent on the ionic radii of cations and anions. By SVM computation, some criteria obtained are useful for the assessment of the relevant phase diagrams. For the binary systems of Me_2XO_4-Me'_2XO_4 (Me, Me'= alkali metals, XO_4 =SO_4, CrO_4, etc.), most of them obey the criterion of formation of continuous or extended solid solution expressed as follows:

$$9.795R_s - 9.522R_l + 5.793 > 0 \qquad (6.6)$$

where R_l denotes the ionic radius of larger cation, and R_s denotes the ionic radius of smaller cation.

For binary systems of alkali halides with common cations or common anions, most of systems obey the criterion of formation of extended solid solution which can be expressed as follows:

$$12.75R_s - 12.39R_1 + 3.117R_{common} + 1.132 > 0 \qquad (6.7)$$

However, it can be found that there are a few published phase diagrams, such as Na_2MoO_4-K_2MoO_4, Na_2WO_4-K_2WO_4, RbF-RbCl and CsF-CsCl systems, appear as the *outliers* in SVM computation for solid solubility prediction. Therefore these published phase diagrams should be subjected to computerized work and experimental work for assessment.

6.4.2 *The assessment of some phase diagrams containing alkali molybdates or tungstates*

More than ten phase diagrams of alkali molybdate-containing or alkali tungstate-containing systems are in controversies between different authors. Since alkali molybdates or tungstates are intermediate products in metallurgical processes, and additives of molten salt fuel cell. It is desirable to do assessment work to settle down these controversies and confirm the reliable phase diagrams for these systems. The phase diagrams involving in the controversy can be classified into two categories: one is the common anions systems (such as Na_2MoO_4-K_2MoO_4 and Na_2WO_4-K_2WO_4 system), the other is the common cation systems (such as Na_2MoO_4-Na_2SO_4 system). According to literature records, the phase diagrams involving controversies are listed in Table 6.4.

Table 6.4   Phase diagrams involving controversies [106; 78].

Systems	Result supporting extended solid solution	Result negate extended solid solution
Na_2MoO_4-K_2MoO_4	Author:Amadori,M. forming continuous solid solution at high temperature.	Author: Bukhanova,G.A. No solid solution,but has two incongruently melting compounds.
Na_2SO_4-Na_2MoO_4	Author:Boeke,H.E. Forming continuous solid	Author:Bergman,A.G. No solid solution but

	solution at high temperature.	two incongruently melting compounds.
Na,K|Cl,MoO$_4$		Author:Bukhanova,G.A. No solid solution but two incongruently melting compounds.
Na,K|MoO$_4$,SO$_4$		Author:Bukhanova,G.A. No solid solution but two incongruently melting compounds.
Na,K|MoO$_4$,F		Author:Bukhanova,G.A. No solid solution but two incongruently melting compounds.
Na,K|MoO4,P2O7	Author:Amadori,M. Continuous solid solution at high temperature.	
Li,Na,K|MoO4		Author:Bergman,A.G No solid solution but two incongruently melting comounds.
Na,K|MoO$_4$.WO$_4$		Author:Bukhanova,G.A. No solid solution but two incongruently melting compounds.
Na$_2$WO$_4$-K$_2$WO$_4$		Author:Bukhanova,G.A. No solid solution but two incongruently melting compounds.
Na,K|Cl,WO$_4$	Author:Amadori,M. Forming continuous solid solution at high temperature.	Author:Bukhanova,G.A. No solid solution but two incongruently melting compounds.
Na,K|WO$_4$,P$_2$O$_7$		Author: Bukhanova,G.A. No solid solution but two incongruently melting compounds.
Na,K|WO$_4$,F		Author:Bukhanova,G.A. No solid solution but two incongruently melting compounds.
Na,K|F,CO3,MoO4		Kochikarov,Z.A. No solid solution but two incongruent binary compounds.

From Table 6.4 it can be seen that the problems of the formation of extended solid solution of three binary systems lead to the problems of the existence of extended regions of solid solution in the phase diagrams of 10 ternary systems, because these 10 ternary phase diagrams all include these binary systems as one or two edges.

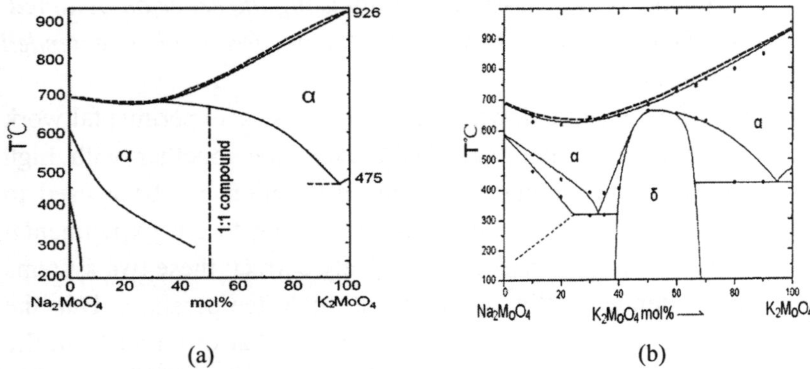

(a)   (b)

Fig. 6.6   Phase diagram of Na_2MoO_4-K_2MoO_4 system.
a. The phase diagram published by Amadori.
b. The phase diagram determined in our laboratory by Ding Yimin and Chi Liang using DTA and high temperature X-ray diffraction.

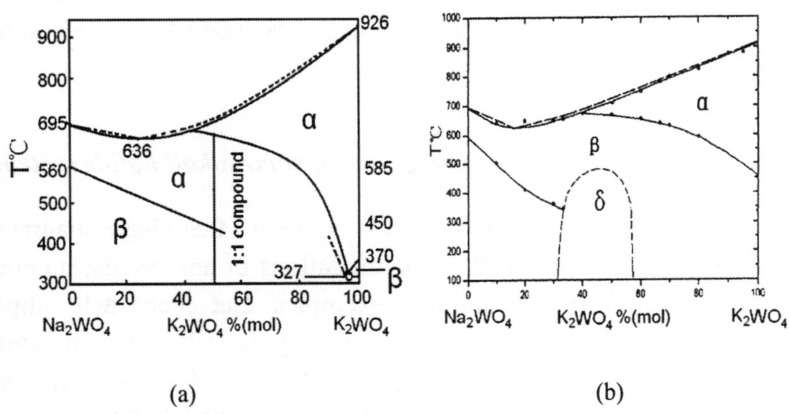

(a)   (b)

Fig. 6.7   Phase diagram of Na_2WO_4-K_2WO_4 system.
a. The phase diagram published by Drobasheva.
b. The phase diagram determined in our laboratory by Ding Yimin and Chi Liang, using DTA and high temperature X-ray diffraction.

In order to find some criteria about the existence of extended or continuous solid solutions, a data file consisting of 27 binary systems of M_2XO_4-M'_2XO_4 type has been built. All these systems are belonging to two classes: class "1" forms extended solid solutions at high temperature; and the members of class "2" do not exhibit significant solid solubility. It has been found that a good separation of these 27 samples can be realized by support vector classification. *By using the criterion obtained, K_2MoO_4-Na_2MoO_4 and K_2WO_4-Na_2WO_4 can be predicted as extended solid solution formers*

In order to finally settle down these controversies, experimental work has been done in our laboratory. DTA technique together with high temperature and room temperature X-ray diffraction has been used to re-determine these two phase diagrams. According to our experimental results, the view point of Amadori is relatively correct: these two systems indeed form extended solid solutions at high temperature. But the concrete phase relations that we found are somewhat different from the diagrams published by Amadori. These phase diagrams are illustrated in Fig. 6.6 and 6.7. The phase diagrams published by Amadori and Drobasheva are also presented for comparison.

Based on the above-mentioned results, it can be concluded that the phase diagrams of 9 ternary systems listed in Table 6.4 should also be re-investigated due to the ignorance of the existence of extended solid solubility in the relevant binary systems.

6.4.3 *The assessment of phase diagrams of some alkali halide systems*

For the phase diagrams of molten salt systems, the phase diagrams between two alkali halides with common cations or anions (the number of them is seventy) are the simplest examples. But even such simple systems the information about solid solubility is still not completely settled. One obvious example is the solid solubility of KF-KCl system. As early as the year of 1907, Plato had published that KF had limited solid solubility (up to 8 mol%) in solid KCl at high temperature, but other authors did not confirm this conclusion in later years. In his thermodynamic assessment work, Sangster assumed that this solid solubility was negligible [119]. Based on this assumption, the results of calculation gave the conclusion that the mixing of molten KF and KCl was strongly endothermic at KCl end. This conclusion is in contradiction with the result of the experimental measurement of heat of mixing by accurate calorimetry made by Kleppa. At the same time, the neglect of solid solubilities of RbF in RbCl and that of CsF in CsCl also lead to large difference between the enthalpy of mixing calculated by Sangster and that measured by Kleppa. These disagreements are shown in Table 6.5.

Table 6.5　Enthalpy of mixing of KF-KCl, RbF-Cl,CsF-CsCl of molten salts.

Systems	ΔH_{mixing} calculated by Sangster ($J \cdot mol^{-1}$)	ΔH_{mixing} measured by Kleppa ($J \cdot mol^{-1}$)
KF-KCl (at KCl end)	+3066	-420
RbF-RbCl (at RbCl end)	+5207	-840.8
CsF-CsCl (at CsCl end)	+2700	-2549

In order to find the origin of this disparity, comprehensive work with DTA measurement, limiting slope calculation and atomic parameter-pattern recognition method with SVM has been used in our research work. The result of re-determination of phase diagram of KF-KCl system is just the same as that of Plato published. Both the result of DTA and the limiting slope at the KCl end confirm the significant solid solubility of KF in KCl. According to the mathematical model built by SVM, both RbF-RbCl system and CsF-CsCl system should also exhibit significant solid solubilities at the chloride end. Since there is no data of these solid solubilities in the previously published papers, experimental work has been done in our laboratory. Both the DTA results and the limiting slope calculation show that there are significant solid solubilities at the chloride end (Fig. 6.8 and Fig. 6.9).

After taking account of the solid solubilities of KF in KCl, RbF in RbCl and CsF in CsCl at high temperatures, the thermodynamic calculation results show that the mixing of molten salt solutions near the chloride end are exothermic. It is in agreement with the experimental results made by Kleppa.

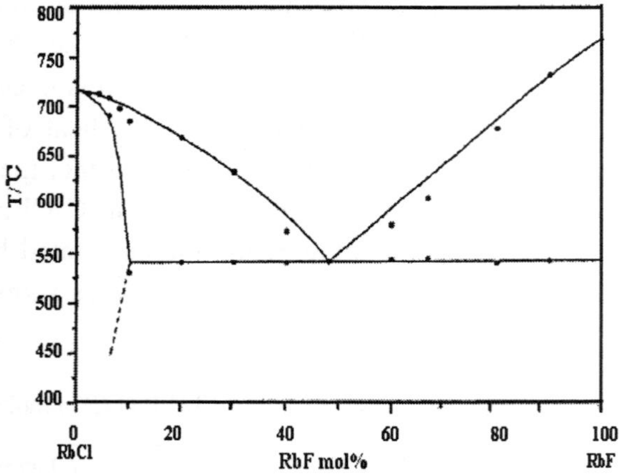

Fig. 6.8   The phase diagram of RbF-RbCl system.

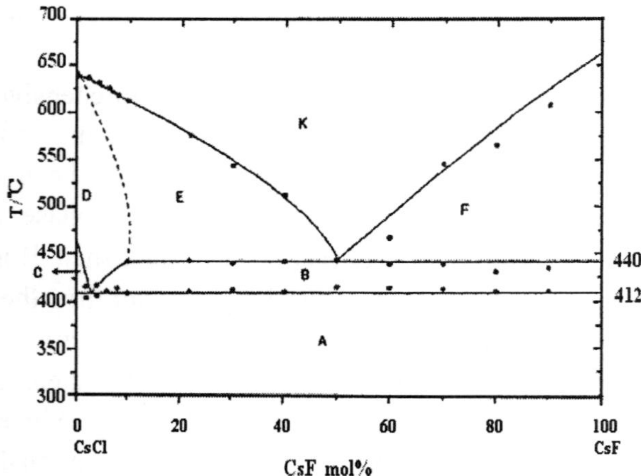

Fig. 6.9   The phase diagram of CsF-CsCl system.

6.5   Prediction of Melting Types of Intermediate Compounds

The intermediate compounds can be classified into two different classes according to their melting types: congruently melting compounds and incongruently melting compounds. Different melting types of intermediate compounds give rise to different geometry of phase diagrams (Fig. 6.10). Therefore the prediction of the melting type of intermediate compound is also a necessary step for the computerized prediction of phase diagrams by atomic parameter-pattern recognition method.

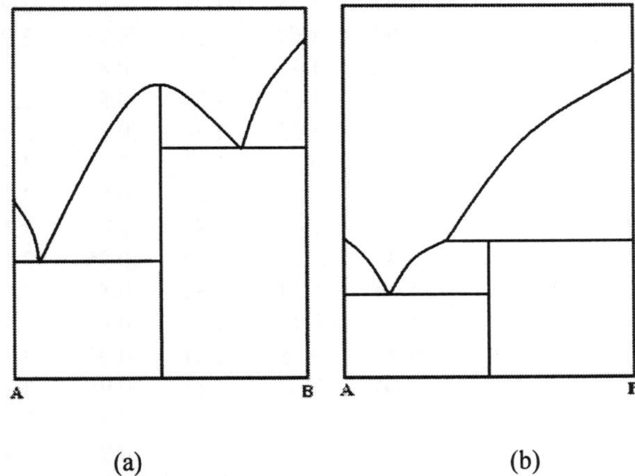

Fig. 6.10   Phase diagrams with congruently melting or incongruently melting intermediate compounds.
(a) phase diagram with congruently melting intermediate compound.
(b) phase diagram with incongruently melting intermediate compound.

Many binary systems consisting of halides of monovalent metals and halides of trivalent metals form intermediate compounds of $Me_3Me'X_6$ (X=F, Cl, Br, I) type. Some of them are congruently melting compounds and the others are incongruently melting compounds. Table 6.6 lists the melting types and relevant atomic parameters of 89 intermediate compounds of this valence type.

Table 6.6   Melting types of $Me_3Me'X_6$ type compounds.

Compound	Melting type[*]	$R_{Me'}$	R_{Me}	R_x	X_{Me}	X_x	N_d
Li_3AlF_6	1	0.50	0.60	1.36	0.95	3.9	0
Na_3AlF_6	1	0.50	0.95	1.36	0.9	3.9	0
K_3AlF_6	1	0.50	1.33	1.36	0.8	3.9	0
Cs_3AlF_6	1	0.50	1.69	1.36	0.75	3.9	0
Rb_3BiF_6	1	1.08	1.48	1.36	0.8	3.9	0
Na_3ScF_6	1	0.81	0.95	1.36	0.9	3.9	3
Cs_3ScF_6	1	0.81	1.69	1.36	0.75	3.9	0

Cs_3YF_6	1	0.93	1.69	1.36	0.75	3.9	0
K_3LaF_6	2	1.15	1.33	1.36	0.8	3.9	0
K_3CeF_6	2	1.14	1.33	1.36	0.8	3.9	0
Cs_3CeF_6	1	1.14	1.69	1.36	0.75	3.9	0
Cs_3PrF_6	1	1.12	1.69	1.36	0.75	3.9	0
K_3SmF_6	1	1.07	1.33	1.36	0.8	3.9	0
Rb_3SmF_6	1	1.07	1.48	1.36	0.8	3.9	0
Cs_3SmF_6	1	1.07	1.69	1.36	0.75	3.9	0
K_3DyF_6	1	1.00	1.33	1.36	0.8	3.9	0
K_3ErF_6	1	0.98	1.33	1.36	0.8	3.9	0
K_3TbF_6	1	0.96	1.33	1.36	0.8	3.9	0
Na_3InCl_6	2	0.81	0.95	1.81	0.9	3.1	10
K_3InCl_6	2	0.81	1.33	1.81	0.8	3.1	10
Tl_3SbCl_6	2	0.92	1.44	1.81	1.4	3.1	10
Cs_3SbCl_6	1	0.92	1.69	1.81	0.75	3.1	10
K_3BiCl_6	1	1.08	1.33	1.81	0.8	3.1	10
Rb_3BiCl_6	1	1.08	1.48	1.81	0.8	3.1	10
Na_3RhCl_6	2	0.72	0.95	1.81	0.9	3.1	6
Li_3TiCl_6	2	0.64	0.60	1.81	0.95	3.1	1
Na_3VCl_6	2	0.72	0.95	1.81	0.9	3.1	2
Na_3CrCl_6	2	0.62	0.95	1.81	0.9	3.1	3
K_3CrCl_6	1	0.62	1.33	1.81	0.8	3.1	3
Rb_3CrCl_6	1	0.62	1.48	1.81	0.8	3.1	3
Cs_3CrCl_6	1	0.62	1.69	1.81	0.75	3.1	3
Li_3ScCl_6	2	0.81	0.60	1.81	0.95	3.1	0
Cs_3ScCl_6	1	0.81	1.69	1.81	0.75	3.1	0
Na_3YCl_6	2	0.93	0.95	1.81	0.9	3.1	0
K_3YCl_6	1	0.93	1.33	1.81	0.8	3.1	0
Rb_3YCl_6	1	0.93	1.48	1.81	0.8	3.1	0
Cs_3YCl_6	1	0.93	1.69	1.81	0.75	3.1	0
K_3LaCl_6	1	1.15	1.33	1.81	0.8	3.1	0
Cs_3LaCl_6	1	1.15	0.95	1.81	0.9	3.1	0
K_3CeCl_6	1	1.14	1.33	1.81	0.8	3.1	0
Cs_3CeCl_6	1	1.14	1.69	1.81	0.75	3.1	0
Cs_3PrCl_6	1	1.12	1.69	1.81	0.75	3.1	0
K_3NdCl_6	1	1.10	1.33	1.81	0.8	3.1	0
Cs_3NdCl_6	1	1.10	1.69	1.81	0.75	3.1	0
K_3SmCl_6	1	1.07	1.33	1.81	0.8	3.1	0
Rb_3SmCl_6	1	1.07	1.48	1.81	0.8	3.1	0
Cs_3SmC	1	1.07	1.69	1.81	0.75	3.1	0
Na_3EuCl_6	2	1.05	0.95	1.81	0.9	3.1	0
K_3EuCl_6	1	1.05	1.33	1.81	0.8	3.1	0
K_3GdCl_6	1	1.03	1.33	1.81	0.8	3.1	0
Na_3DyCl_6	2	1.00	0.95	1.81	0.9	3.1	0
K_3DyCl_6	1	1.00	1.33	1.81	0.8	3.1	0

Cs_3DyCl_6	1	1.00	1.69	1.81	0.75	3.1	0
Na_3HoCl_6	2	0.99	0.95	1.81	0.9	3.1	0
K_3HoCl_6	1	0.99	1.33	1.81	0.8	3.1	0
K_3ErCl_6	1	0.98	1.33	1.81	0.8	3.1	0
Cs_3ErCl_6	1	0.98	1.69	1.81	0.75	3.1	0
Na_3TbCl_6	2	0.96	0.95	1.81	0.9	3.1	0
Na_3YbCl_6	2	0.95	0.95	1.81	0.9	3.1	0
K_3YbCl_6	1	0.95	1.33	1.81	0.8	3.1	0
Cs_3YbCl_6	1	0.95	1.69	1.81	0.75	3.1	0
Na_3LuCl_6	2	0.93	0.95	1.81	0.9	3.1	0
K_3PuCl_6	1	1.00	1.33	1.81	0.8	3.1	0
Cs_3PuCl_6	1	1.00	1.69	1.81	0.75	3.1	0
Na_3InBr_6	2	0.81	0.95	1.95	0.9	2.9	10
K_3InBr_6	2	0.81	1.33	1.95	0.8	2.9	10
Cs_3InBr_6	2	0.81	1.69	1.95	0.75	2.9	10
K_3SbBr_6	2	0.92	1.33	1.95	0.8	2.9	10
K_3ScBr_6	1	0.81	1.33	1.95	0.8	2.9	0
Na_3HoBr_6	2	0.99	0.95	1.95	0.9	2.9	0
Cs_3HoBr_6	1	0.99	1.69	1.95	0.75	2.9	0
Na_3InI_6	2	0.81	0.95	2.15	0.9	2.6	10
K_3InI_6	2	0.81	1.33	2.15	0.8	2.6	10
Cs_3InI_6	2	0.81	1.69	2.15	0.75	2.6	10
Ag_3BiI_6	1	1.08	1.27	2.15	1.9	2.6	10
K_3ScI_6	1	0.81	1.33	2.15	0.8	2.6	0
Rb_3ScI_6	1	0.81	1.48	2.15	0.8	2.6	0
K_3LaI_6	1	1.15	1.33	2.15	0.8	2.6	0
Cs_3LaI_6	1	1.15	1.69	2.15	0.75	2.6	0
K_3PrI_6	1	1.12	1.33	2.15	0.8	2.6	0
Cs_3PrI_6	1	1.12	1.69	2.15	0.75	2.6	0
K_3NdI_6	1	1.10	1.33	2.15	0.8	2.6	0
Cs_3NdI_6	1	1.10	1.69	2.15	0.75	2.6	0
K_3SmI_6	1	1.07	1.33	2.15	0.8	2.6	0
Cs_3SmI_6	1	1.07	1.69	2.15	0.75	2.6	0
K_3HoI_6	1	0.99	1.33	2.15	0.8	2.6	0
Rb_3HoI_6	1	0.99	1.48	2.15	0.8	2.6	0
Cs_3HoI_6	1	0.99	1.69	2.15	0.75	2.6	0
Cs_3TmI_6	1	0.96	1.69	2.15	0.75	2.6	0

*Class "1" denotes congruently melting compounds, and class "2" denotes incongruently melting compounds. N_d denotes the number of d electrons in the next outmost shell in the atom of Me'.

By SVC with Gaussian kernel and C = 100, the samples listed in Table 6.6 can be classified with the rate of correctness of 100%. In LOO cross-validation test, the rate of correctness of prediction is 97.8%.

6.6 Modeling of Melting Points or Decomposition Temperature of Intermediate Compounds

Many binary or ternary systems have intermediate compounds. The congruently melting intermediate compound formation has great influence on the geometry of phase diagrams. If we wish to make prediction of the unknown phase diagram completely, it is necessary to have some method for the prediction of the melting points of unknown intermediate compounds. Here we will describe some examples of the melting point estimation by SVM and atomic parameter method:

(1) Computerized Modeling of Melting Points of 1-1 Type Intermetallic Compounds between Rare Earth and Nontransition Metals

Since the 1-1 type compounds between rare earth and nontransition metals include two categories: some of them melt congruently, and the others are incongruently melting compounds. At first it is necessary to make some mathematical model to differentiate these two categories of compounds, and then try to predict the melting points of the compounds of the first category, or try to predict the decomposition temperatures (peritectic temperature) of the compounds of the second category. We use support vector classification method to make the mathematical model for the differentiation of the intermediate compounds of these two categories.

Forty one known 1-1 type intermediate compounds between rare earth and nontransition metals have been used as the training set. By support vector classification method using the kernel function of second degree or Gaussian type of kernels, the separation of these two categories of compounds is rather good. By LOO cross-validation method and support vector classification, the rate of correctness of computerized prediction is more than 92%.

After the classification of the compounds of these two categories, we can start to make mathematical modeling for the quantitative prediction of the melting points and decomposition points of intermediate compounds.

Table 6.7 is a data file for the modeling of the melting points of 1-1 type congruently melted intermetallic compounds between rare earth and nontransition metals. And Table 6.8 is a data file for the modeling of the decomposition temperature of incongruently melted intermetallic compounds between those metals.

Table 6.7   Data file for the modeling of the melting point of intermetallic compounds between rare earth and non-transition metals.

Compounds	$T_{m.p.}$ °C	$T_{m.p.RE}$ °C	$T_{m.p.nontr}$ °C	$\Delta\phi$	R_{RE}/R_{NTR}	Z_{nontr}^{*}
LaBi	1615	918	271	0.98	1.104	3
NdBi	1900	1021	271	0.96	1.071	3
LaCd	946	918	321	0.88	1.197	3
CeZn	825	798	419	0.92	1.309	2
ErGa	1340	1529	30	0.88	1.245	3
EuGa	1030	822	30	0.90	1.275	3
HoGa	1280	1474	30	0.88	1.252	3
SmGa	1390	1074	30	0.90	1.277	3
GdIn	1250	1313	156	0.70	1.084	3
YIn	1220	1522	156	0.70	1.083	3
YbIn	1067	819	156	0.68	1.046	3
LaZn	815	918	419	0.93	1.346	2
PrZn	882	931	419	0.91	1.311	2
LaIn	1125	918	156	0.73	1.129	3
NdIn	1230	1021	156	0.71	1.095	3
LaHg	1100	918	-38	1.03	1.193	2
GdTl	1270	1313	304	0.70	1.050	3
NdGa	1488	1021	30	1.00	1.291	3
GdGa	1400	1313	30	0.90	1.277	3
HoIn	1270	1474	156	0.68	1.062	3
LaTl	1220	918	304	0.73	1.094	3
NdZn	923	1021	419	0.91	1.306	2
YbPb	1116	619	327	0.88	0.994	4
SmZn	960	1074	419	0.90	1.293	2
TbTl	1300	1356	304	0.69	1.058	3
YZn	1105	1522	419	0.90	1.292	2

The mathematical model for the estimation of the melting points of the compounds listed in Table 6.7 can be expressed as follows:

$$T_{mp} = 0.526\Delta\phi + 0.368T_{mp.RE} + 0.031T_{mp.nontr} - 0.337(R_{RE}/R_{NTR}) + 0.481Z_{nontr} - 0.091 \quad (6.8)$$

Here $T_{m.p.}$ is the melting point of intermediate compound, $\Delta\phi$ is the difference of Miedema electronegativities, $T_{m.p.RE}$ and $T_{m.p.nontr}$ are the melting points of the rare earth and nontransition metals respectively. R_{RE}/R_{NTR} is the ratio of atomic radii of rare earth and nontransition metals. Z_{nontr} is the number of valence electrons of the atoms of transition metals.

Using LOO cross-validation method, the predicted values of melting point can be obtained. The average relative absolute error of this computation is 7.7%. Figure 6.11 illustrates the comparison between the experimental values and the predicted values by SVR computation.

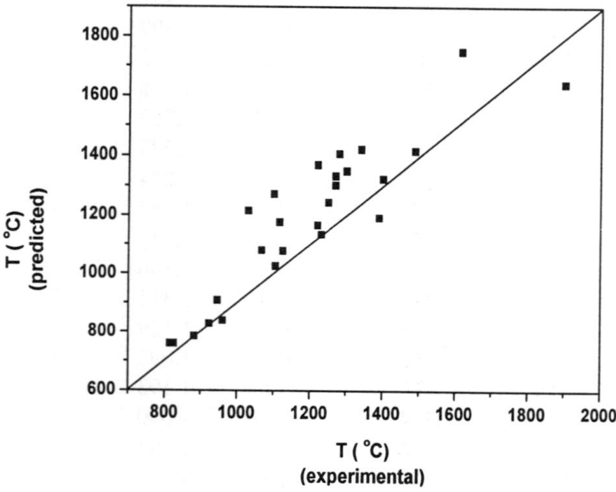

Fig. 6.11   Results of melting points prediction by SVR.

Table 6.8   Data file for the modeling of the decomposition temperature of incongruent intermediate compounds between rare earth and nontransition metals.

Compounds	$T_{d.p.}$ °C	$T_{m.p.RE}$ °C	$T_{m.p.nontr}$ °C	$\Delta\phi$	R_{RE}/R_{NTR}	Z_{nontr}^{*}
LaAl	873	918	660	1.23	1.311	3
CeAl	845	798	660	1.02	1.274	3
PrAl	905	931	660	1.01	1.277	3
NdAl	940	1021	660	1.01	1.272	3
GdAl	1075	1313	660	1.00	1.258	3
HoAl	1115	1474	660	0.98	1.233	3
YAl	1130	1522	660	1.00	1.258	3
EuMg	463	822	650	0.25	1.123	2
YMg	934	1522	650	0.25	1.124	2
LaGa	1010	918	30	0.93	1.380	3
PrGa	1015	931	30	0.91	1.296	3

LuIn	1080	1663	156	0.68	1.043	3
LaSn	1295	918	231	0.98	1.157	4
PrSn	1253	931	231	0.96	1.126	4
NdSn	1257	1021	231	0.96	1.122	4

[*] $T_{d.p.}$ denotes the decomposition temperature of incongruent intermediate compounds, $T_{m.p.RE}$ denotes the melting point of rare earth metals, and $T_{m.p.nontr}$ denotes the melting points of nontransition metals. $\Delta\phi$ denotes the difference of Miedema's electronegativity. R_{RE}/R_{NTR} denotes the ratio of metallic radii. Z_{nontr} denotes the number of valence electron of atom of nontransition metals.

By a similar procedure, the mathematical model obtained for the decomposition temperature of incongruently melted intermediate compounds between rare earth and nontransition metals obtained by SVR computation can be expressed as follows:

$$T_{d.p.} = 0.451\,T_{mp.RE} + 0.024\,T_{mp.nontr} - 0.065\,\Delta\phi + 0.198\,(R_{RE}/R_{NTR}) + 0.794\,Z_{nontr} + 0.017 \quad (6.9)$$

The comparison between the values of prediction in LOO cross-validation and the real values obtained by experimental measurements is shown in Fig. 6.12.

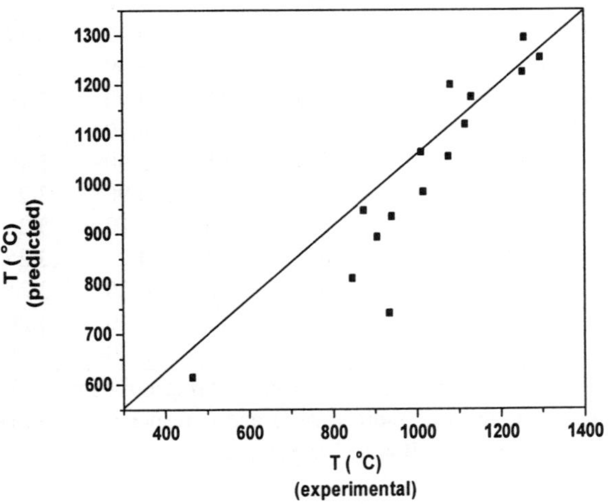

Fig. 6.12   Results of prediction of decomposition temperature by SVR.

6.7 Prediction of Crystal Types of Intermediate Compounds

Atomic parameter-pattern recognition method is rather effective for the computerized prediction of the crystal types of unknown intermediate phases.

In previous literatures, it has been reported that 10 ternary intermetallic compounds of A_2BC type have BiF_3-type structure, while other 10 compounds exist with Co_2GaPr-type structure. The data of those 20 compounds are listed in Table 6.9.

Table 6.9　The crystal types of A_2BC-type intermetallic compounds.

Compound	Z_a^*	R_a^*	R_b^*	R_c^*	ϕ_a^*	ϕ_b^*	ϕ_c^*	Crystal type
Co_2InDy	9	1.252	1.663	1.773	5.10	3.90	3.21	Co_2GaPr
Co_2InGd	9	1.252	1.663	1.802	5.10	3.90	3.20	Co_2GaPr
Co_2InPr	9	1.252	1.663	1.828	5.10	3.90	3.19	Co_2GaPr
Co_2InNd	9	1.252	1.663	1.821	5.10	3.90	3.19	Co_2GaPr
Co_2GaPr	9	1.252	1.411	1.828	5.10	4.10	3.19	Co_2GaPr
Co_2InSm	9	1.252	1.663	1.802	5.10	3.90	3.20	Co_2GaPr
Co_2GaLa	9	1.252	1.411	1.877	5.10	4.10	3.17	Co_2GaPr
Ni_2InPr	10	1.246	1.663	1.828	5.20	3.90	3.19	Co_2GaPr
Ni_2InNd	10	1.246	1.663	1.821	5.20	3.90	3.19	Co_2GaPr
Ni_2InLa	10	1.246	1.663	1.877	5.20	3.90	3.17	Co_2GaPr
Cu_2InTb	11	1.278	1.663	1.782	4.45	3.90	3.21	BiF_3
Cu_2InY	11	1.278	1.663	1.801	4.45	3.90	3.20	BiF_3
Cu_2AlSc	11	1.278	1.432	1.641	4.45	4.20	3.25	BiF_3
Cu_2InNd	11	1.278	1.663	1.821	4.45	3.90	3.19	BiF_3
Cu_2InSm	11	1.278	1.663	1.802	4.45	3.90	3.20	BiF_3
Ni_2GaSc	10	1.246	1.411	1.641	5.20	4.10	3.25	BiF_3
Ni_2AlSc	10	1.246	1.432	1.641	5.20	4.20	3.25	BiF_3
Pd_2AlSc	10	1.376	1.432	1.641	5.45	4.20	3.25	BiF_3
Pd_2InY	10	1.376	1.663	1.801	5.45	3.90	3.20	BiF_3
Pd_2InSc	10	1.376	1.663	1.641	5.45	3.90	3.25	BiF_3

$^*Z_a$ denotes the total number of valence electrons and d electrons in unfilled shell of atom A. R_a, R_b, R_c denote the metallic radii of A, B and C respectively. ϕ_a, ϕ_b, ϕ_c denote the Miedema electronegativity of A, B and C respectively.

By SVC with linear kernel function, the samples listed in Table 6.9 can be separated into two classes completely. The rate of correctness of prediction in LOO cross-validation is also 100%. The criterion of Co_2GaPr-type structure can be expressed by the following inequality:

$$92.10\text{-}101.46\,R_a + 16.14\,R_c - 3.753\phi_b + 6.837\phi_c > 0 \qquad (6.10)$$

By using this criterion, it can be predicted that a new compound, Co_2ZnLa, should exist with Co_2GaPr structure. The prediction has been confirmed by a recently published paper [146].

6.8 Modeling of Liquid-Liquid Immiscibility of Inorganic Systems

6.8.1 *Regularity of liquid-liquid immiscibility of molten salt systems*

Some molten salt systems form two immiscible layers. Since the free energy of mixing of molten salts should be related to the electrostatic potential energy of ions, and the liquid-liquid immiscibility can occur only when the temperature of the system is lower than the temperature of critical mixing. For the molten salt systems with the same valence type, the occurrence of immiscibility should be related to the ionic radii and the melting points of pure salts. For example, some $AX\text{-}BX_3$ type systems exhibit liquid-liquid immiscibility in the AX_3 rich regions of these systems (Here X is halogen, A and B are monovalent and trivalent metallic elements respectively). Table 6.10 lists the salt systems of $AX\text{-}BX_3$ type used as training set for modeling.

Table 6.10   Liquid-liquid immiscibilty of binary systems of molten salts.

Salt system	Melting point of AX_3 (K°)	R_A	R_B	R_X	Immiscibility
$AgBr\text{-}AlBr_3$	371	0.50	1.00	1.95	Yes
$KBr\text{-}AlBr_3$	371	0.50	1.33	1.95	Yes
$KCl\text{-}AlCl_3$	466	0.50	1.33	1.81	Yes
$NaCl\text{-}AlCl_3$	466	0.50	0.95	1.81	Yes
$NaBr\text{-}AlBr_3$	371	0.50	0.95	1.95	Yes
$CuI\text{-}InI_3$	483	0.81	0.96	2.15	Yes
$TlBr\text{-}AlBr_3$	371	0.50	1.40	1.95	Yes
$LiCl\text{-}AlCl_3$	466	0.50	0.60	1.81	Yes
$AgI\text{-}InI_3$	483	0.81	1.00	2.15	Yes
$AgCl\text{-}AlCl_3$	466	0.50	1.00	1.81	Yes

$CsCl\text{-}AlCl_3$	466	0.50	1.69	1.81	Yes
$TlCl\text{-}AlCl_3$	466	0.50	1.40	1.81	Yes
$NaI\text{-}AlI_3$	464	0.50	0.95	2.15	No
$AgBr\text{-}InBr_3$	709	0.81	1.00	1.95	No
$NaCl\text{-}CrCl_3$	1425	0.62	0.95	1.81	No
$NaCl\text{-}ErCl_3$	1047	0.98	0.95	1.81	No
$NaCl\text{-}EuCl_3$	896	1.05	0.95	1.81	No
$NaCl\text{-}FeCl_3$	577	0.64	0.95	1.81	No
$NaCl\text{-}BiCl_3$	502	1.08	0.95	1.81	No
$LiF\text{-}PrF_3$	1643	1.12	0.60	1.36	No
$LiF\text{-}HoF_3$	1630	0.99	0.60	1.36	No
$LiF\text{-}InF_3$	1443	0.81	0.60	1.36	No
$LiF\text{-}ScF_3$	1500	0.81	0.60	1.36	No
$LiF\text{-}YF_3$	1660	0.93	0.60	1.36	No
$LiF\text{-}LaF_3$	1700	1.15	0.60	1.36	No
$TlCl\text{-}GdCl_3$	882	1.03	1.40	1.81	No
$RbCl\text{-}YCl_3$	973	0.93	1.48	1.81	No
$TlCl\text{-}BiCl_3$	502	1.08	1.40	1.81	No
$NaCl\text{-}YCl_3$	973	0.93	0.95	1.81	No
$NaF\text{-}YF_3$	1660	0.93	0.95	1.36	No
$RbCl\text{-}CrCl_3$	1425	0.62	1.48	1.81	No
$RbCl\text{-}GaCl_3$	351	0.62	1.48	1.81	No
$RbCl\text{-}NdCl_3$	1033	1.10	1.48	1.81	No
$NaF\text{-}YbF_3$	1650	0.95	0.95	1.36	No
$RbBr\text{-}TiBr_3$	1200	0.64	1.48	1.95	No
$RbCl\text{-}SmCl_3$	951	1.07	1.48	1.81	No
$NaCl\text{-}GaCl_3$	351	0.62	0.95	1.81	No
$NaCl\text{-}TbCl_3$	861	1.02	0.95	1.81	No
$NaCl\text{-}SmCl_3$	1013	1.07	0.95	1.81	No
$NaCl\text{-}GaCl_3$	351	0.62	0.95	1.81	No
$LiCl\text{-}GaCl_3$	351	0.62	0.60	1.81	No
$NaCl\text{-}HoCl_3$	991	0.99	0.95	1.81	No
$NaCl\text{-}InCl_3$	859	0.81	0.95	1.81	No
$NaF\text{-}PrF_3$	1643	1.12	0.95	1.36	No
$NaF\text{-}SmF_3$	1670	1.07	0.95	1.36	No
$NaF\text{-}ScF_3$	1500	0.81	0.95	1.36	No
$NaF\text{-}GdF_3$	1650	1.03	0.95	1.36	No
$KCl\text{-}YbCl_3$	1000	0.95	1.33	1.81	No
$CsCl\text{-}ScCl_3$	1233	0.81	1.69	1.81	No
$CsCl\text{-}SmCl_3$	1013	1.07	1.69	1.81	No
$CsBr\text{-}SmCl_3$	1013	1.07	1.69	1.81	No
$CsI\text{-}SmI_3$	1098	1.07	1.69	2.15	No
$CsBr\text{-}TiBr_3$	1200	0.64	1.69	1.95	No
$CsCl\text{-}NdCl_3$	1033	1.10	1.69	1.81	No
$CsCl\text{-}DyCl_3$	927	1.00	1.69	1.81	No

CsCl-NdCl$_3$	1033	1.10	1.69	1.81	No
CsCl-CrCl$_3$	1425	0.62	1.69	1.81	No
AgI-GaI$_3$	485	0.62	1.00	2.15	No
CuCl-FeCl$_3$	577	0.64	0.96	1.81	No
CsCl-ErCl$_3$	1047	0.98	1.69	1.81	No
CsCl-GaCl$_3$	351	0.62	1.69	1.81	No
CsCl-FeCl$_3$	577	0.64	1.69	1.81	No
CsCl-LaCl$_3$	1125	1.15	1.69	1.81	No
CsBr-InBr$_3$	709	0.81	1.69	1.95	No
CuCl-GaCl$_3$	351	0.62	0.96	1.81	No
CuI-GaI$_3$	485	0.62	0.96	2.15	No
CuCl-InCl$_3$	859	0.81	0.96	1.81	No
KI-NdI$_3$	1048	1.10	1.33	2.15	No
KCl-NdCl$_3$	1033	1.10	1.33	1.81	No
KCl-LaCl$_3$	1125	1.15	1.33	1.81	No
KCl-InCl$_3$	859	0.81	1.33	1.81	No
KCl-SmCl$_3$	951	1.07	1.33	1.81	No
KBr-TiBr$_3$	1200	0.64	1.33	1.95	No
KCl-TbCl$_3$	861	1.02	1.33	1.81	No
KCl-YCl$_3$	973	0.93	1.33	1.81	No
TlCl-InCl$_3$	859	0.81	1.40	1.81	No
LiF-CeF$_3$	1733	1.14	0.60	1.36	No
KCl-HoCl$_3$	991	0.99	1.33	1.81	No
KCl-EuCl$_3$	896	1.05	1.33	1.81	No
KI-AlI$_3$	464	0.50	1.33	2.15	No
KF-AlF$_3$	1545	0.50	1.33	1.36	No
KCl-CrCl$_3$	1425	0.62	1.33	1.81	No
KCl-BiCl$_3$	502	1.08	1.33	1.81	No
KCl-ErCl$_3$	1047	0.98	1.33	1.81	No
KCl-DyCl$_3$	927	1.00	1.33	1.81	No
TlI-InI$_3$	483	0.81	1.40	2.15	No

Using support vector classification with Gaussian kernel, the data set listed in Table 6.10 can be classified with clear-cut boundaries in the feature space. The LOO cross-validation test gives the rate of correctness of prediction equal to 98.9%.

6.8.2   *Regularity of liquid-liquid immiscibility of liquid alloy systems*

SVM and atomic parameter method can be also used to study the liquid-liquid immiscibility of alloy systems. Table 6.11 lists the data of

    Support Vector Machine in Chemistry

36 binary alloys (16 of them exhibit liquid-liquid immiscibility, and the rest of them form homogeneous liquid solutions). The following four atomic parameters are used in SVC computation for the classification of the phase diagrams with or without liquid-liquid immiscibilities:

T_{mp}:   the higher melting temperature of components

$\Delta\phi$:   difference of Miedema's electronegativities

$\Delta n_{ws}^{1/3}$: difference of valence electron densities in Wagner-Seitz cell calculated by Miedema

(R_1/R_2): ratio of atomic radii proposed by Teatum.

Table 6.11   Liquid-liquid immiscibility of binary metallic systems.

Number	System	T_{mp} °C	$\Delta\phi$	$\Delta n_{ws}^{1/3}$	R_1/R_2	Liquid-Liquid immisciblity
1	Bi-Zn	419	0.05	-0.16	1.22	Yes
2	Ca-Na	850	-0.15	0	1.03	Yes
3	Ga-Cd	321	0.05	0.07	0.90	Yes
4	Pb-Zn	419	0	-0.17	1.26	Yes
5	Pb-Ga	327	0	-0.16	1.24	Yes
6	Li-K	180	0.60	0.33	0.657	Yes
7	Li-Na	180	0.15	0.16	0.817	Yes
8	Mg-Na	650	0.75	0.35	0.838	Yes
9	Bi-Al	659	-0.05	-0.23	1.18	Yes
10	Pb-Al	659	-0.10	-0.24	1.22	Yes
11	Al-In	659	0.30	0.22	0.861	Yes
12	Al-K	659	1.95	0.74	0.603	Yes
13	Al-Na	659	1.50	0.57	0.749	Yes
14	Bi-Ga	659	0.05	-0.13	1.21	Yes
15	Al-Cd	659	0.15	0.15	0.913	Yes
16	Mg-K	650	1.20	0.52	0.674	Yes
17	Bi-Hg	271	-0.05	-0.08	1.08	No
18	Bi-In	271	0.25	-0.01	1.02	No
19	Bi-K	271	1.90	0.51	0.715	No
20	Bi-Li	271	1.30	0.18	1.09	No
21	Bi-Sr	770	1.75	0.32	0.790	No
22	Cd-Ca	850	1.50	0.33	0.394	No
23	Ga-Ca	850	1.55	0.40	0.715	No
24	Ca-Li	850	-0.30	-0.07	1.26	No
25	Mg-Ca	850	0.90	0.26	0.812	No
26	Pb-Ca	850	1.55	0.24	0.887	No
27	Sn-Ca	850	1.60	0.33	0.783	No
28	Tl-Ca	850	1.35	0.21	0.869	No
29	Sn-Cd	321	0.10	0	0.985	No

30	Tl-Cd	304	-0.15	-0.12	1.09	No
31	Zn-Cd	419	0.05	0.08	0.889	No
32	Pb-Na	327	1.40	0.33	0.916	No
33	Sn-Na	321	1.45	0.42	0.808	No
34	Tl-Na	304	1.20	0.30	0.898	No
35	Sn-Pb	327	-0.05	-0.09	1.13	No
36	Pb-Tl	327	0.20	0.03	1.02	No

The order of elements in this table is according to the following rules: (1) the element with higher valency is selected as the first element, (2) for the binary systems with elements of the equal valency, the element with higher electronegativity is selected as the first element.

By using Gaussian kernel and C=100, the representative points of the samples listed in Table 6.11 can be completely separated in the high dimensional feature space. By LOO cross-validation test, the rate of correctness of prediction is 91.6%

6.9   SVM Applied to Intelligent Database of Phase Diagrams

Since atomic parameter-pattern recognition method has been proved to be a powerful tool for the assessment and computerized prediction of the phase diagrams of inorganic systems, it is desirable to build some intelligent databases of phase diagrams both for the retrieval of known phase diagrams and the computerized prediction of unknown systems. For example, we have built an intelligent database for the phase diagrams of molten salt systems (IMSPDB) [32]. Over three thousands of binary and ternary known phase diagrams have stored into this phase diagram database for retrieval. And at the same time, it also includes a data processing plateform. It includes a knowledge base consisting of tables of atomic parameters and mathematical models for the computerized prediction of unknown phase diagram. It can be also used for the computerized assessment for newly published phase diagrams. Figure 6.13 and Fig. 6.14 illustrate some of its interfaces.

Since many computerized prediction problems involved in this intelligent database are problems of small sample size, support vector machine plays an important role in the model-building for this intelligent database [7].

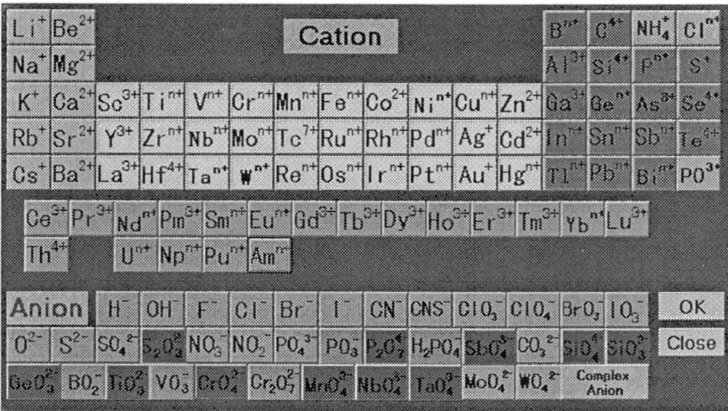

Fig. 6.13   An interface of IMSPDB (for phase diagram retrieval).

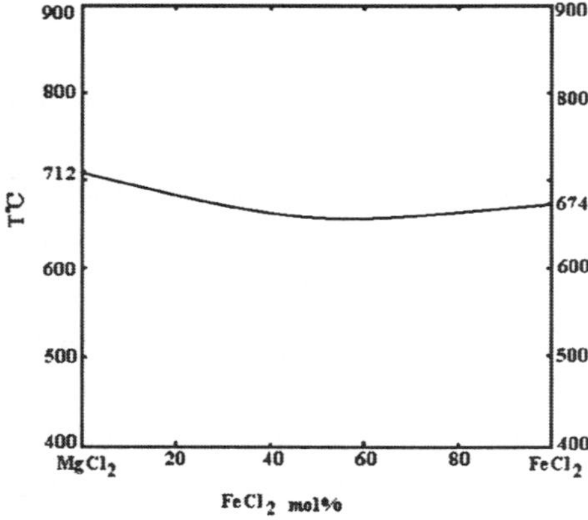

Fig. 6.14   An interface of IMSPDB (a result of liquidus curve prediction).

Chapter 7

SVM Applied to

Thermodynamic Property Prediction

7.1　Significance of Estimation of Thermodynamic Properties of Chemical Substances

Since the experimental determination of thermodynamic quantities is usually a rather tedious work, and up to now the known thermodynamic data are far from complete, it is desirable to have some semi-empirical methods for the estimation of unknown thermodynamic quantities of chemical substances. Because the thermodynamic quantities of a compound or a mixture should be related to the microscopic structure of this compound or this mixture, and the microscopic structure of this compound or this mixture can be approximately described by its atomic parameters or molecular descriptors, it is possible to estimate these thermodynamic quantities by some functions of the atomic parameters or molecular descriptors of this compound or this mixture. So we can have a semi-empirical method for the prediction of unknown thermodynamic data by transduction, i.e., to find the mathematical model describing the relationship between the thermodynamic quantities and the atomic parameters or molecular descriptors based on the SVM computation of the known data firstly, and then use this mathematical model obtained to predict the unknown thermodynamic quantities of some target substances.

145

From the practical point of view, the most important thermodynamic quantities include the enthalpy of formation of compounds, the free energy of mixing and the activity coefficient of solutions. In this chapter, we will give some typical examples of the prediction of these thermodynamic quantities by using SVM and atomic parameters or molecular descriptors in computation.

7.2　Modeling of Enthalpy of Formation of Compounds

The enthalpy of formation of compounds is one of the most basic thermodynamic quantities. Here we will describe two examples of the estimation of the enthalpies of formation. One of the examples is dealing with inorganic compounds by using atomic parameters, and the other is dealing with organic compounds by using molecular descriptors.

7.2.1　*Modeling of the enthalpy of formation of intermetallic compounds*

The enthalpy of formation of intermetallic compounds is a kind of fundamental data in alloy chemistry. In the years of seventies of last century, Miedema proposed a semi-empirical model for the estimation of the enthalpy of formation of intermetallic compounds. According to this model, the energy change in the intermetallic compound formation process is dependent chiefly on two parameters. One parameter is the difference of Miedema's electronegativities (denoted by $\Delta\phi$) of two metallic elements. It can be considered as a measure of the charge transfer in compound formation. Since charge transfer is the result of electron transfer from higher energy level to lower energy level, so $|\Delta\phi|$ can be considered as the driving force of intermetallic compound formation. Another parameter is $n_{ws}^{1/3}$, which represents the valence electron density in the Wagner-Seitz cell of metals. The difference of $n_{ws}^{1/3}$ (denoted by $\Delta n_{ws}^{1/3}$) is considered as the resistance of intermetallic compound formation. So some linear equation involving these two parameters has been proposed to estimate the enthalpy of formation of intermetallic compounds [98]. Although Miedema proposed that the enthalpy of formation of intermetallic compounds can be calculated as the function of $\Delta\phi$ and $\Delta n_{ws}^{1/3}$ only, it is reasonable to expect that there should be some *other* atomic parameters affecting the enthalpy of formation of intermetallic compounds. For example, it is well-known that the atomic radii of constituent atoms are influential factors for the crystal type of intermetallic compounds, and that the crystal type should be an influential factor for the lattice energy of intermetallic compounds. Hence it is reasonable to consider the atomic radii of constituent atoms as the atomic parameters affecting the enthalpy of formation of intermetallic compounds. Based on this consideration, we have tried to use Miedema's parameters together with some additional atomic parameters as features to make mathematical modeling for more accurate prediction of the enthalpy of formation of intermetallic compounds. Here we demonstrate the results of this method for the modeling of the enthalpy of formation of the compounds formed by rare earth metals and post-transition metals.

Kleppa and his coworkers measured the enthalpies of formation of a series of intermetallic compounds between rare earth metals and transition metals by accurate high temperature calorimetric method [62]. Based on these data, we have used SVR-based feature selection method (this method is described in Chapter 4 of this book) to search the optimal set of atomic parameters for support vector regression. It has been found that an effective feature set suitable for the computerized prediction of the enthalpy of formation of these compounds includes only two parameters: Miedema's $\Delta\phi$ and the ratio of atomic radii R_1/R_2 (here the values of the atomic radius proposed by Teatum are used). By SVR and leave-one-out (LOO) cross-validation, the results of prediction are rather good. Table 7.1 shows the comparison between the experimental values and the predicted values of the enthalpy of formation of these compounds. The results of calculation by Miedema's model [104] are also listed for comparison.

It can be seen that the results of SVR computation are rather satisfactory. Moreover, it has been found by SVR computation that the inclusion of radius ratio of metallic atoms significantly improves the accuracy of enthalpy prediction. A reasonable explanation of this fact is that the rare earth-transition metal compounds listed here do not have the same crystal structure. Depending on the radius ratio and electronegativity difference, this compound series includes compounds with FeB-type, CrB-type and CsCl-type crystal structures. Different crystal structures should have different lattice energies. Therefore the influence of atomic radius on the enthalpy of formation is evident and cannot be neglected.

Table 7.1 Enthalpy of formation of 1:1 intermetallic compounds between rare earth and post-transition metals ($kJ \cdot mol^{-1}$).

Compounds	Experimental values	Predicted values by SVR	Predicted values by Miedema model
ScNi	-44.7	-54.4	-57
LaNi	-24.8	-28.5	-38
CeNi	-30.3	-34.9	-40
PrNi	-28.1	-33.6	-42
NdNi	-25.0	-38.2	-42
SmNi	-36.4	-35.2	-44
GdNi	-29.9	-36.3	-44
TbNi	-36.8	-38.5	-46
DyNi	-35.2	-39.3	-46
HoNi	-41.7	-39.5	-45
ErNi	-42.1	-40.7	-48
TmNi	-46.4	-42.4	-48
LuNi	-47.3	-43.7	-52
ScRh	-94.5	-88.7	-91
YRh	-76.1	-73.0	-79
LaRh	-56.3	-67.5	-73
NdRh	-64.2	-71.4	-77
GdRh	-72.4	-72.9	-79
TbRh	-72.3	-74.6	-82
DyRh	-76.5	-75.6	-81
HoRh	-87.2	-75.5	-79
ErRh	-87.4	-74.6	-83
TmRh	-88.9	-75.8	-83
ScPd	-89.3	-91.9	-128
YPd	-94.9	-82.8	-123
LaPd	-76.2	-76.0	-119
CePd	-78.3	-82.7	-121
PrPd	-78.8	-81.4	-122
NdPd	-77.2	-81.3	-122
SmPd	-82.4	-82.7	-124
GdPd	-82.6	-82.7	-123
TbPd	-85.2	-84.4	-125

DyPd	-83.3	-85.5	-124
HoPd	-91.5	-85.6	-122
ErPd	-91.1	-86.3	-126
TmPd	-92.6	-88.1	-125
LuPd	-95.0	-89.6	-128
ScIr	-89.7	-108.7	-92
YIr	-65.9	-90.3	-78
HoIr	-80.7	-92.7	-78
ErIr	-82.9	-93.8	-83
LuIr	-85.5	-96.6	-87
ScPt	-104.8	-120.2	-132
YPt	-104.0	-102.9	-123
LaPt	-99.7	-96.0	-117
CePt	-103.8	-100.7	-119
PrPt	-103.4	-100.3	-121
NdPt	-104.4	-101.2	-121
SmPt	-108.7	-102.7	-123
GdPt	-109.3	-102.7	-123
TbPt	-115.7	-109.2	-125
DyPt	-109.4	-105.5	-125
HoPt	-121.8	-105.4	-122
ErPt	-118.7	-106.3	-127
TmPt	-121.0	-107.4	-125
LuPt	-118.4	-109.5	-130

7.2.2 *Modeling of the enthalpy of formation of polycyclic aromatic hydrocarbons*

Table 7.2 lists the data of the enthalpy of formation and molecular descriptors of benzene and some polycyclic aromatic hydrocarbons [2; 3].

Table 7.2 Enthalpy of formation and molecular descriptors of some aromatic hydrocarbons.

Compound	ΔH_f^*	W^*	X_v^*	X_e^*	Length*	Volume*	Area*
Benzene	82.8	27	3.000	3.000	7.046	193.8	104.8
Naphthalene	150.6	109	4.966	5.454	9.915	286.0	141.0
Anthracene	218.3	279	6.933	7.942	11.651	336.4	160.7
Phenanthrene	209.1	271	6.950	7.926	11.752	366.9	171.2
Naphthacene	286.1	569	8.899	10.430	14.116	408.3	188.8
Benz(a)anthracene	276.9	557	8.916	10.414	13.942	472.3	209.6

Chrysene	267.7	545	8.993	10.397	13.939	447.3	1.99.8
Triphenylene	258.5	513	8.950	10.414	11.682	411.5	196.6
Pyrene	230.5	362	7.933	9.409	11.662	420.7	189.6
Perylene	279.9	654	9.933	11.897	11.800	426.9	191.4
Benzo(a)pyrene	279.9	680	9.916	11.897	13.883	502.1	219.2
Benzo(e)pyrene	289.1	652	9.933	11.897	11.765	481.0	210.3
Picene	326.3	963	10.916	12.868	16.112	504.6	223.6
Dibenzo(a,h)anthracene	335.5	975	10.889	12.885	15.898	539.6	234.5
Dibenzo(a,j)anthracene	335.5	973	10.889	12.885	14.532	537.0	231.6
Dibenzo(a,c)anthracene	326.3	911	10.916	12.902	13.922	608.9	254.4
Dibenzo(a,h)pyrene	347.7	1142	11.899	14.385	16.084	581.8	248.1
Anthanthrene	310.5	839	10.899	13.397	12.898	480.3	211.0
Benzo(ghi)perylene	301.3	815	10.916	13.380	11.779	480.0	211.0
Coronene	322.7	1002	11.899	14.863	11.722	533.1	228.2
Dibenzo(a,e)pyrene	338.5	1082	11.916	14.385	13.904	653.5	261.1
Dibenzo(a,l)pyrene	351.2	1066	11.916	14.390	13.693	827.5	291.3
Dibenzo(a,i)pyrene	347.7	1142	11.899	14.385	16.089	581.9	248.3

[*] ΔH_f denotes the enthalpy of formation ($kJ \cdot mol^{-1}$). W denotes Wiener topological index. X_v denotes vertex connectivity. X_e denotes edge connectivity. Length, Volume, Area denotes the length, volume and area of molecules respectively.

Figure 7.1 illustrates the comparison of the values of experimental enthalpy of formation of polycyclic aromatic hydrocarbons and the predicted values by SVR in LOO cross-validation test.

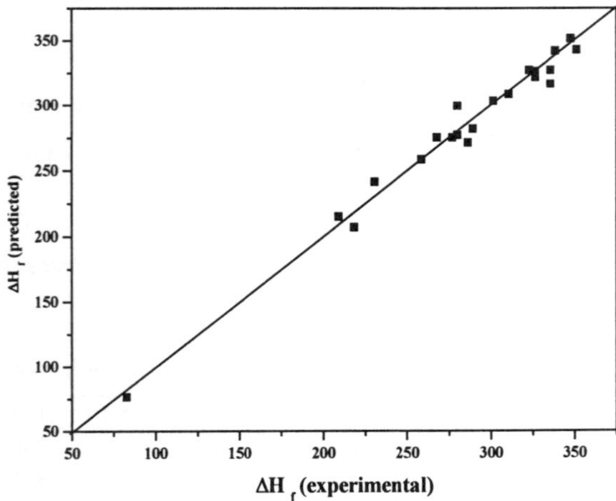

Fig. 7.1	Comparison between the values of experimental enthalpy of formation of polycyclic aromatic hydrocarbons and the predicted values by SVR in LOO cross-validation.

Based on the same data set listed in Table 7.2, the errors of prediction of SVR in cross-validation are significantly lower than those of ordinary linear regression.

7.3   Modeling of Free Energy of Mixing of Liquid Alloy Systems

The data of excess free energy of mixing and activity coefficient of components of liquid alloys are very useful for the process design in pyrometallurgy. Table 7.3 lists some data of excess free energies (at 50/50 composition) and atomic parameters of 33 binary liquid alloys [72]. The correlation between these data is investigated by SVM.

Table 7.3   Data of integral excess free energy and atomic parameters of liquid alloys.

System	Class[*]	ΔF^{XS} [**]	R_1/R_2 [**]	ϕ_1	ϕ_2	$(n_{ws}^{1/3})_1$	$(n_{ws}^{1/3})_2$
K-Tl	1	-17050	1.388	2.25	3.90	0.65	1.12
Na-Hg	1	-13290	1.215	2.70	4.20	0.82	1.24
Na-Sn	1	-13000	1.177	2.70	4.15	0.82	1.24
K-Pb	1	-12460	1.358	2.25	4.10	0.65	1.15
Na-Pb	1	-12000	1.092	2.70	4.10	0.82	1.15
K-Hg	1	-11620	1.510	2.25	4.20	0.65	1.24
Na-Tl	1	-2070	1.114	2.70	3.90	0.82	1.12
Mg-Pb	1	-6813	0.915	3.45	4.10	1.17	1.15
Cd-Hg	1	-2048	0.997	4.05	4.20	1.24	1.24
Mg-Al	1	-1672	1.119	3.45	4.20	1.17	1.39
Pb-Bi	1	-1254	1.029	4.10	4.15	1.15	1.16
In-Sn	1	-1045	1.025	3.90	4.15	1.17	1.24
Na-Cd	1	-460	1.219	2.70	4.05	0.82	1.24
Cd-Bi	1	-355	0.922	4.05	4.15	1.24	1.16
Zn-Hg	2	878	0.886	4.10	4.20	1.32	1.24
Zn-Sn	2	1630	0.859	4.10	4.15	1.32	1.24
Sn-Bi	2	272	0.909	4.15	4.15	1.24	1.16
In-Pb	2	585	0.950	3.90	4.10	1.17	1.15
K-Na	2	669	1.243	2.25	2.70	0.65	0.82
Hg-Bi	2	669	0.925	4.20	4.15	1.24	1.16
Zn-Ga	2	836	0.988	4.10	4.10	1.32	1.31
Hg-Pb	2	878	0.899	4.20	4.10	1.24	1.15
Hg-Sn	2	920	0.969	4.20	4.15	1.24	1.24
Cd-Sn	2	920	1.015	4.05	4.15	1.24	1.24
Cd-In	2	1003	0.943	4.05	3.90	1.24	1.17
Cd-Tl	2	1672	0.914	4.05	3.90	1.24	1.12
Cd-Pb	2	2090	0.896	4.05	4.10	1.24	1.15

Zn-Cd	2	2174	0.889	4.10	4.05	1.32	1.24
Al-Sn	2	2215	0.927	4.20	4.15	1.39	1.24
Zn-In	2	2362	0.838	4.10	3.90	1.32	1.17
Cd-Ga	2	2675	1.111	4.05	4.10	1.24	1.31
Zn-Bi	2	2675	0.820	4.10	4.15	1.32	1.16
Zn-Pb	2	4953	0.797	4.10	4.10	1.32	1.15

[*]Class "1" includes the binary systems with negative excess free energy and class "2" includes that with positive excess free energy.

[**] ΔF^{XS} denotes the integral excess free energy ($kJ \cdot mol^{-1}$). R_1/R_2 denotes the radius ratio. The order of elements listed in this table is according to ascending order of valence (if the two elements have same valence number, the order is according to ascending order of Basanov electronegativity).

By SVC with linear kernel, the classification of class "1" and "2" is rather good, and the rate of correctness of prediction is 91% in LOO cross-validation. The relative averaged absolute error of the prediction of ΔF^{XS} by SVR in LOO cross-validation is lower than 10%.

7.4 Prediction of Activity Coefficient of Concentrated Electrolyte Solutions

Concentrated electrolyte solutions are the intermediate materials of many industrial chemical and hydrometallurgical processes. The water of salt lakes and some underground water are also concentrated electrolyte solutions. Therefore both the chemical or metallurgical engineers and geochemists concern the thermodynamic properties of concentrated electrolyte solutions.

The regularities and estimation methods of the activity coefficients of concentrated electrolyte solutions have been investigated for many years, but up to now these problems are not completely solved yet. In this field, Pitzer's ion-ion interaction model [109], as a semi-empirical model, has been most widely used. In Pitzer's method, the thermodynamic function (such as logarithm value of activity coefficient, $\log \gamma_{\pm}$) is expanded into a series, and the coefficients of the terms of this series, $\beta^{(0)}$ and $\beta^{(1)}$ and C, are used to calculate the activity coefficients of solutions of different concentrations. The coefficients $\beta^{(0)}$, $\beta^{(1)}$ and C have to be calculated from the experimental data of activity coefficients of concentrated electrolyte solutions. In our previous work, it has been found that the values of $\beta^{(0)}$, $\beta^{(1)}$ and C exhibit some approximate relationships with ionic radius and ionic charge of cations and anions of the electrolytes, since these coefficients actually reflect the values of log $\gamma_\pm$ itself. So it appears possible to correlate the values of known data of log $\gamma_\pm$ with the ionic radii and ionic charges directly by SVR, and then the mathematical model obtained from known data can be used for the computerized prediction of unknown log$\gamma_\pm$ of other electrolytic solutions.

By using SVR-based feature selection method, the following features are selected for log $\gamma_\pm$ modeling of concentrated electrolyte solutions:

(R_+/R_-): ratio of cationic and anionic radii

R_+: cationic radius

R_-: anionic radius

(Z_+/Z_-): ratio of the charge numbers of cation and anion

It has been found that the following equation can be used to estimate the log $\gamma_\pm$ of electrolyte solutions within the range of concentration from 0.5 to 4m [149]:

$$\ln \gamma_\pm = -6.9100 + 7.4933\,(R_+/R_-) - 4.8652\,R_+ + 3.9587\,R_- + 0.04786\,(Z_+/Z_-) + 0.3641\,m$$

where m is the molality of electrolyte solution.

7.5　Regularity of the Solubility of C_{60} in Organic Solvents

C_{60} is a new material having many possible applications. Up to now, C_{60} is prepared by solvent extraction from the volatile dust of the vaporization of graphite by arc heating, and then chromatographic method is used to purify it by separation of its allied compounds. Since this method of preparation is rather expensive, it is desirable to find some cheaper way for C_{60} production. Based on the requirements of new method design, many data of solubilities of C_{60} in various solvents have been measured and published in literatures. Some authors used these data to investigate the relationship between the solubility data and the molecular descriptors of solvents. But it is now widely recognized that C_{60} combines with many solvent molecules to form C_{60} solvates. For example, C_{60} combines with benzene to form $C_{60} \cdot 4C_6H_6$, and combines with pentane to form $C_{60} \cdot C_5H_{12}$. Hence many published solubility data of C_{60} in organic solvents are actually not the data of the phase equilibria with pure C_{60}, but the data of equilibria with various C_{60} solvates with different stoichiometries. In order to investigate this problem in a more strict way, Korobov proposed that it was necessary to measure the free energy of formation of various C_{60} solvates, and then calculate the 'hypothetical solubilities' based on the experimental solubility data and the free energy of solvate formation. Table 7.4 illustrates the calculated data of hypothetical solubilities [80].

Table 7.4   Hypothetical solubilities of C_{60} in solvents.

Solvent	Composition of solvates	ΔF^0 * ($\text{kJ} \cdot \text{mol}^{-1}$)	$10^4 \, x^*$	$10^4 \, x'^*$
Benzene	1:4 compound	3.0	2.0	6.0
Toluene	no solvate		4.2	4.2
o-xylene	1:2 compound	2.3	14.2	37.7
m-xylene	1:2 compound	2.3	3.6	10.4
bromobenzene	1:2 compound	6.2	4.4	54.6
o-dichlorobenzene	1:2 compound	1.4	38.3	68.3
m-dichlorobenzene	1:2 compound	4.2	3.8	19.7
heptane	1:1 compound	0.6	0.098	0.12
octane	1:1 compound	3.9	0.045	0.22

* ΔF^0 is the free energy of formation of C_{60} solvates. x is the experimental solubility of C_{60}. x' is the hypothetical solubility of C_{60}. The unit of x or x' is mole fraction.

Table 7.5 lists the values of molecular parameters of these solvents.

Table 7.5   Molecular parameters of solvents.

Solvent	ε^*	δ^*	γ^*	π^*	μ^*	E_t^*
Benzene	2.27	18.8	28.2	0.55		
Toluene	2.38	18.8	27.9	0.49		
o-xylene	2.57	18.0	29.5	0.51		
m-xylene	2.37	18.0	28.1	0.47		
bromobenzene	5.40	20.2	35.5	0.77		
o-dichlorobenzene	9.93	20.5	36.2	0.77		
m-dichlorobenzene	5.04	20.0	35.5	0.65	1.34	36.7
heptane	1.92	15.2	19.7	-0.06	0	31.1
octane	1.95	15.5	21.2	0.01	0	31.1

* ε denotes the relative permittivity , δ denotes the solubility parameter, γ denotes the surface tension, π denote the polarizability, μ denotes the dipole moment of solvent molecules ,and E_t denotes polarity parameter.

Since the number known data of hypothetical solubilities is quite limited, SVR should be the suitable method to find the mathematical model describing the relationship between the hypothetical solubilities and the molecular parameters of solvents.

By SVR-based feature selection, it has been found that a feature set consisting of δ, π and μ can be used for mathematical modeling of the hypothetical solubility of C_{60} in various solvents [144]. Figure 7.2 illustrates the correlation between the values of hypothetical solubility and the values calculated by SVR using the above-mentioned three parameters.

It can be seen from the above-mentioned examples that support vector regression together with atomic parameters or molecular descriptors can provide a useful tool for the estimation of the thermodynamic functions of compounds and their mixtures.

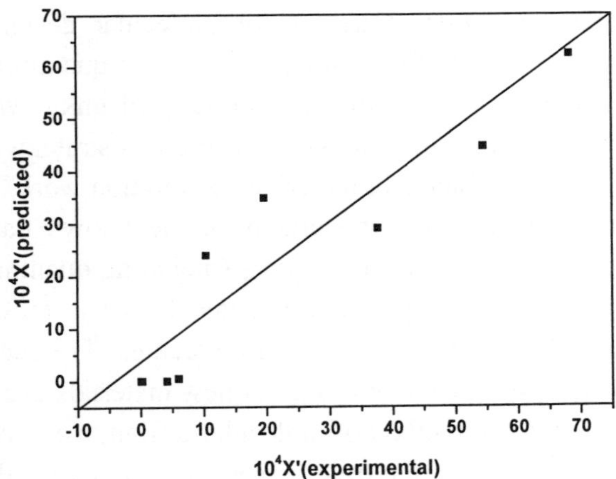

Fig. 7.2   Correlation between the values of hypothetical solubility and the values calculated by SVR.

Chapter 8

SVM Applied to Molecular and Materials Design

8.1 Concepts of Molecular Design and Materials Design

In principle, there are two strategies for molecular design. The first strategy is to start from the first principle, i.e., from quantum mechanics and statistical mechanics to predict the property of unknown materials. Up to now, however, it is still impossible to use this strategy to solve the most of complicated problems in materials exploration work. The second way is a semi-empirical one. It starts from the known data of some molecules and some molecular descriptors (including quantum chemical parameters) to find semi-empirical rules, and then use these empirical rules to predict the property of unknown molecules. The second way is already practicable for molecular design or new materials exploration.

In the semi-empirical method of molecular design, we have to use the data of known molecules as the training set. In many cases, the numbers of available data of known molecules used in the training sets are rather limited. Therefore the data processing tasks are usually dealing with the problem of small sample size, so support vector machine should be suitable for solving these problems.

In new materials exploration work, there are two problems having general significance: The first problem is "what is the chemical composition of the substance having desirable properties?", and the second problem is "what are the optimal conditions of preparation or production for this material?" Since both of these two problems are

156

dealing with very complicated systems or very complicated processes, we have to solve these problems by using some semi-empirical methods. For solving the first problem, we have to find the relationships between the microscopic structure of materials and their properties. We usually call this relationship as QSPR (quantitative structure-property relationship). For solving the second problem, we have to find some mathematical models for the optimization of the processes.

In this chapter, some typical problems for new materials exploration work will be described to demonstrate the applications of SVM and other methods of pattern recognition techniques in molecular design or materials design.

The tasks of molecular design or materials design can be classified into several different categories:

The first type of task is to solve the "formability problems", i.e., to find some mathematical model or criterion for the stability of some unknown molecules or chemical substances. The second type of task is the "property prediction", i.e., to make mathematical models for the structure-property relationships and use these models to predict the property of new materials (or the inverse problem: to search the unknown new materials with some pre-assigned property). The third type of task is to solve the "optimization problems", i.e., to find the conditions for optimizing some properties of certain materials. The fourth type of task is to solve the "problem of control", i.e., is to find the mathematical model to control some index of materials within a desired range. And the fifth type of task is to find the multivariate relationships between the conditions of preparation and the properties of materials. Different SVM techniques should be used for these different purposes. In the following sections, we will use different examples of materials design tasks to demonstrate various strategies of solution by SVM technique.

8.2   SVM Applied to New Compound Synthesis Problems

The most exciting achievement of materials research is to find some new compound (or new phases) with specified structure and outstanding properties. In this chapter, the materials design problems of compounds with perovskite-type and perovskite-like type structures will be discussed.

8.2.1 *Materials design problems for materials with perovskite structure*

Although perovskite is the name of a rare mineral of no great importance, the compounds having perovskite-type crystal structure with general formula ABX_3 (X = oxygen or halogen) include numerous complex oxides or halides with outstanding functional properties [54; 145; 102]. Since 1945, when the ferroelectric properties of barium titanate were discovered, a series of complex oxides and complex halides with perovskite-type structure have been found to be valuable functional materials. In recent years, searching new complex oxides and complex halides with perovskie-type structure has become an active research field of new materials exploration.

The crystal structure of compounds with ideal perovskite structure is illustrated in Fig. 8.1. It is the structure of a unit cell of $SrTiO_3$ crystal. In this structure, tetravalent Ti^{4+} cation is surrounded by six oxygen anions to form octahedral structure, and bivalent Sr^{2+} cation is surrounded by twelve oxygen anions to form cubo-octahedral structure. Based on the understanding of such type of crystal structure, Goldschmidt proposed a famous crystal-chemical criterion of the formability or the stability of perovskite structure for ABX_3-type compounds:

$$t = \frac{R_a + R_x}{\sqrt{2}\left(R_b + R_x\right)}$$

(8.1)

here t is called *tolerance factor*, and R_a, R_b are cationic radii of A, B ions respectively, R_x is the ionic radius of anion of X. According to Goldschmidt, *the cubic perovskite structure is stable only if the value of tolerance factor has an approximate range of 0.8 < t < 0.9, and the distorted perovskite structure can be stable in a somewhat larger range of tolerance factor.* This criterion is widely used in the exploration work of new compounds with perovskite-type or perovskite-like type structure.

Owing to the accumulation of the crystallographic data of compounds with perovskite structure, it is now widely recognized that the range of tolerance factor for the stability of perovskite structure should be 0.75 < t <1.00.

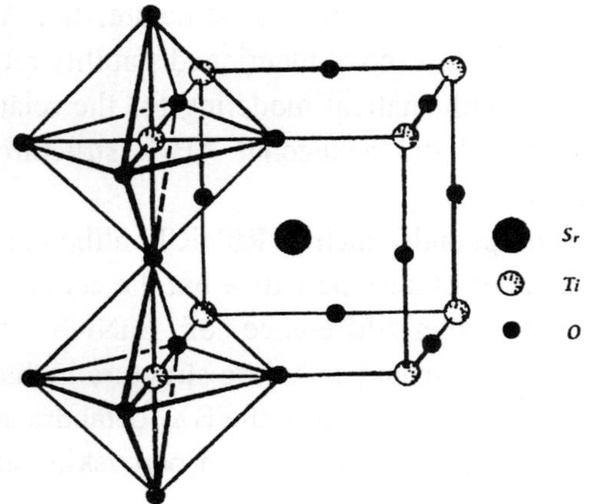

Fig. 8.1   Crystal structure of $SrTiO_3$, a typical compound with ideal perovskite structure*.

Although Goldschmidt's tolerance factor t is indeed very useful for the exploration of new compounds with perovskite structure, *it is only a necessary condition but not the sufficient condition for the formation or the stability of perovskite structure* [145]. Many systems having t in the range of 0.75—1.0 do not form perovskite-type compound.

For example, $MnSiO_3$ has t = 0.856, but it has $CdGeO_3$ structure; $RbMnCl_3$ has t = 0.88, but it has hexagonal $BaTiO_3$ structure. NaI-MgI_2 system has t = 0.826, but it has no intermediate compound at all. Therefore it is desirable to investigate the complementary conditions for the formability of perovskite structure, in order to help the computerized materials design for new materials with perovskite structure. Atomic parameters and SVM technique can be used for this purpose.

---

*Originally $CaTiO_3$ crystal was considered as having cubic structure, but in later work it has been found that its crystal lattice has slightly distorted structure. So $SrTiO_3$ should be considered as the typical compound with ideal perovskite structure.

Since BX_6 octahedra and AX_{12} cubo-octahedra are the basic sub-structures of perovskite lattice, as shown in Fig. 8.1, it is easy to see that the stability of BX_6 octahedra and AX_{12} cubo-octahedra are also necessary conditions for the stability of perovskite lattice. It is obvious that the condition $0.75<t<1.00$ is not enough to assure the stability of the BX_6 octahedral and AX_{12} cubo-octahedral structure. It is necessary to find the suitable criteria for the above-mentioned stability requirements.

(1) SVM applied to mathematical modeling for the relative stability of perovskite structure and the hexagonal ABX_3 structures involving face-shared octahedra:

Some ABX_3 type compounds, such as $RbNiCl_3$, although having $0.75 < t <1.00$, do not form perovskite-type lattice but the crystal lattices with $BaNiO_3$ structure. The chief difference of $BaNiO_3$ structure (or hexagonal $BaTiO_3$ structure) from perovskite structure is that in $BaNiO_3$ structure (or hexagonal $BaTiO_3$ structure) the BX_6 octahdra are shared by their face with each other (Fig. 8.2), while in perovskite structure they are shared with each other by their corners.

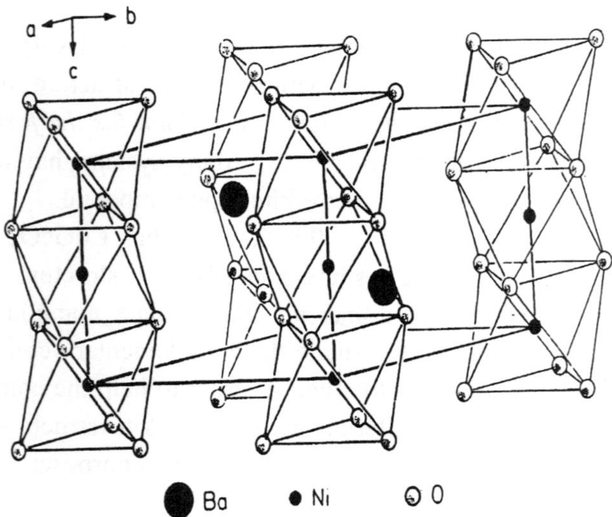

Fig. 8.2   Face-shared BX_6 structure in $BaNiO_3$ lattice.

In order to find the criterion of relative stability between $BaNiO_3$ structure and perovskite structure, the data listed in Table 8.1 are used for data processing.

Table 8.1   Crystal-types of some complex chlorides*[102].

Compound	R_a*	R_b*	X_a*	X_b*	N_d*	Crystal-type
$CsPbCl_3$	1.88	1.19	0.75	1.6	0	perovskite
$CsHgCl_3$	1.88	1.02	0.75	1.9	0	perovskite
$CsCaCl_3$	1.88	1.00	0.75	1.0	0	perovskite
$CsCdCl_3$	1.88	0.95	0.75	1.7	0	both type**
$CsSrCl_3$	1.88	1.18	0.75	1.0	0	perovskite
$KMnCl_3$	1.64	0.83	0.8	1.4	5	perovskite
$KCaCl_3$	1.64	1.00	0.8	1.0	0	pervskite
$KMgCl_3$	1.64	0.72	0.8	1.2	0	perovskite
$RbCaCl_3$	1.72	1.00	0.8	1.0	0	perovskite
$RbSrCl_3$	1.72	1.18	0.8	1.0	0	perovskite
$RbCdCl_3$	1.72	0.95	0.8	1.7	0	perovskite
$TlMnCl_3$	1.70	0.83	1.4	1.4	5	perovskite
$RbMnCl_3$	1.72	0.83	0.8	1.4	5	Hexagonal $BaTiO_3$-type
$RbNiCl_3$	1.72	0.69	0.8	1.8	8	$BaNiO_3$
$RbCoCl_3$	1.72	0.745	0.8	1.7	7	$BaNiO_3$
$RbFeCl_3$	1.72	0.78	0.8	1.7	6	$BaNiO_3$
$RbCrCl_3$	1.72	0.80	0.8	1.6	4	$BaNiO_3$
$CsMgCl_3$	1.88	0.72	0.75	1.2	0	$BaNiO_3$
$CsNiCl_3$	1.88	0.69	0.75	1.8	8	$BaNiO_3$
$CsCoCl_3$	1.88	0.745	0.75	1.7	7	$BaNiO_3$
$CsFeCl_3$	1.88	0.78	0.75	1.7	6	$BaNiO_3$
$CsVCl_3$	1.88	0.79	0.75	1.5	3	$BaNiO_3$
$CsCrCl_3$	1.88	0.80	0.75	1.6	4	$BaNiO_3$

*R_a and R_b denote the Shanon-Preweit ionic radii of A and B in $ABCl_3$ respectively. X_a and X_b denote the Basanov electronegativity of A and B respectively. N_d denotes the number of d electrons in the unfilled shell of d electrons of B ions.

**Both the $CsCdCl_3$ with perovskite structure and $CsCdCl_3$ with hexagonal $BaTiO_3$ structure are reported in literature, so this compound is a borderline case in Table 8.1.

By using SVC with linear kernel function, 100% of separation of the chlorides with perovskite structure and the chlorides with face-shared structures can be achieved. And the sample point of the borderline case, $CsCdCl_3$, is located close to the optimal hyperplane of classification. In leave-one-out (LOO) cross-validation test, the rate of correctness of prediction is 91%.

The criterion for the formation of face-shared structure found by SVC can be expressed as follows:

$$4.52R_b - 1.83R_a + 2.23X_a - 0.142X_b - 4.10N_d + 0.589 < 0 \qquad (8.2)$$

It implies that large A^+ cation, small B^{2+} cation and large electronegativity of B favor the formation of face-sharing of BX_6 octahedra. This fact can be explained as follows: In perovskite-type lattice the network of corner shared BX_6 octahedra form *cages* of A^+ ions. The repulsive force due to the large A^+ and small cage formed by small B^{2+} and X^- will make perovskite-type lattice unstable, while the BX_6 octahedra in face-shared structure form parallel chains, and A^+ ions are located between these chains without strict confinement. So large A^+ and small B^{2+} favor the face-shared structure. On the other hand, according to Pauling's rule [107] about the relative stability of corner-shared polyhedra and face-shared polyhedra, stronger of B^{2+}-B^{2+} inter-ionic electrostatic repulsion will destabilize the face-shared structure. Large elctronegativity of B tends to induce the covalency of B-X bond and the smaller effective charge on B^{2+} ion, so the electrostatic repulsion will be weaker. This will make the face-shared structure more stable.

(2) SVM applied to mathematical modeling to differentiate the perovskite-type compound forming systems and the $AX-BX_2$ system with $0.75 < t < 1.00$ but without 1:1 intermediate compound formation:

Table 8.2 lists the data of some $AX-BX_2$ systems (X = Cl, Br, I) having tolerance factors in the range of 0.75—1.00. Some of them form perovskite-type compounds, and the others have no 1:1 intermediate compound. SVM has been used to differentiate these two types of systems.

Table 8.2   Data of some perovskite-type compound forming systems and the systems without 1:1 compound formation.

System	R_a^{*}	R_b^{*}	R_x^{*}	Compound formation
Cs,Pb|Cl	1.88	1.19	1.81	Perovskite-type compound
Cs,Hg|Cl	1.88	1.02	1.81	Perovskite-type compound
Cs,Cd|Cl	1.88	0.95	1.81	Perovskite-type compound
Tl,Mn|Cl	1.70	0.83	1.81	Perovskite-type compound
K,Mg|Cl	1.64	0.72	1.81	Perovskite-type compound
Cs,Sr|Cl	1.88	1.18	1.81	Perovskite-type compound
Rb,Cd|Cl	1.72	0.95	1.81	Perovskite-type compound

K,Mn|Cl	1.64	0.83	1.81	Perovskite-type compound
Cs,Pb|Br	1.88	1.19	1.96	Perovskite-type compound
Cs,Hg|Br	1.88	1.02	1.96	Perovskite-type compound
Cs,Cd|Br	1.88	0.95	1.96	Perovskite-type compound
K,Ca|Cl	1.64	1.00	1.81	Perovskite-type compound
Rb,Sr|Cl	1.72	1.18	1.81	Perovskite-type compound
Rb,Ca|Cl	1.72	1.00	1.81	Perovskite-type compound
Na,Mg|Br	1.39	0.72	1.96	No intermediate compound
Na,Mg|I	1.39	0.72	2.20	No intermediate compound
K,Mg|I	1.64	0.72	2.20	No intermediate compound
Na,Co|I	1.39	0.745	2.20	No 1:1 intermediate compound
K,Co|I	1.64	0.745	2.20	No 1:1 intermediate compound
Cs,Hg|I	1.88	1.02	2.20	No 1:1 intermediate compound

[*] R_a, R_b, R_x denote the Shanon-Preweit radii of A, B and X ions respectively.

It can be found that most of the systems without 1:1 compound formation have the radius ratio (R_b/R_x) << 0.414. This fact can be explained by Pauling's first rule about the structure of complex ionic crystalline compounds: (R_b/R_x) << 0.414 means the BX_6 octahedron unstable, while BX_6 octahedron is one of the basic parts of perovskite structure. *Therefore compounds ABX_3 with (R_b/R_x) << 0.414 can not form stable perovskite-type lattice, even if the value of tolerance factor t is in the favorable range for perovskite formation.*

(3) SVM applied to mathematical modeling of distortion of perovskite structure:

Most of perovskite-type compounds exhibit some degree of lattice distortion. It is the asymmetry induced by lattice distortion that gives rise to many valuable physical properties of perovskite-type compounds. Therefore lattice distortion of perovskite-type compounds is also concerned by materials scientists.

Table 8.3 lists the data of some perovskite-type compounds with or without distortion, and the relevant atomic parameters of these compounds.

By using SVC with linear kernel function, 100% separation of the sample points with and without distortion can be achieved. The criterion obtained can be expressed by the following inequality:

$$6.48R_a - 9.06R_b - 2.19X_a - 3.60X_b + 4.38 < 0 \qquad (8.3)$$

for the occurrence of lattice distortion. It implies that small A^{2+} ion, large B^{4+} ion and large electronegativity of cationic elements favor the occurrence of lattice distortion of perovskite structure. By SVC with linear kernel function, the rate of correctness of prediction in LOO cross-validation test is more than 95%.

Table 8.3   The lattice distortion of perovskite-type compounds.

Compound	R_a^*	R_b^*	X_a^*	X_b^*	Distortion
$BaMoO_3$	1.61	0.59	0.9	1.6	No
$BaNbO_3$	1.61	0.68	0.9	1.7	No
$BaSnO_3$	1.61	0.69	0.9	1.9	No
$BaHfO_3$	1.61	0.71	0.9	1.4	No
$BaZrO_3$	1.61	0.72	0.9	1.5	No
$BaPuO_3$	1.61	0.86	0.9	1.3	No
$BaNpO_3$	1.61	0.87	0.9	1.4	No
$SrTiO_3$	1.44	0.605	1.0	1.6	No
$SrMnO_3$	1.44	0.53	1.0	2.1	No
$SrCeO_3$	1.44	0.87	1.0	1.2	Yes
$SrUO_3$	1.44	0.89	1.0	1.9	Yes
$SrPbO_3$	1.44	0.775	1.0	1.8	Yes
$SrZrO_3$	1.44	0.72	1.0	1.5	Yes
$SrIrO_3$	1.44	0.625	1.0	2.1	Yes
$SrRuO_3$	1.44	0.62	1.0	2.0	Yes
$CaSnO_3$	1.34	0.69	1.0	1.9	Yes
$CaZrO_3$	1.34	0.72	1.0	1.5	Yes
$CaHfO_3$	1.34	0.71	1.0	1.4	Yes
$CaNbO_3$	1.34	0.68	1.0	1.7	Yes
$CaMoO_3$	1.34	0.59	1.0	1.6	Yes
$CaRuO_3$	1.34	0.62	1.0	2.0	Yes
$CaTiO_3$	1.34	0.605	1.0	1.6	Yes
$CaMnO_3$	1.34	0.53	1.0	2.1	Yes

$^*R_a$ and R_b denote Shanon-Preweit ionic radii of A and B respectively. X_a and X_b denote Basanov electronegativities of A and B respectively.

The physical meaning of the empirical rule mentioned above is very easy to be understood. Large B ion gives rise to large unit cell, and small A ion cannot keeps 12 neighboring X ions in contact simultaneously. So that the unit cell must to distort to make the decrease of coordination number of A ion.

8.2.2 *Materials design problems of perovskite-like structures*

Since the discovery of the La-Ba-Cu-O superconductor by Bednorz and Muller, a series of high T_c superconducting compounds has been synthesized. All of them have two-dimensional CuO_2 sheets as current carriers, and between two adjacent CuO_2 sheets there is always an insulating atomic or ionic layer. Some important high T_c superconducting compounds, such as La_2CuO_4, are formed by alternative stacking of perovskite-like layers and layers with rock salt structure (This structure is also called K_2NiF_4-type structure).

Since the discovery of La_2CuO_4, a high temperature superconductor, the searching of perovskite-like structure has become an active research field in materials chemistry. One of the most important perovskite-like structures is the layered structure formed by stacking two-dimensional perovskite structure with two-dimensional layer of other crystal lattices, such as rock salt or fluorspar lattices. In chapter 6 of this book, the regularity of the formation of compounds of these types has been analyzed by atomic parameter-pattern recognition method. We will not repeat it here, but only describe an example to demonstrate how to use these principles in materials design for high temperature superconductor research work.

As we have mentioned in chapter 6 of this book, when the perovskite layer and the rock salt layer have been stacked together, their bond distance must be so adjusted to make the bond length of two kinds of layers matched with each other. Since the bond distance is an influential factor affecting many properties of functional materials, we can change the bond length in one kind of layer by making atom-atom substitution to adjust the bond length of the layer of other kind, in order to modify the properties of functional materials.

Since the T_c of oxide superconductor is dependent on the bond length and bond angle in CuO_2 sheets, which are the current carrier for superconductivity. It is understandable that the change of the kinds of atoms in adjacent layers may be useful for adjusting the values of T_c of oxide superconductors. It has been well-known that large cations in the adjacent layer of CuO_2 sheet will promote superconductivity (Up to now, all elements with large ionic radii, such as Hg, Tl, Pb, Sr, Ba and rare earth elements have been used in superconductor preparation), but the influence of the atoms of *next adjacent layers* seems not to be noticed by materials scientists yet. In this respect, the application of support vector machine to find the relevant regularities is probably a new way to get useful concepts for superconductor exploration work. Here an example about the influence of the ions in the next adjacent layer to CuO_2 sheets on the Cu-O-Cu bond angle and T_c of 1222-type oxide superconductors will be described [95].

Figure 8.3 illustrates the structure of a half unit cell of typical 1222 type compounds. In this unit cell the ions "M" are the ions in next adjacent layers of CuO_2 sheets.

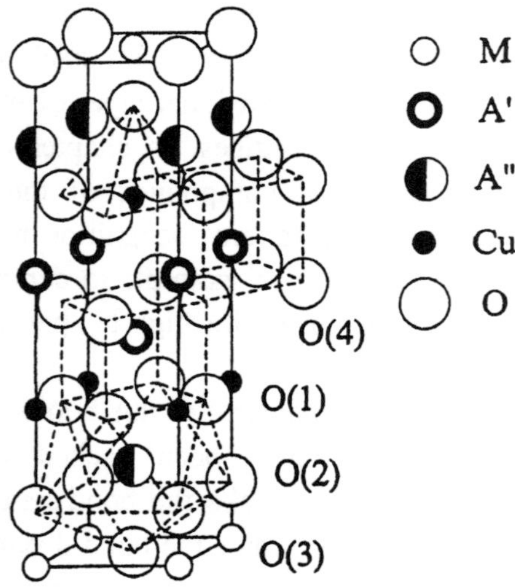

Fig. 8.3   Structure of half unit cell of 1222 compound.

Figure 8.4 illustrates the correlation of the three kinds of parameters, R_M, Z_M and X_M (Here R_M denotes the averaged ionic radii, Z_M denotes the averaged charge number, and X_M denotes the averaged electronegativities of elements of "M") of the atom M in next adjacent layers and the Cu-O-Cu angle of 1222-type oxide supercomputers. Figure 8.5 illustrates the comparison between the experimental T_c values and the T_c values predicted by SVR in LOO cross-validation test.

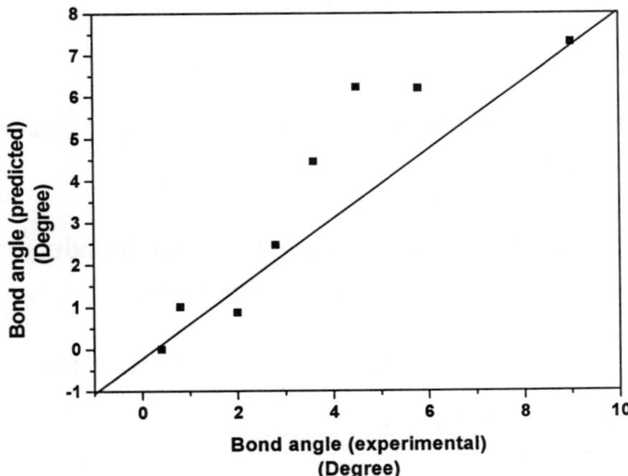

Fig. 8.4   Comparison between the experimental values and SVR-predicted values of Cu-O-Cu bond angles of 1222-type superconductors.

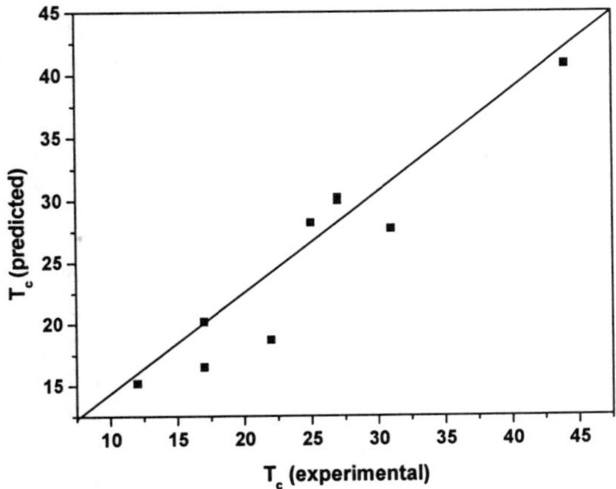

Fig. 8.5   Comparison between the experimental T_c values and the T_c values predicted by SVR for 1222-type superconductors.

From Fig. 8.4 and Fig. 8.5 it can be seen that the correlations are rather good. It appears possible to improve T_c by adjusting the atomic parameters of the M ion, or the atoms in next adjacent layers of CuO_2 sheets.

8.3 SVM Applied to the Computerized Prediction of Properties of Materials

8.3.1 *Computerized estimation of the energy gaps of semiconductor compounds*

Table 8.4 lists the data of the energy gaps of semiconductor compounds, and some atomic parameters of these compounds.

Table 8.4 Energy gaps of compound semiconductors.

No.	Compound	Gap (ev)	$Z1^*$	$Z2^*$	ΔX^*	$\Sigma(Z/Rcov.^*)$
1	InSb	0.23	3	5	0.2	5.6
2	GeSn	0.30	4	4	0.3	6.14
3	InAs	0.36	3	5	0.3	6.15
4	GaSb	0.72	3	5	0.3	5.95
5	SiGe	0.90	4	4	0	6.70
6	InSe	1.25	3	6	0.7	7.18
7	InP	1.35	3	4	0.4	6.60
8	GaAs	1.40	3	5	0.4	6.50
9	CdTe	1.50	2	6	0.4	5.78
10	CdSe	1.80	2	6	0.7	6.53
11	GaSe	2.02	3	6	0.8	7.53
12	HgS	2.10	2	6	0.6	7.15
13	GaP	2.30	3	5	0.5	6.95
14	ZnSe	2.30	2	6	0.5	5.95
15	CdS	2.48	2	6	0.8	7.17
16	GaS	2.50	3	6	0.9	8.17
17	ZnSe	2.70	2	6	0.8	6.70
18	AgI	2.80	1	7	0.6	5.96
19	CuBr	2.91	1	7	0.9	6.94
20	CuI	2.95	1	7	0.6	6.06
21	CuCl	3.17	1	7	1.1	7.80
22	ZnS	3.90	2	6	0.9	7.34
23	SiC	6.00	4	4	0.7	8.61

*Z1 and Z2 denote the valencies of elements. ΔX denotes the difference of Basanov electronegativity of elements. Z/Rcov denotes of the ratio of valency to covalent radius of atom.

It can be shown that the relationship between the energy gap and the atomic parameters listed in Table 8.4 is not a linear one. By using nonlinear regression with second order polynomial equation the result of data fitting is relatively good, but in LOO cross-validation method the error of prediction is rather large, while the results of support vector regression not only give good correlation, but also give acceptable results by LOO cross-validation test. Figure 8.6 shows the comparison between the prediction results of LOO cross-validation test and the experimental values of the energy gap of these semiconductors.

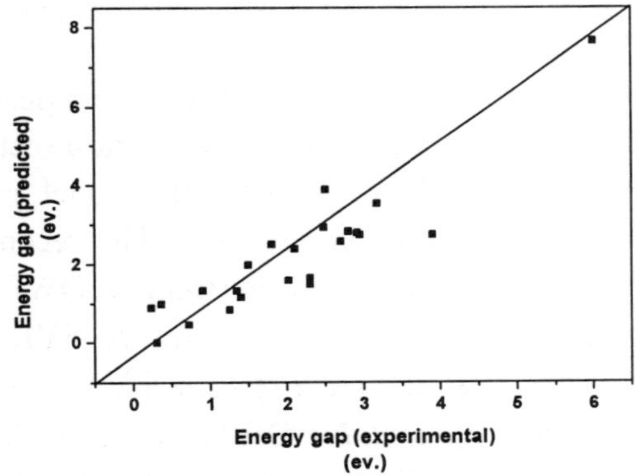

Fig. 8.6   Comparison between the experimental values and the predicted values of energy gaps by SVR in LOO cross-validation.

8.3.2   *SVR applied to mathematical modeling of maximum absorption wavelength of azo dyes*

The most important property of azo dyes is their maximum absorption wavelength (λ_{max}) because it determines the color of azo dyes. Theoretically speaking, λ_{max} of azo dyes is dependent on the electronic structure of azo dye molecules. Quantum chemical calculation has been applied to estimate λ_{max} of some azo dyes. For planar dye molecules, PPP method of quantum chemistry can be used for λ_{max} estimation. But PPP method is not suitable for the estimation of the λ_{max} of azo dyes with non-planar molecules. For azo dyes with non-planar molecules, the application of more sophisticated quantum chemical algorithms has been investigated. Although the calculation of the λ_{max} of some other dye by ZINDO/S algorithm has certain degree of success, the application of

ZINDO/S to azo dye is not very successful. In ZINDO/S algorithm, an adjustable parameter, OWF has to be pre-assigned. Its default value is 0.585. Adachi and his coworkers used this method to calculate the λ_{max} of azobenzene, and found that the results were seriously under-estimated. It means that ZINDO/S with its default value is not suitable for the estimation of the λ_{max} of azo dyes [1; 82].

In order to find some more effective estimation method for the λ_{max} of azo dyes, we have tried to combine the quantum chemical algorithm with the data processing technique together. This is performed by two following ways [150].

The first method is to correlate the quantum chemical parameters of azo dye molecules with the experimental λ_{max} data directly, to build the mathematical model by SVR, and then this model can be used to estimate the unknown λ_{max} of other azo dyes. The second method is to use the experimental λ_{max} data to find the *suitable* OWF value for each dye molecule by an inversed calculation, and then SVR is used to find the mathematical model describing the relationship between the quantum chemical parameters and the *suitable* OWF. Based on the mathematical model found, the *suitable* OWF value of the dye molecule with unknown λ_{max} can be estimated. Both of these two methods have been tested in our laboratory. It has been found that the results of second method are slightly better than that of first one, but both of these methods can give the results much better than that obtained by using ZINDO/S with default value of OWF.

Table 8.5 lists the experimental data of λ_{max} and the molecular descriptors of 37 azo dyes. The molecular descriptors used are obtained as follows: firstly the optimal configuration of 37 azo dye molecules are calculated by molecular mechanics algorithm, then the quantum chemical parameters obtained by PM3 algorithm together with other molecular descriptors are selected by a SVR-based feature selection method. By this way, a feature set including three parameters: ΔE (the difference of HOMO and LUMO), HOMO and the O/N ratio (the ratio of the numbers of oxygen atoms to the number of nitrogen atoms), has been selected for SVR computation.

Table 8.5 λ_{max} and molecular descriptor of azo dye molecules.

No.	R1	R2	HOMO (eV)	ΔE_1 (eV)	O/N ratio	$\lambda^*_{Exp.}$ (nm)	$\lambda^*_{Pred.1}$ (nm)	$\lambda^*_{Pred.2}$ (nm)
1	H_2NO_2S, OCH_3, H_3C (substituted benzene)	HO, tert-butyl substituted phenol	-8.933	8.301	0.75	335.8	391.2	320.2
2	OCH_3, H_3C (substituted benzene)	HO, tert-butyl substituted phenol	-8.734	7.858	1	391.8	435.1	354.2
3	H_3C benzothiazole	HO, tert-butyl substituted phenol	-8.948	7.54	3	400.2	439.2	385.8
4	OC, $(H_3CH_2C)_2NO_2S$ (substituted benzene)	HO, OH (dihydroxy benzene)	-9.034	7.874	0.6	405.6	423.7	352.4
5	H_2NO_2S, OCH_3, H_3C (substituted benzene)	HO, N, OH, HO (pyrimidine)	-9.433	8.068	0.83	407.6	389.1	342.7
6	OCH_3, O_2N (substituted benzene)	HO, N, OH, HO (pyrimidine)	-9.562	7.762	0.83	415.2	409.2	362.1
7	quinoline	HO, tert-butyl substituted phenol	-8.728	7.539	3	418.4	448.6	358.1
8	OCH_3, $(H_3CH_2C)_2NO_2S$ (substituted benzene)	HO, tert-butyl substituted phenol	-9.041	7.84	0.75	421.2	425.3	363.1
9	H_3C, N, S (thiazole)	HO, tert-butyl substituted phenol	-9.071	7.673	3	424.6	422.9	382.3
10	O_2N (phenyl thiazole)	HO, tert-butyl substituted phenol	-9.221	7.418	1.33	428.2	449.1	391.9
11	N–N, S (thiadiazole)	HO, tert-butyl substituted phenol	-9.261	7.678	4	428.6	407.7	357.8

No.	R₁	R₂						
12	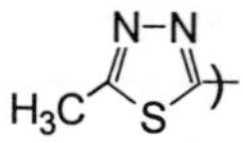 H₃C–thiadiazole (N–N, S)	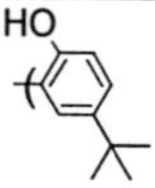 HO–phenol (4-*tert*-butyl)	-9.192	7.434	4	429.8	431.1	367.5
13	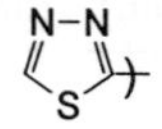 thiadiazole (N–N, S)	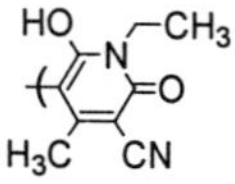 HO, N–CH₃, =O, H₃C, CN	-9.161	7.167	3	431.6	461.4	376.7
14	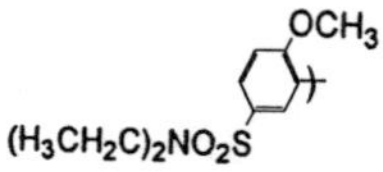 OCH₃, (H₃CH₂C)₂NO₂S–benzene	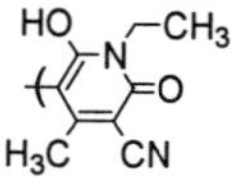 HO, N–CH₃, =O, H₃C, CN	-9.051	7.649	1	439.8	439.2	363.1
15	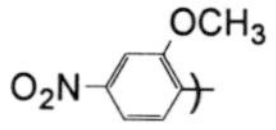 OCH₃, O₂N–benzene	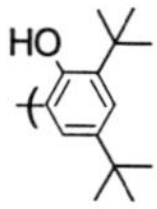 HO–di-*tert*-butylphenol	-9.167	7.457	0.75	449.8	452.0	376.8
16	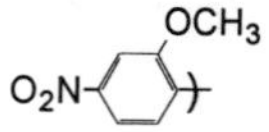 OCH₃, O₂N–benzene	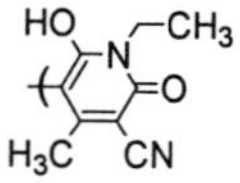 HO, N–CH₃, =O, H₃C, CN	-9.166	7.272	1	452.2	465.8	382.3
17	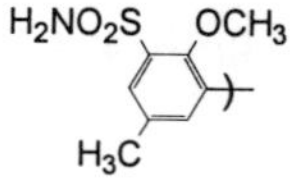 H₂NO₂S, OCH₃, H₃C–benzene	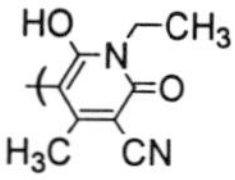 HO, N–CH₃, =O, H₃C, CN	-9.055	7.581	1	455.6	444.7	363.3
18	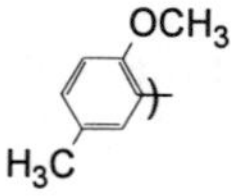 OCH₃, H₃C–benzene	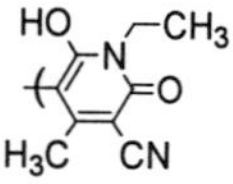 HO, N–CH₃, =O, H₃C, CN	-8.849	7.607	1.33	458.6	449.0	351.2
19	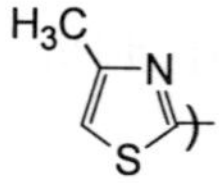 H₃C, N, S (thiazole)	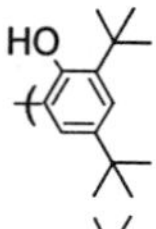 HO–di-*tert*-butylphenol	-8.971	7.594	3	458.8	433.7	384.8
20	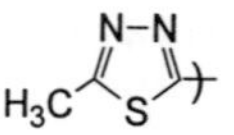 H₃C–thiadiazole (N–N, S)	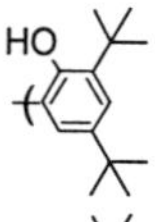 HO–di-*tert*-butylphenol	-9.086	7.347	4	459.2	442.8	375.3
21	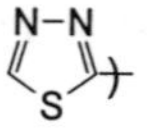 thiadiazole (N–N, S)	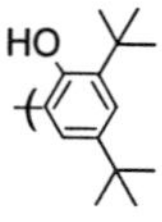 HO–di-*tert*-butylphenol	-9.135	7.346	4	459.8	440.8	377.9
22	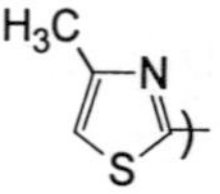 H₃C, N, S (thiazole)	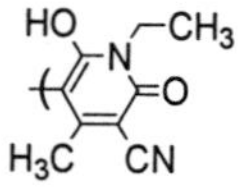 HO, N–CH₃, =O, H₃C, CN	-8.987	7.392	2.5	465.6	453.3	398.4
23	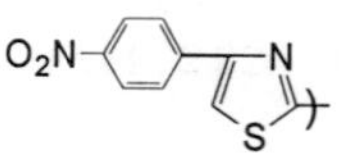 O₂N–phenyl–thiazole (N, S)	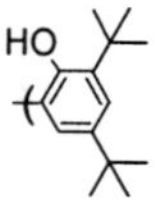 HO–di-*tert*-butylphenol	-9.164	7.332	1.33	466	458.7	391.6
24	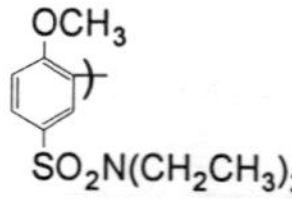 OCH₃, SO₂N(CH₂CH₃)₂–benzene	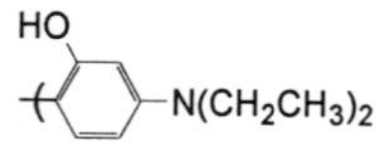 HO–benzene–N(CH₂CH₃)₂	-8.441	7.404	1	466.6	485.4	365.3

No.	Structure A	Structure B	$\lambda_{Exp.}$ col 1	col 2	col 3	col 4	col 5	col 6
25	O$_2$N–C$_6$H$_4$–thiazole(S)	pyridone (HO, N–CH$_3$, =O, H$_3$C, CN)	-9.231	7.166	1.5	469.4	468.6	434.5
26	thiadiazole (N–N, S)	HO–naphthyl	-8.878	7.039	4	477.4	477.4	399.9
27	O$_2$N–, OCH$_3$ (phenyl)	HO–C$_6$H$_3$–N(CH$_2$CH$_3$)$_2$	-8.64	7.062	1	484.6	505.6	372.8
28	H$_3$C–thiazole (N, S)	HO–naphthyl	-8.716	7.24	3	491.2	474.1	410.4
29	OCH$_3$, (H$_3$CH$_2$C)$_2$NO$_2$S (phenyl)	HO–naphthyl	-8.765	7.512	0.75	491.8	464.4	373.4
30	H$_2$NO$_2$S, OCH$_3$, H$_3$C (phenyl)	HO–naphthyl	-8.782	7.456	0.75	499.6	468.3	379.0
31	OCH$_3$, O$_2$N (phenyl)	HO–naphthyl	-8.938	7.171	0.75	500.2	485.6	399.3
32	quinoline (N)	HO–naphthyl	-8.403	7.294	3	501.8	482.8	383.4
33	OCH$_3$, H$_3$C (phenyl)	HO–naphthyl	-8.527	7.423	1	502	480.2	379.4
34	N–N, H$_3$C, S (thiadiazole)	HO–C$_6$H$_3$–N(CH$_2$CH$_3$)$_2$	-8.267	6.736	5	502.6	521.8	381.6
35	H$_3$C–thiazole (N, S)	HO–C$_6$H$_3$–N(CH$_2$CH$_3$)$_2$	-8.167	6.991	4	504.4	511.4	394.4
36	H$_3$CO–benzothiazole (N, S)	HO–C$_6$H$_3$–N(CH$_2$CH$_3$)$_2$	-8.45	7.161	2	517.8	498.7	398.2
37	O$_2$N–(phenyl)–OCH$_3$	pyrazole (CH$_3$, HO, N–N, phenyl)	-9.026	7.639	1.25	525.6	439.4	372.9

*$\lambda_{Exp.}$ denotes the experimental value of λ_{max}, $\lambda_{Pred.1}$ denotes the predicted value by using SVM and $\lambda_{Pred.2}$ denotes the calculated value by using Hyperchem with OWF=0.585.

Figure 8.7 illustrates the comparison between the experimental values and the SVR predicted values of λ_{max} of 37 azo dyes. The calculation results by ZINDO/S with default value of OWF are also plotted for comparison.

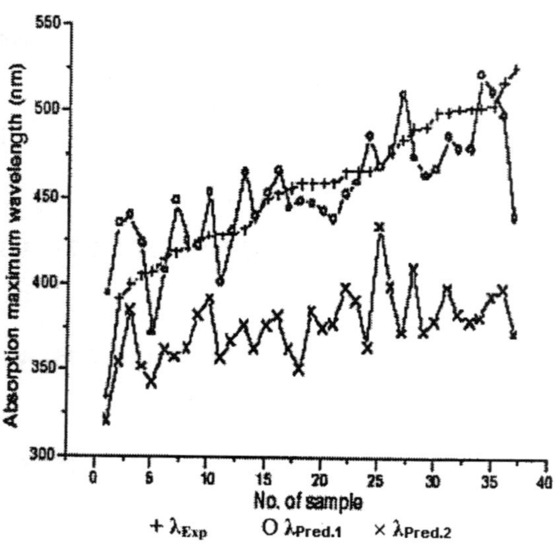

Fig. 8.7   Comparison between the calculated and experimental values of λ_{max} for 37 azo dyes.

8.3.3   *SVR applied to mathematical modeling of cellulose affinity of dyestuff*

The cellulose affinity in dyeing process is another important property of dyestuff. The dyeing of textile fibers led to many theories, which tried to explain the mechanism of dye uptake. One of the most interesting topics is the relationships between the molecular structure and the cellulose affinity of dyestuffs [129]. Table 8.6 lists the data of cellulose affinity of 21 diazo dyes and the following kinds of molecular parameters of these dye molecules: the length of the conjugated chain (n), the van der Waals volume (V_w), the dipole moment (μ), the HOMO by PM3 method of quantum chemical calculation and six steric parameters (λ_1, λ_2, λ_3 and a, b, c) describing the molecular conformation determined by molecular mechanics.

Table 8.6   Cellulose affinity and molecular descriptor for diazo dyes.

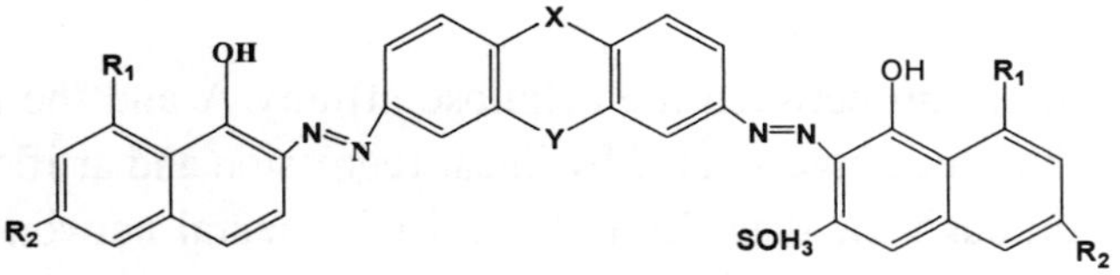

R_1	R_2	X	Y	A	λ_1	λ_2	λ_3	n	V_W	E_{HOMO}	logP	a	b	c
NH_2	SO_3H	-	none	24.60	1.88	1.02	0.10	23	382.12	-8.81	0.29	14.61	9.52	3.76
NH_2	SO_3H	-	CONH	24.10	1.93	0.97	0.11	23	373.88	-8.79	1.42	14.63	9.82	4.21
NH_2	SO_3H	-	CO	23.70	1.86	0.98	0.16	22	366.62	-8.85	0.73	15.17	8.67	4.06
NH_2	SO_3H	NHCONH	none	23.70	2.53	0.41	0.06	26	409.98	-8.70	0.60	10.00	13.87	6.65
NH_2	SO_3H	-	NH	23.60	2.37	0.44	0.19	22	363.00	-8.78	0.99	12.94	11.45	4.45
NH_2	SO_3H	CONH	none	21.00	2.64	0.29	0.07	25	401.90	-8.73	0.61	14.94	10.89	5.95
NH_2	SO_3H	NH	-	19.00	1.66	0.95	0.32	21	363.00	-8.66	0.66	12.15	9.01	3.9
NH_2	SO_3H	NH	none	18.90	2.44	0.46	0.09	24	390.20	-8.67	1.00	13.76	7.10	8.15
NH_2	SO_3H	CO	none	18.30	1.81	1.09	0.10	24	393.82	-8.83	0.99	14.42	6.40	6.99
NH_2	SO_3H	none	CONH	17.60	1.61	0.83	0.56	23	382.96	-8.68	-0.11	9.26	10.42	4.82
NH_2	SO_3H	none	NHCONH	16.40	2.20	0.62	0.18	24	340.50	-8.81	0.95	8.14	9.33	9.83
NH_2	SO_3H	none	CO	13.90	2.21	0.57	0.22	22	356.82	-8.83	0.99	8.31	10.79	8.16
H	NH_2	-	none	30.40	1.93	0.95	0.13	25	332.42	-8.63	2.25	13.46	9.23	3.15
H	NH_2	-	CONH	28.90	1.86	0.91	0.24	25	348.74	-8.61	1.71	13.65	9.27	3.97
H	NH_2	NHCONH	none	27.40	2.53	0.41	0.06	28	356.82	-8.52	1.24	9.23	12.67	6.21
H	NH_2	-	CO	27.10	1.74	1.11	0.15	24	332.42	-8.66	0.75	14.22	8.34	3.65
H	NH_2	CONH	none	27.10	2.66	0.28	0.07	27	348.74	-8.53	0.90	14.07	9.53	6.17
H	NH_2	none	NHCONH	22.90	1.56	1.00	0.44	26	356.82	-8.66	1.24	7.95	8.65	9.22

H	NH$_2$	none	CO	20.40	2.41	0.50	0.08	24	340.66	-8.69	1.28	10.48	11.68	5.54
H	NH$_2$	none	CONH	22.40	1.61	0.87	0.52	25	348.74	-8.51	0.90	8.72	9.99	4.78
H	NH$_2$	CO	none	21.00	2.43	0.49	0.08	26	340.66	-8.68	1.15	9.03	12.03	5.69

The relationship between the cellulose affinity A and the molecular descriptors has been investigated by linear regression and artificial neural network. Although the data fitting by artificial neural network is rather good, the prediction ability in LOO cross-validation is not so good due to overfitting. Using SVR and the polynomial kernel of the second degree, the result of LOO cross-validation test is relatively satisfactory. The comparison between the experimental values and the predicted values of cellulose affinity (A) by using the data set after KL transformation is shown in Fig. 8.8.

Using SVR-based feature selection, the most important parameters affecting the cellulose affinity of diazo dyes selected are: the length of conjugate chain (n), HOMO and steric parameters.

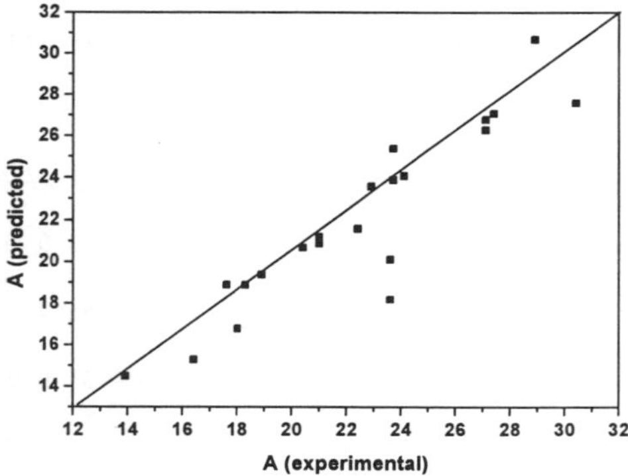

Fig. 8.8 Comparison between the values of experimental cellulose affinity and the values predicted by SVR.

8.4   SVM Applied to Process Design for Materials Preparation

Although the discovery of new materials is very exciting in materials research, the most part of tasks of materials research everyday is to try to improve the preparation technology of known materials. The economic effect of such kind of improvement is very significant because these efforts eventually determine the cost and the quality of products or the competitive ability in international market. Here some examples of the applications of SVM for the modeling of materials preparation processes will be described.

8.4.1 *SVM applied to intelligent control of In_2O_3 semiconductor film preparation (problem of thickness control)*

In_2O_3 semiconductor nanometer film is a new material for combustible gas detector uses. It can be prepared by sol-gel method. How to control the thickness of the semiconductor film is one of the crucial problems in the preparation work. There are several factors influencing the thickness of film: the mass percentage of In_2O_3 and PVA in the bath, the viscosity of coating liquids, the drawing rate and the drawing number in preparation. So it is desirable to have a mathematical model for the automatic control in the film production. SVM and other data processing methods have been used for this purpose. Table 8.7 lists the experimental data used as training set.

Table 8.7 Factors affecting the thickness of In_2O_3 films.

No.	Thickness (nm)	Drawing rate ($cm \cdot min^{-1}$)	Drawing number	Viscosity (centipoise)	In_2O_3 content (%)	PVA content (%)
1	34.2	3.45	4	6.97	0.03	0.03
2	332.3	3.91	4	77.9	0.05	0.06
3	52.7	4.36	4	18.8	0.03	0.04
4	38.9	2.43	3	18.8	0.03	0.04
5	77.8	6.10	5	18.8	0.03	0.04
6	69.6	6.50	5	18.8	0.03	0.04
7	77.3	6.19	5	18.8	0.03	0.04
8	21.5	2.95	3	6.97	0.03	0.03
9	89.0	4.13	4	10.28	0.05	0.03
10	81.1	4.17	4	10.28	0.05	0.03
11	79.9	4.41	4	10.28	0.05	0.03
12	84.5	4.46	4	10.28	0.05	0.03
13	163.8	3.64	4	83.2	0.03	0.07
14	154.0	3.39	4	83.2	0.03	0.07

15	140.2	2.99	4	83.2	0.03	0.07
16	257.8	2.77	3	83.2	0.03	0.07
17	131.6	2.64	3	83.2	0.03	0.07
18	129.8	2.61	3	83.2	0.03	0.07
19	134.5	2.85	3	83.2	0.03	0.07
20	145.1	3.29	5	83.2	0.03	0.07
21	170.5	3.63	5	83.2	0.03	0.07
22	155.9	3.45	5	83.2	0.03	0.07
23	146.4	4.29	5	83.2	0.03	0.07
24	75.0	3.14	4	38.1	0.03	0.05
25	74.8	3.31	4	38.1	0.03	0.05
26	79.5	3.14	4	38.1	0.03	0.05
27	79.4	2.35	3	38.1	0.03	0.05
28	390.6	6.48	5	53.7	0.07	0.05
29	40.0	3.87	4	6.97	0.03	0.03
30	324.0	6.40	5	53.7	0.07	0.05
31	46.3	4.12	4	6.97	0.03	0.03
32	53.6	5.92	5	6.97	0.03	0.03
33	39.7	5.21	5	6.97	0.03	0.03
34	33.6	2.33	3	6.97	0.03	0.03
35	20.0	2.67	3	6.97	0.03	0.03
36	20.0	2.57	3	6.97	0.03	0.03
37	412.7	6.41	5	77.9	0.05	0.06
38	373.1	5.84	5	77.9	0.05	0.06
39	78.9	2.62	3	77.9	0.07	0.03
40	86.4	2.86	3	77.9	0.07	0.03
41	102.5	2.75	3	77.9	0.07	0.03
45	97.8	2.76	3	22.3	0.05	0.04
46	57.7	2.61	3	22.3	0.05	0.04
47	39.0	2.39	3	18.8	0.03	0.04
48	38.5	2.44	3	18.8	0.03	0.04
49	41.2	4.08	4	6.97	0.03	0.03
50	280.5	2.97	3	77.9	0.05	0.06
51	387.0	5.98	5	77.9	0.05	0.06
52	392.5	6.58	5	77.9	0.05	0.06
53	238.1	2.67	3	53.7	0.07	0.05
54	259.0	3.31	3	53.7	0.07	0.05
55	268.9	3.63	4	53.7	0.07	0.05
56	283.2	3.83	4	53.7	0.07	0.05
57	316.8	3.85	4	53.7	0.07	0.05
58	300.3	4.28	4	53.7	0.07	0.05
59	337.2	6.35	5	18.8	0.07	0.05
60	378.2	6.56	5	53.7	0.07	0.05
61	44.7	2.62	3	18.8	0.03	0.04

It has been found that the support vector regression with polynomial kernel of second degree can make the mathematical model for the thickness control of the semiconductor films. Figure 8.9 shows the comparison between the experimental thickness data and the predicted thickness in LOO cross-validation test [9].

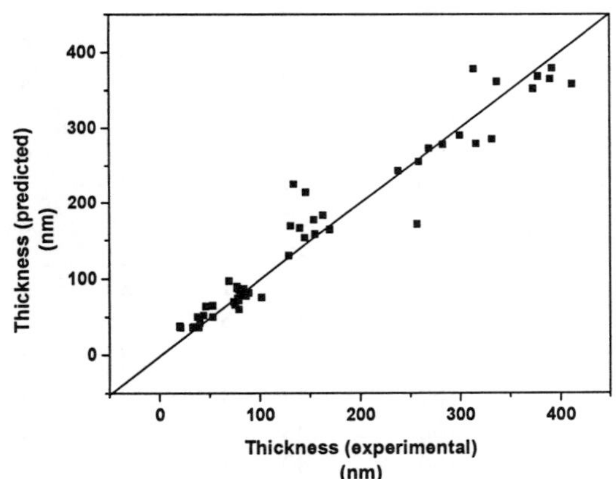

Fig. 8.9　The result of prediction of thickness of In_2O_3 film by SVR in LOO cross-validation test.

8.4.2　*SVM applied to modeling for control of AlN semiconductor film growth (modeling preparation condition for materials with specified structure)*

Aluminium nitride thin film, having 100 or 002 crystal orientation, is useful for piezo-electric devices. The most important factors influencing the crystal orientation are: sputtering pressure (P_s), target-base distances (D) and power of operation (W) in preparation process.

(1) Mathematical model for the preparation of the films with 100-type crystal orientation:

The experimental data set for training is listed in Table 8.8.

By support vector classification with linear kernel, the following criterion can be obtained for the separation of 100 structure forming region from 002 structure forming region:

$$P_s + D - 5.0 > 0 \tag{8.4}$$

Table 8.8   Factors affecting crystal orientation of AlN films.

No.	$P_s(P_a)$	D (cm)	P_w(W)	Crystal orientation	No	$P_s(P_a)$	D (cm)	P_w(W)	Crystal orietation
1	3	4	2	100	31	7	3	4	100
2	3	4	3	100	32	8	1	4	100
3	3	4	4	100	33	1	2	4	100
4	3	4	5	100	34	1	2	3	002
5	1`	5	4	100	35	1	2	4	002
6	1	6	4	100	36	1	2	5	002
7	1	7	4	100	37	1	2	6	002
8	1	8	4	100	38	1	1	4	002
9	2	4	4	100	39	3	1	4	002
10	2	5	4	100	40	2	1	4	002
11	2	6	4	100	41	1	2	1	m*
12	2	7	4	100	42	1	2	2	m
13	3	3	4	100	43	1	2	7	m
14	3	4	4	100	44	1	4	4	m
15	3	5	4	100	45	2	2	4	m
16	3	6	4	100	46	2	3	4	m
17	4	2	4	100	47	3	2	4	m
18	4	3	4	100	48	4	1	4	m
19	4	4	4	100	49	1	3	4	m
20	4	5	4	100	50	3	4	1	m
21	5	1	4	100	51	3	4	5	m
22	5	2	4	100	52	3	4	6	m
23	5	3	4	100	53	2	8	4	a**
24	5	4	4	100	54	3	7	4	a
25	6	1	4	100	55	4	6	4	a
26	6	2	4	100	56	5	5	4	a
27	6	3	4	100	57	6	5	4	a
28	6	4	4	100	58	7	4	4	a
29	7	1	4	100	59	8	3	4	a
30	7	2	4	100					

*m denotes mixed crystal orientation.
**a denotes amorphous structure.

In order to assure the orientation of crystalline film to be 100 type, it is also necessary to find the criteria to avoid the formation of mixed orientation or amorphous state.

The 100 orientation structure-forming region and mixed orientation structure-forming region is not linearly separable. But using Gaussian kernel the separation is good and the rate of correctness of prediction in

LOO cross-validation test is 95.4%. By support vector classification, and select the mutually connected "good sample points" located far away from optimal plane of separation, we can find some optimal region to avoid the formation of mixed orientation. For example, the following region can be used as optimal regions:

$$7 \leq P_s \leq 8; 1 \leq D \leq 3; \text{ and } P_w = 4 \tag{8.5}$$

or $\qquad 1 \leq P_s \leq 4; 3 \leq D \leq 8; P_w = 4; \text{ and } D > 7.51\text{-}0.97P_s$

By using support vector classification, it can be shown that the criterion to avoid the formation of amorphous structure can be expressed as follows:

$$88.9 - 7.99P_s\text{-}9.98D > 0 \tag{8.6}$$

(2) Mathematical model for the preparation of the films with 002-type crystal orientation:

The separation of 002 orientation-forming region and mixed orientation region is not clear. But by using support vector classification an optimal region to assure 002 orientation structure formation can be found as follows:

$$1 \leq P_s \leq 2; 1 \leq D \leq 2; 4 \leq P_w \leq 6; P_w > 2.5D + 0.39$$

By support vector classification, it can be shown that the criterion to avoid the formation of amorphous structure for 002 type film preparation is as follows [29]:

$$2.12 - 0.249P_s\text{-}0.375D > 0 \tag{8.7}$$

8.4.3 *SVM applied to modeling of properties of cathode materials for Ni/H battery production (multivariate modeling problem)*

Ni/H battery is a newly developed battery with high electrochemical capacity. The property of cathode materials is one of the most important factors affecting the quality of battery products. The chief indices of the cathode materials are the electrochemical capacity and its rate of declination in the charging-discharging processes (these factors determine the capacity and life of the battery). There are many factors affecting the quality of the cathode materials: the content of metallic elements (Ni, La, Co, Mn, Nd, Al, Ti and Si in $LaNi_5$ based alloys), the phase composition and the preparation conditions. Here an example of application of SVM in the modeling for the relationships between the C_0 (electrochemical capacity of cathode materials) or the ratio C_{400}/C_0 (C_{400} is the electrochemical capacity after 400 times of charging-discharging processes, so C_{400}/C_0 represents the rate of declination of electrochemical capacity) and the composition of cathode materials will be described. The experimental data used as training set are given in Table 8.9.

Table 8.9   Comparison between C_0 and C_{400}/C_0 of cathode alloys and their composition (atom%).

No.	La	Nd	Ni	Co	Al	Si	Ti	C_{400}/C_0	C_0
1	1	0	5	0	0	0	0	0.16	372
2	1	0	5	0	0.1	0	0	0.12	355
3	1	0	4	1	0	0	0	0.25	372
4	1	0	3	2	0	0	0	0.30	342
5	1	0	4	1	0.1	0	0	0.30	365
6	1	0	4	1	0	0.1	0	0.33	350
7	1	0	2	2.5	0	0	0	0.45	324
8	0.8	0.2	3	2	0	0	0	0.48	302
9	1	0	3	2	0.1	0	0	0.59	315
10	1	0	2	3	0	0	0	0.61	292
11	1	0	1.7	3.3	0	0	0	0.64	273
12	1	0	3	2	0	0.1	0	0.65	310
13	1	0	2.5	2.5	0	0.1	0	0.81	296
14	1	0	2.5	2.5	0.1	0	0	0.82	313
15	0.2	0.3	2.5	2.4	0.1	0	0	0.87	293
16	0.8	0.2	2.5	2.4	0	0.1	0	0.88	293
17	0.7	0.2	2.5	2	0.5	0	0.1	0.90	254
18	1	0	2	3	0.1	0	0	0.94	289
19	1	0	2	3	0	0.1	0	0.96	280

[*]The unit of electrochemical capacity of cathode materials is $mA \cdot hr \cdot g^{-1}$.

(1) Modeling for C_0:

By using linear kernel and support vector regression with $\varepsilon = 0.1$., the mathematical model of C_0 found can give rather good prediction results in LOO cross-validation. Figure 8.10 illustrates the comparison between the experimental data and the results of prediction [92].

(2) Modeling for C_{400}/C_0:

By using linear kernel and support vector regression with $\varepsilon = 0.1$, it can be shown that the mathematical model found can give good prediction results in LOO cross-validation for C_{400}/C_0 values, as shown in Fig. 8.11.

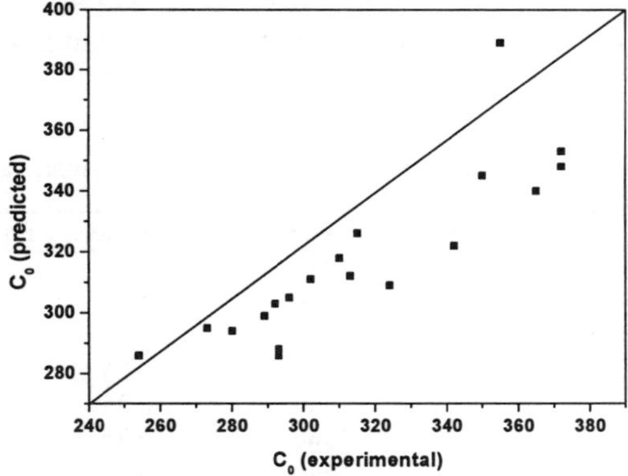

Fig. 8.10   Results of computerized prediction of electrochemical capacity C_0 by SVR.

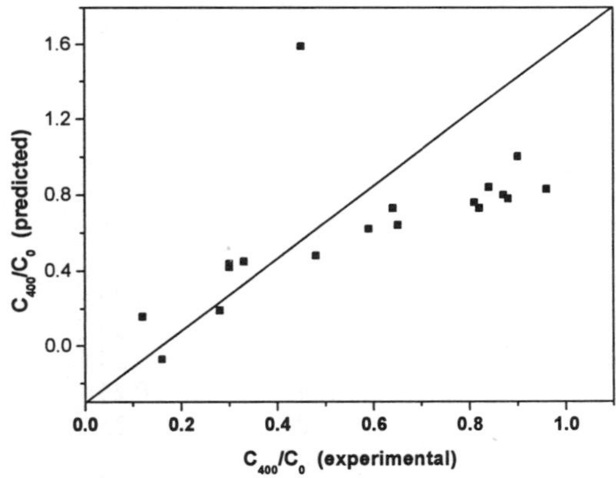

Fig. 8.11   Result of computerized prediction of C_{400}/C_0 by SVR.

8.4.4   *SVM applied to optimization of VPTC semiconductors (optimization problems)*

VPTC materials are a kind of ceramic semiconductors for electronic uses. The task of the research work of VPTC materials is to search the optimal composition and the optimal preparation conditions for high value of

ρ_0/ρ_{min}.(the ratio of the electric resistance at zero degree centigrade to the minimum electric resistance) of these materials. Table 8.10 lists the data used as the training set. The five influencing factors are: Yb_2O_3 content, excess TiO_2, sintering temperature, sintering time, and relative cooling rate.

Table 8.10   Factors affecting of ρ_0/ρ_{min}. of VPTC materials.

No.	ρ_0/ρ_{min}	$Yb_2O_3\%$	Exccess TiO_2	Sintering T °C	Sintering time	Cooling rate
1	20	0.4	1	1360	4	0.5
2	15	0.4	0	1360	4	0.5
3	14.5	0.3	1	1360	4	0.5
4	14.4	0.4	0	1340	0.25	0.5
5	13.1	0.3	0	1360	0.25	0.5
6	13	0.4	0	1360	0.25	0.5
7	12.2	0.4	1	1340	0.25	0.5
8	12.1	0.4	1	1360	0.25	0.5
9	11.3	0.3	0	1360	1	0.5
10	9.6	0.3	1	1360	0.25	0.5
11	9.1	0.3	1	1280	0.25	0.5
12	8.5	0.3	1	1340	0.25	0.5
13	8.1	0.14	0	1360	0.25	0.1
14	6.9	0	0	1360	1	0.5
15	6.4	0	0	1340	0.25	0.5
16	6.1	0	0	1360	1	0.5
17	5.8	1	1	1360	1	0.5
18	5.6	0	0	1280	0.25	0.5
19	5.3	1	1	1340	0.25	0.5
20	5.1	1	1	1340	0.25	0.5
21	4.7	0.15	0	1360	0.25	0.5
22	4.2	0.11	1	1340	0.25	0.5
23	3.9	0.14	1	1340	0.25	0.5
24	3.6	0.15	1	1300	0.25	0.5
25	3.4	0.09	0	1320	0.25	0.5
26	3.1	0.15	1	1360	0.25	0.5
27	2.8	0.13	1	1360	0.25	0.9
28	2.6	0.15	0	1360	0.25	0.9
29	2.5	0.15	1	1360	0.25	0.9
30	2.4	0.11	1	1360	0.25	0.9

It can be shown that the relationship between the target value, ρ_0/ρ_{min}, and the five influencing factors is nearly linear one. Figure 8.12 shows the comparison between the experimental values and the predicted values of ρ_0/ρ_{min} by SVR in LOO cross-validation. It has been found that the averaged absolute error of prediction by SVR is smaller than that obtained by linear regression.

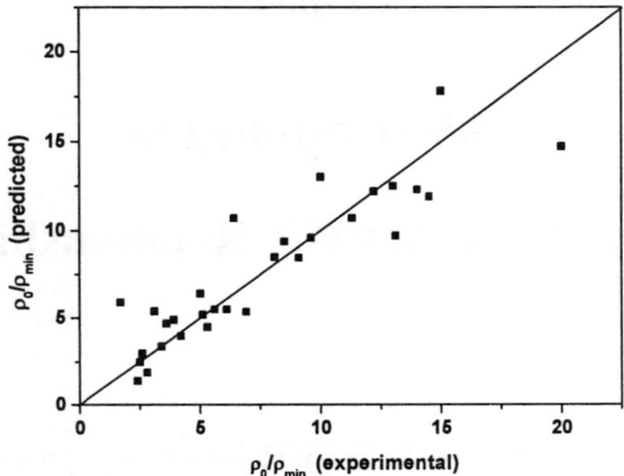

Fig. 8.12 Comparison between the experimental values and predicted values of ρ_0/ρ_{min} by SVR in LOO cross-validation.

Chapter 9

SVM Applied to

Structure-Activity Relationships

9.1 Concept of Structure-Activity Relationships (SAR)

The major tasks of drug development are to characterize medical condition and determine receptor targets, achieve active site complementarity, consider biochemical mechanism of receptor, adhere to laws of chemistry, synthetic feasibility, biological considerations, etc. Combinatorial chemistry and high through-put screening are exciting techniques that are being adopted by the pharmaceutical industries in an effort to reduce costs and shorten discovery and optimization time. Computer chemists are contributing to this effort through combinatorial chemistry, library analysis, diversity analysis, and structure-activity relationship (SAR) studies. It has been realized that SAR plays a very important role in the process of drug development.

SAR work can be classified into two categories: QSAR (quantitative structure-activity relationships) and qSAR (qualitative structure-activity relationships). In QSAR analysis, biological activity is quantitatively expressed as a function of physico-chemical properties of molecules. QSAR involves modeling a continuous activity for quantitative prediction of the activity of new compounds. qSAR aims to separate the compounds into a number of discrete types, such as active and inactive or good and bad. It involves modeling a discrete activity for qualitative prediction of the activity of new compounds.

The seminal work in the field of QSAR was reported by Hansch who had demonstrated the use of regression analysis for model building. It was reported that there existed the analytic relationship between the bioactivities of growth hormone of the plant and the effect of substituents of their compounds. Later, QSAR model was built for the biological activity of plant growth regulators and chloromycetin derivatives with Hammett constants and partition coefficients, using linear regression method [67]. The success of regression analysis in QSAR model building depends upon an assumed linear relationship between the biological activity and one or more descriptors. As the number of descriptors increases, however, regression analysis becomes problematic in many cases. One problem likely to occur in large descriptor sets, for example, is information redundancy when descriptors are correlated. Latent variable techniques have become accepted methods of addressing this issue [115; 49]. These techniques include the use of principal components regression (PCR) and partial least squares regression (PLSR). A second problem encountered in using regression analysis is the assumption of a model form (i.e. quadratic, cubic, use of cross terms, etc.) beforehand. In order to address this issue, variable selection techniques such as stepwise forward and stepwise backward multiple linear regression analysis (MLR) are introduced. One recurrent problem in all of these methods is the fact that by using computational methods to generate descriptors, a modern data set may contain more descriptors than compounds, that is, more columns of parameters than rows of compounds. This results in the introduction of the insidious problem described by Topliss and Edwards [130].

In order to overcome the above-mentioned difficulties, more sophisticated data processing techniques, including cluster analysis, factor analysis, and principle component analysis, have been used in SAR work to study the relationships between biological activity and physico-chemical parameters. With the increasing demand for the research of SAR, the number of multivariate methods introduced into the literature for SAR has increased. Many new algorithms, including artificial neural network, genetic algorithm and decision tree, have been also introduced into the field of SAR and drug design.

The need for a more refined search of potential drugs makes the statistical methods and intelligent computation becoming an integral part of the drug production process. SAR analysis is now one technique used to shorten the searching time for new drugs. It is used to guide optimization of lead compound and investigate mechanisms of chemical-biological interactions. The underlying assumption behind SAR analysis is that there is a relationship between the variation of biological activity within a group of molecular compounds with the variation of their respective structural and chemical features. The analyst searches for a rule or function that predicts the biological activities of new compounds from the values of their physico-chemical descriptors.

In the SAR research work, one has to deal with two difficult problems: feature selection and model selection for a particular data set with finite number of samples and multiple descriptors.

On the one hand, it is a tough work to select appropriate physico-chemical descriptors of molecules, for there are large number of descriptors concerning with electrostatic effects, geometrical parameters and hydrophobic roles. Usually, the analyst tries different kinds of descriptors to examine if the tentative descriptors are related to a known target activity and attempts to model the relationship between them within a finite number of compounds. *A priori* knowledge of the underlying chemistry may also be considered. The methods used to represent molecular features when producing a SAR model vary greatly and a lot of work is carried out on this sub-problem alone.

On the other hand, it is a hard work to select a proper model with high accuracy of prediction for the activities of new compounds, since there are a lot of statistical or chemometric methods for data processing. The most commonly used methods are multiple linear regression (MLR), partial least square regression (PLSR), nonlinear regression with hypothesis of nonlinear items including quadratic ones, artificial neural networks (ANN) including back-propagation (BP) ANN, self-organized ANN and neural trees (NT), genetic algorithms (GA), traditional pattern recognition methods including principal component analysis (PCA), multiple discriminate vector (MDV) including Fisher vector to discriminate between two classes, nonlinear mapping (NLM), decision trees (DT), etc.

Generally speaking, empirical data modeling of SAR application concerns a process of induction to build up a model of the system, from which it is hoped to deduce responses of the system that have yet to be observed. Ultimately the quantity and quality of the observations govern the performance of this empirical model. By its observational nature data obtained is finite and sampled; typically this sampling is non-uniform and due to the high dimensional nature of the problem the data will form only a sparse distribution in the input space. Besides, the accuracies of the data of the biological activities and the physico-chemical parameters measured or calculated are limited and the influence of noise is usually cannot be neglected. Consequently the problem is actually ill-posed in the sense of Hadamard [111; 65]. It means that the problem of overfitting is inevitable if the traditional regression or artificial neural network are used. Since the size of the training sets in the most of SAR work are usually relative small, the problem of overfitting in artificial neural network and nonlinear regression methods will become a rather serious problem.

As an effective way to overcome the problem of overfitting, support vector machine (SVM) [61; 132], as a newly developed method, has been introduced in the field of SAR. The paper of Burbidge reported the pioneering work in this field. In this work, it was reported that the prediction ability of support vector classification (SVC) was significantly better than that of artificial neural network and decision tree in the SAR computation for the prediction of the inhibition of dihydrofolatase by pyrimidines [16]. In this chapter, more SAR work using both support vector classification (SVC) and support vector regression (SVR) methods will be described.

As we have mentioned before, the success of SVM does not mean the fact that the classical data processing methods would be useless in SAR. On the contrary, a clever strategy is to integrate SVM together with other data processing methods together to make problem-solving for SAR. In this chapter, we firstly describe various kinds of chemometric methods, which are widely adopted computational approaches in drug design in section 2. And the useful molecular descriptors are introduced in section 3. Then, some applications of SAR using SVC or SVR are demonstrated in the following sections. In these examples of application, the results of

SVC or SVR are compared with that of some other algorithms, and it can be seen that the prediction abilities of SVM in these examples are all better than those of other algorithms.

9.2   Brief Introduction to Some of Chemometric Methods Used in SAR

Now a series of chemometric methods have been used in SAR. Generally speaking, different chemometric methods are suitable for different cases for each kind of these methods has its own advantages and disadvantages, and the best approach should be the combination of several chemometric techniques that complement each other, as that we have discussed in chapter one. So it is helpful to introduce briefly the different chemometric methods widely used in SAR. Here chemical pattern recognition methods will be specially emphasized since different projections can provide plentiful information from different projection maps from the hyperspaces. In SAR practice, some chemical pattern recognition methods are often used as complementary methods in the application of SVM.

9.2.1   *Chemical pattern recognition in SAR*

Starting from the pioneering work of Kowalskii published in the year of 1972, various chemical pattern recognition techniques have been the powerful tools in SAR work.

All chemical pattern recognition methods are based on the computerized recognition of the multidimensional figures (or the figures of their two-dimensional projections) of the distribution of the sample points of different classes in feature spaces. In SAR research, the molecular descriptors (independent variables, often called features) influencing the biological activity (target variables, often called dependent variables) are used to span some multidimensional spaces. And the representative points of molecules with different biological activities as plotted into these spaces with different symbols. Then various pattern recognition methods can be used to recognize the figure of the distribution zones of different sample points. By this way, the mathematical model for the classification can be obtained. If we adjust the criterion of classification, some semi-quantitative models describing the regularities of target can be found, provided that the noise is not too serious.

In SAR, chemical pattern recognition methods can be categorized into unsupervised methods and supervised pattern recognition methods.

The unsupervised pattern recognition normally consists of cluster analysis. Using a couple of dozen descriptors of molecules, it is possible to see which activities of molecules are most similar and draw a picture of these similarities, called a dendrogram, in which more closely related activities are closer to each other

The methods of unsupervised pattern recognition can be directly applied to SAR. It can be used to determine the similarities in structures of molecules with a variety of activities. The more similar the molecular structures, the closer the activities of molecules: chemical similarity mirrors biological similarity. Sometimes the amount of information is very large, for example in large crystallographic databases, so that cluster analysis is the only practicable way of searching for similarities between molecules. Unsupervised pattern recognition differs from exploratory data analysis in that the aim of the methods is to detect similarities, without particular prejudice as to whether or how many groups will be found.

The supervised pattern recognition requires a training set of known groupings to be available in advance, and tries to answer a precise question as to the class of an unknown sample. Of course, firstly, it is always necessary to establish whether chemical measurements are actually good enough to fit into the predetermined groups. There are a large number of methods for supervised pattern recognition, mostly aimed at classification with discriminant functions.

In SAR, the descriptors of molecules are used to determine whether their activities are high or not. Using data set of known groupings, a supervised pattern recognition model can be set up. Finally, the activity of an unknown sample can be predicted.

The supervised pattern recognition methods include K nearest neighbor method (KNN), principal component analysis (PCA), Fisher method and partial least square method (PLS). While some newly developed methods such as sphere-linear mapping (SLM), optimal projection recognition (OPR), hyper-polyhedron, box methods are also very useful for the data processing of complicated data sets in SAR [30].

KNN methods are conceptually rather simple, and do not require elaborate statistical computations. The basic concept of this method is to discriminate the class of an unknown sample point by assuming that its class is equal to the class of the majority of its nearest neighbors. The method is implemented by calculating the distance between the unknown sample point to representative points of all members of the training set, and then taking the *majority vote* for classification of the unknown sample based on known classes of K samples with smallest distances closest to it.

PCA is a method based on the Karhunen-Loeve transformation (KL transformation) of the data points in the feature space. In KL transformation, the data points in the feature space are rotated such that the new coordinates of the sample points become the linear combination of the original coordinates. And the first principal component is chosen to be the direction with largest variation of the distribution of sample points. After the KL transformation and the neglect of the components with minor variation of coordinates of sample points, we can make dimension reduction without significant loss of the information about the distribution of sample points in the feature space. Up to now PCA is probably the most widespread multivariate statistical technique used in chemometrics. Within the chemical community the first major application of PCA was reported in 1970s, and form the foundation of many modern chemometric methods. Conventional approaches are univariate in which only one independent variable is used per sample, but this misses much information for the multivariate problem of SAR, in which many descriptors are available on a number of candidate compounds. PCA is one of several multivariate methods that allow us to explore patterns in multivariate data, answering questions about similarity and classification of samples on the basis of projection based on principal components.

Fisher method can be derived from multiple discriminant vector method, for the case of sample set with two classes only. The first discriminant P_1 called Fisher vector is $\alpha W^{-1}[m_1-m_2]T$. Here α is the standardized coefficient. W is the within class scatter matrix, and m_1, m_2 are the average vectors of two different kinds of classes respectively. The second vector P_2 to plot samples on the two dimensional plane can be available by using maximizing the following criterion:

$$\frac{P_2^T B P_2}{P_2^T W P_2} - \lambda P_2^T P_1 \qquad (9.1)$$

where B is the between class scatter matrix, the λ is Lagrange coefficient.

Compared with the regression methods or ANN, the results of pattern recognition methods are usually semi-quantitative or qualitative information about classification. This is of course a limitation of pattern recognition methods. This fact, however, is not always a disadvantage, because many data files exhibit strong noise and a quantitative calculation may be *too precise* for these data processing problems. Besides, in many cases practical problems are *yes or no* problems (for example, the problem may be *whether a compound is active or inactive*). It is easily understandable that pattern recognition is especially suitable for these data processing problems.

9.2.2 *Linear or nonlinear regression in SAR*

The Hansch-Fujita method is a classical QSAR technique with traditional statistics, which constructs the quantitative relationship between physico-chemical descriptors of substituents and biological activities by using classical regression analysis.

In the research of QSAR with classical statistics, it is assumed that the data between the activities and the descriptors obey linear relationship, where the noise is very small, the distribution of data is relatively uniform, and the independent variables are linearly independent of each other. These assumptions can greatly simplify the computation. As is well-known, MLR model is just based on these assumptions. Actually, MLR will be a very good method if these assumptions are indeed obeyed. Unfortunately, most of the data in various QSAR problems can not satisfy these conditions. Besides, since the Hansch-Fujita method uses MLR model with two independent features only, the predictive ability and applicable range of this model is more restricted.

In regression work, if some of the features used are linearly dependent of each other, then collinearity problem happens. In order to solve the collinearity problem, principal component regression (PCR) and partial least squares regression (PLSR) methods are proposed.

PCR is one of linear regression methods using some of principal components instead of original features so that the collinearity problem is eliminated since the selected principal components are orthogonal to each other.

PLSR is an extension of the multiple linear regression model. It is probably the least restrictive of the various multivariate extensions of the multiple linear regression model. This flexibility allows it to be used in situations where the use of traditional multivariate methods is severely limited, such as the case that when there are fewer observations than predictor variables. Furthermore, PLSR can be used as an exploratory analysis tool to select suitable predictor variables and to identify outliers before classical linear regression. Especially in chemometrics, PLSR has become a standard tool for modeling linear relationships between multivariate measurements.

PLSR method not only can avoid the collinearity problem, but also can filter off a part of noise by using predicted error sum of square (PRESS) computation. But it should be emphasized that PLSR is not very effective to solve nonlinear problems, since the addition of nonlinear terms can be only tried by trial and error method.

In principle, the addition of nonlinear terms can extend the linear regression method to nonlinear problems. In some cases it is indeed feasible to solve nonlinear problems. One of the difficulties is how to guess the exact form of suitable nonlinear function. Usually, The addition of quadratic terms is tried out if there exists a reasonable combination of quadratic terms by using polynomial regression method, sometimes the addition of quadratic terms indeed decrease the PRESS of the regression model.

9.2.3 *Artificial neural network in SAR*

Artificial neural network (ANN) has been applied to SAR analysis since the late 1980s, mainly in response to accuracy demands. In QSAR research, ANN can be used to construct a quantitative model between structural descriptors and biological activities. Then, biological activities of unknown compounds are predicted from this model. One of the advantages of ANN is that it can be used for the modeling of nonlinear relationships. The classical QSAR technique such as Hansch-Fujita method is appropriate only for the modeling of linear relationship between physico-chemical descriptors of substituents and biological activities using the multiple linear regression (MLR) method. However, there maybe exist nonlinear relationships in SAR problems, and ANN is just an effective technique for the modeling of nonlinear relationships in QSAR study. For example, it was reported that the Hopfield neural network could be used to construct the 3D QSAR such as comparative molecular field analysis (CoMFA), which successfully reproduced the real molecular alignments obtained from X-ray crystallography [96].

Compared with linear and nonlinear regression methods, the advantage of ANN is its ability to correlate a nonlinear function without assumption about the form of this function beforehand. And the trained ANN can be used for unknown prediction. Therefore, ANN has been widely used in data processing of SAR. But if we use ANN solely, sometimes the results of prediction may be not very reliable. Experimental results indicate that some of the test samples predicted by ANN as optimal samples are really not true optimal samples. This is a typical example of so-called *overfitting* that makes the prediction results of trained ANN not reliable enough. Since the data files in many practical problems usually have strong noise and non-uniform sample point distribution, the overfitting problem may lead to more serious mistake in these practical problems.

Based on the concept of statistical learning theory, some modification of the algorithms of ANN has been proposed. For example, it has been argued that the artificial neural networks with minimized weight can be used to make the overfitting depressed. On the other hand, an early stopping algorithm of ANN sometimes can be also used to depress overfitting. The weight decay ANN and early stopping ANN are now widely used in SAR work. However, even by weight decay or early stopping method, the overfitting problem of ANN cannot completely avoided yet.

Other limitations of ANN method are the convergence problem and the danger in extrapolation. When the data file exhibits strong noise, the training of ANN usually cannot converge. Since the nonlinear nature of ANN, it is generally accepted that extrapolation is not allowable.

9.2.4 *Genetic algorithms in SAR*

Genetic algorithm (GA) is a novel algorithm rooted in Darwin's theory and attracts much attention as a powerful tool for various optimization problems in QSAR analysis [57]. As ANN, genetic algorithms also rely on a specific representation of the problem to be solved. The genetic algorithm will generate a number of chromosome strings at random, each of them representing an individual in the initial (or parent) population, to which the evolutionary principles of selection and mutation are applied. The algorithm will favor individual with higher fitness (lower error) to be selected and to be propagated into the next generation. Typically, this evaluation process consumes most of the execution time, no matter whether the fitness is determined by calculation or by experiment.

Usually, a new generation is produced via applying genetic operators to the selected individuals. The basic operators are mutation, where one or more digits of the chromosome string are changed, and crossover, where two strings are cut and crosswise recombined to form two new strings, which contain the features of both parents. Other operators like reproduction (copy) or inversion (swapping of substrings) are of minor importance.

In SAR research, this directed stochastic search makes genetic algorithms a very robust and universal tool for global optimization problem of conformation search, which can be expressed in a reasonably small set of parameters.

One of limitations that genetic algorithm suffers is the stochastic nature of both population initializations and the genetic operators used during training, which can make results hard to reproduce.

9.2.5   *Decision tree in SAR*

The simplest kind of decision tree consists of nodes (circles) and the branches (segments) connecting the nodes. Each internal node may grow out two or more branches. Each node corresponds with a certain characteristic and the branches correspond with a range of values. These ranges of values must give a partition of the set of values of the given characteristic. When precisely two branches grow out from an internal node, each of these branches can give a true or false statement concerning the given characteristic. There are, however, more complex kinds of trees, in which each internal node corresponds to more complex statements, not one but several characteristics are given. For example, these statements can be defined by a linear combination of quantitative characteristics, corresponding to various subregions of multivariate space split by a hyperplane.

In SAR research, a decision tree represents a logic model of regularities of the researched bioactivities since it allows to process both quantitative and qualitative characteristics simultaneously. The weakness of decision trees is the fact that in a case where all characteristics are quantitative, the decision trees may represent too rough approximation of the optimum solution. For example, the regression tree is piecewise a constant approximation of the regression function, although it is possible to compensate this weakness by increasing the number of leaves, i.e. by decreasing the length of appropriate "segments" or "steps". On the other hand, the decision tree model had the advantage of easy interpretation, for example, it can be used to demonstrate for classifying and predicting the aquatic toxicity mechanisms of phenols including polar narcosis, respiratory uncoupling, pro-electrophilicity, and soft electrophilicity [117].

9.3   Brief Introduction to Molecular Descriptors Used in SAR

In SAR research, the purpose is to connect the biological activities of a series of compounds with their physico-chemical properties by using regression analysis, pattern recognition methods, or other sophisticated data processing methods. Generally speaking, the activities and properties are connected by a function F as follows for QSAR problems:

$$\text{Biological activity} = F \text{ (Physicochemical Properties)}$$

or a series of inequalities presented as criteria to discriminate between the molecules with high and low biological activities.

In SAR work, the biological activity of compounds is usually expressed by the values of IC_{50} or ED_{50}. Physico-chemical properties used in SAR can be broadly classified into three general types: electronic parameters, steric parameters and hydrophobic parameters. Electronic parameters include Hammett constants ($\sigma, \sigma^-, \sigma^+$), Swain and Lupton field parameter (F), Swain and Lupton resonance parameter (R), etc. Steric parameters include Taft's steric parameter (E_s), molecular volume (V_m), molecular surface area (Area), molecular weight (MW), van der Waals radius (r), molar refractivity (MR), parachor (P_r), etc. Hydrophobic parameters include partition coefficient (LogP), distribution coefficient (LogD), substituent constant (π), solubility parameter (LogS) etc. Very often, physicochemical properties can also be theoretical parameters obtained from the computational work using semi-emperical methods of quantum chemistry such as PM3 and other quantum chemical methods. The theoretical parameters available include atomic net charge (Q), energy of highest occupied molecular orbital (E_{HOMO}), energy of lowest unoccupied molecular orbital (E_{LUMO}), electrostatic potential (EP), superdelocalizability (S_r), bond length (L), bond angle, molecular dipole moment, etc. Some of these parameters are explained as follows.

HOMO is the highest energy level of the molecule that contains electrons. It is crucially important in governing molecular reactivity and other properties. When a molecule acts as a Lewis base (an electron pair donor) in bond formation, the electrons are supplied from the HOMO of molecule. How readily this occurs is reflected in the energy of the HOMO. Molecules with high HOMOs have greater tendencies to donate their electrons and are hence relatively reactive compared to molecules with low-lying HOMOs, thus descriptor HOMO can be considered as a measure of the nucleophilicity of molecules.

LUMO is the lowest energy level in the molecule that contains no electron or only a single electron. This characteristic is important in governing molecular reactivity and properties. When a molecule acts as a Lewis acid (an electron pair acceptor) in bond formation, incoming electron pairs are received in its LUMO. Molecules with low-lying LUMOs have greater tendencies to accept electrons than those with high-energy LUMOs; thus descriptor LUMO can be considered as a measure the electrophilicity of a molecule.

Octanol-water partition coefficient (LogP) is used in QSAR studies and rational drug design as a measure of molecular hydrophobicity. Hydrophobicity affects drug absorption, bioavailability, hydrophobic drug-receptor interactions, metabolism of molecules, as well as their toxicity. LogP has become also a key parameter in studies of the environmental fate of chemicals. The values of LogP of many known compounds have been measured by experimental work. But there is already found some simple empirical equations that can be used to estimate the values of LogP of unknown compounds based on the summation of the contribution values of molecular fragments.

The molecular surface area (Area) descriptor is a 3D spatial descriptor that describes the van der Waals area of a molecule. The molecular surface area determines the extent to which a molecule exposes itself to the external environment. This descriptor is related to the binding, transport properties and the solubility of compounds.

The molecular volume (V_m) is a 3D spatial descriptor that defines the molecular volume inside the contact surface. The molecular volume is calculated as a function of conformation. Molecular volume is related to the binding and transport properties of compounds.

The molar refractivity (MR) can be considered as the sum of either atom or bond refractivities. This sum, which can be estimated directly from the molecular structure, should be equal the value given by the

Lorenz-Lorenz equation when the measured values of density and refractive index have been inserted:

$$MR = \frac{\mu^2 - 1}{\mu^2 + 2} \cdot \frac{M}{d} \tag{9.2}$$

where M is the molecular weight, μ is the refractivity, and d is the density of compound.

The parachor (P_r) is an additive physico-chemical property of a substance related to its molar volume. The value of the parachor is determined by the kind and the number of atoms in a molecule as well as their manner of arrangement and binding.

The superdelocalizability is an index of the reactivity of aromatic hydrocarbons , proposed by Fukui.

$$S_r = 2\sum_{j=1}^{m}\left(\frac{c_{jr}^2}{e_j}\right) \tag{9.3}$$

where S_r is the superdelocalizability at position r, e_j is the bonding energy coefficient in j-th molecular orbital (eigenvalue), c is the molecular orbital coefficient at position r in the HOMO, m is the index of the HOMO (highest occupied molecular orbital). This index is based on the idea that early interaction of the molecular orbitals of two reactants may be regarded as a mutual perturbation, so that the relative energies of the two orbitals change together and maintain a similar degree of overlap as the reactants approach one another. Consequently, summing S for all atomic positions of a molecule gives a metric of electrophilicity, which may be used to predict relative reactivity in a series of molecules.

The van der Waals radius (r) of an atom is the radius of an imaginary hard sphere, which can be used to model the atom for many purposes. Van der Waals radii are determined from measurements of atomic spacing between pairs of unbonded atoms in crystals.

Electrostatic potential (EP) is both a molecular property and a spatial property. It depends on what charges exist in the molecule and how they there are distributed. It also depends on what point (x, y, z) we choose to investigate. If we select a point where the +1 charge is attracted by the molecule, the potential will be negative at this point. On the other hand, if we select a point where the +1 charge is repelled, the potential will be positive.

9.4 SAR of N-(3-Oxo-3,4-dihydro-2H-benzo[1,4]oxazine-6-carbonyl) Guanidines

The Na/H exchanger is a major Na^+ entry pathway in many types of cells and plays an important role in regulation of cell volume and ion concentration. It is rapidly activated at post-ischemic reperfusion and causes a Ca^{2+} overload, which is known to be associated with cellular dysfunction, damage and necrosis. Therefore, Na/H exchange inhibitor is a potentially useful candidate for improvement of ischemia-reperfusion-induced injury. N-(3-Oxo-3,4-dihydro-2H-benzo[1,4]oxazine-6-carbon-yl) guanidines are known to be Na/H exchange inhibitors. We have tried to build SVC model for SAR of N-(3-Oxo-3,4-dihydro-2H-benzo[1,4]o-xazine-6-carbonyl) guanidines as Na/H exchange inhibitory agent. Since SVM is especially suitable to find the regularities of data set with fewer samples, it may be a suitable tool to investigate the SAR of guanidines with a small sample set.

9.4.1 *Data set*

The data set available consisting of 20 compounds is listed in Table 9.1 [142]. The activity (IC_{50}) is evaluated by its ability to inhibit platelet swelling induced by sodium propionate. Samples with $IC_{50} < 0.44$ $\mu mol \cdot L^{-1}$ are assigned to be class 1, while the others are the samples of class 2.

Table 9.1　Na/H exchange inhibitory activities and selected features of 20 compounds.

No.	R_1	R_2	R_3	X	HOMO /eV	LUMO /eV	D /Å	Q	IC_{50} /$\mu mol \cdot L^{-1}$
1	H	H	H	H	-8.93	-0.628	0.99	-0.200	10.0
2	H	H	Et	H	-8.84	-0.575	3.42	-0.197	0.33
3	H	H	Iso-Pr	H	-8.82	-0.516	3.44	-0.198	0.25
4	H	Me	Et	H	-8.80	-0.524	3.42	-0.192	0.29
5	H	Et	Iso-Pr	H	-8.78	-0.478	3.44	-0.189	0.17
6	H	Ph	Iso-Pr	H	-8.79	-0.500	3.44	-0.188	1.00
7	Me	Me	Iso-Pr	H	-8.78	-0.467	3.44	-0.196	0.12
8	Me	Me	H	H	-8.87	-0.558	0.99	-0.196	7.40
9	Me	Me	Me	H	-8.77	-0.508	2.12	-0.196	0.81
10	Me	Me	Et	H	-8.77	-0.498	3.42	-0.195	0.38
11	Me	Me	Pr	H	-8.77	-0.499	4.65	-0.195	0.74
12	Me	Me	Butyl	H	-8.77	-0.498	5.96	-0.195	1.30
13	Me	Me	$(CH_2)_2OEt$	H	-8.82	-0.536	6.44	-0.195	5.60
14	Me	Me	Hexyl	H	-8.78	-0.487	7.63	-0.194	3.70
15	H	H	Me	Cl	-8.95	-0.728	2.12	-0.194	0.33
16	H	H	Et	Cl	-8.96	-0.726	3.42	-0.190	0.22
17	H	H	Iso-Pr	Cl	-8.90	-0.659	3.44	-0.188	0.16
18	H	H	Me	OMe	-8.80	-0.591	2.12	-0.187	0.50
19	H	H	Et	OMe	-8.72	-0.511	3.45	-0.179	0.27
20	Me	Me	Iso-Pr	OMe	-8.75	-0.473	3.44	-0.182	0.37

9.4.2 *Molecular descriptors and feature selection*

The molecular descriptors obtained by computation of molecular mechanics and quantum chemical methods are used to describe the molecular structures of *N*-(3-Oxo-3,4-dihydro-2*H*-benzo[1,4]oxazine-6-carbonyl) guanidines. The three-dimensional structures of the molecules are optimized with the software Hyperchem[1]. Prior to the semi-empirical quantum chemical computation, all structures of the compounds are submitted to MM+ computation of molecular mechanics for energy optimization. The structural descriptors are obtained via the computation of semi-empirical method PM3. The computations are carried out at restricted Hartree-Fock level without configuration interaction.

---

[1] Release 7.0 for Windows Molecular Modeling System, Hypercube Inc. 2002

The feature set used for modeling is selected from the molecular descriptors by using the hyper-polyhedron model [8; 45]. The principle of this algorithm is illustrated in Fig. 9.1. The samples of high activity (class 1) and those of low activity (class 2) can be separated by a hyperpolyhedron with good separability. The principle of feature selection is to find the smallest feature set that can separate the samples of two classes with clear-cut boundaries. By this way, four features, HOMO, LUMO, D (the longest diameter of the substituent R_3), and Q (net charge of the oxygen atom of oxazine) have been selected out as features for modeling.

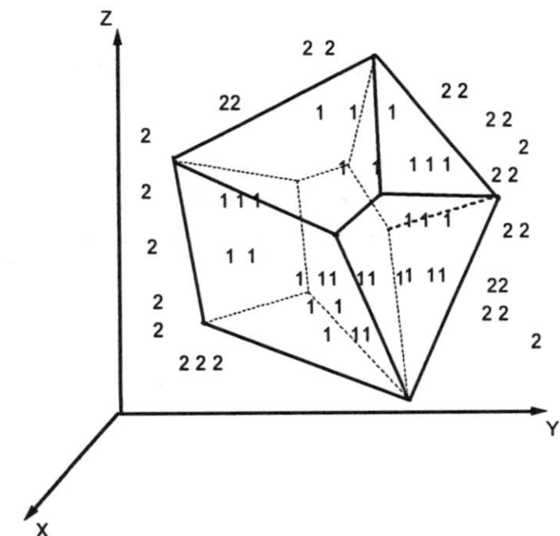

Fig. 9.1　Conceptual hyper-polyhedron in three dimensional space.

9.4.3　*Selection of the kernel function and capacity parameters*

In order to assure the best result of modeling, the leave-one-out (LOO) cross-validation method is used to find the suitable parameter C and the appropriate kernel function for SVC modeling. In this computation, the rate of correctness (P_A) is used as the criterion for the selection of kernel function and parameter C.

Based on the results found by Fig. 9.2 and Fig. 9.3, it has been found that the highest P_A can be obtained by using radial basis kernel function with the parameters C = 80, and σ = 1.0.

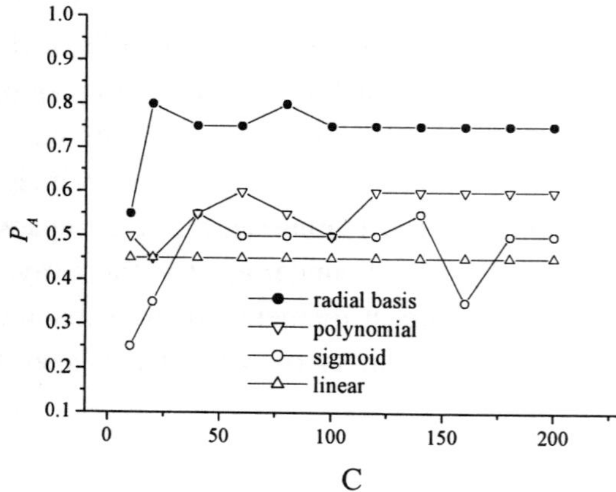

Fig. 9.2   The P_A of different models versus C from 10 to 200.

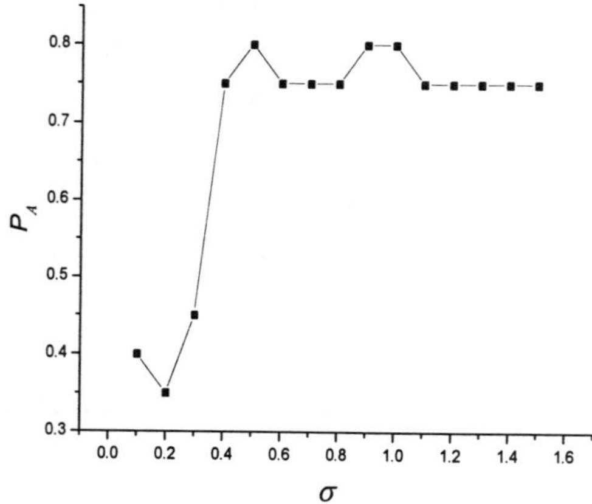

Fig. 9.3   The P_A of model with radial basis function versus σ with
C = 80.

9.4.4   *Modeling by SVC*

Using the radial basis kernel function with capacity parameter C = 80,
the following criterion for the bioactivities has been obtained:

$$g(\mathbf{x}) = \sum_{i \in SV} \alpha_i y_i \exp\{-\frac{\|\mathbf{x} - \mathbf{x}_i\|^2}{\sigma^2}\} + b \qquad (9.4)$$

where $b = 2.242$, $\sigma = 1.00$, α_i (listed in Table 9.2) are corresponding to the Lagrange multipliers of support vectors. $y_i=1$ corresponds to the samples of class 1 ($IC_{50} < 0.44$ $\mu mol \cdot L^{-1}$); while $y_i = -1$ corresponds to the samples of class 2 ($IC_{50} > 0.44$ $\mu mol \cdot L^{-1}$). $\mathbf{x}$ is a vector (pattern of sample) with unknown activity to be discriminated, $\mathbf{x}_i$ is one of the support vectors. $g(x) \geq 0$ corresponds to class 1.

Table 9.2   The sample numbers of support vectors and their corresponding coefficients.

Sample	α_i	Sample	α_i	Sample	α_i	Sample	α_i
No.1	7.545	No.6	80.000	No.10	80.000	No.18	8.450
No.2	10.000	No.7	37.183	No.11	80.000	No.19	15.888
No.4	80.000	No.8	1.863	No.13	4.652	No.20	1.761
No.5	26.420	No.9	75.150	No.15	6.008		

9.4.5   *Results of LOO cross-validation of SVM*

Table 9.3 lists the rate of correctness of different classifiers obtained by using LOO cross-validation method. It can be found that the result of SVC model is somewhat better than those of PCA and Fisher methods in prediction ability.

Table 9.3   The rate of correctness of prediction by PCA, Fisher, KNN and SVC.

Algorithm	PCA	Fisher	KNN (K=5)	SVC
P_A	70%	65%	70%	80%

9.5   SAR of Triazole-Derivatives

Triazole-derivatives are antimycotic compounds widely used in both human and veterinary therapy and in agriculture. Interference with steroid biosynthesis and with formation of fungal cell walls has been identified as the mechanism of the activity of these compounds. Because of their action, these compounds are successfully used as systemic agricultural fungicides (against mildews and rusts of cereal grains, fruits, vegetables, and ornamentals) and against both systemic and topical fungal diseases in humans and domestic animals [50]. SVC method has been used to investigate SAR of 1-(1H-1,2,4-triazole-1-yl)-2-(2,4-difluorophenyl)-3-substituted-2-propan-ols (triazole-derivatives). PCA, Fisher and K-Nearest neighbor (KNN) are also studied for comparison.

9.5.1 *Data set*

The data set consists of 23 triazole-derivatives [148]. The molecular formula investigated of these compounds is shown in Fig. 9.4. According to the bioactivities MIC (Candida albicans931103) of samples, the data set can be divided into two classes. Herein samples of class 1 denote the compounds with high activity, i.e., the molecules with $MIC < 80\mu g \cdot ml^{-1}$, while samples of class 2 denote the compounds with low activity, i.e., the molecules with $MIC \geq 80\mu g \cdot ml^{-1}$.

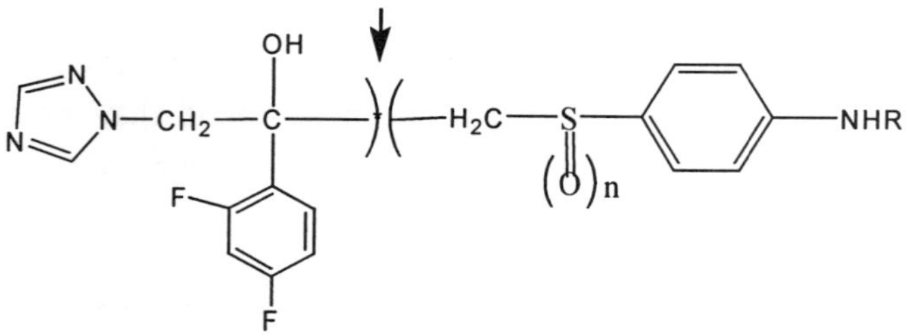

Fig. 9.4   Structure of 1-(1H-1, 2, 4-triazole-1-yl)-2-(2, 4-difluorophenyl) -3-substituted-2-propanols.

9.5.2 *Simplification of a series of molecular descriptors*

In view of the complexity of the molecules studied, the method of calculation is simplified in the following way: a series of sub-structures with different substituted group R (right part in Fig. 9.7) are used for the computation of molecular mechanics and quantum chemical methods. This kind of simplification is similar to the process of Hansch method considering the effects of various substituted groups. The difference of our method with Hansch method is that we did not treat the contribution of atomic groups separately, but treat the every substructure as a whole unit, and calculate its structural parameters after adding a hydrogen atom to the dangling bond.

9.5.3　*Computation of descriptors*

The three-dimensional structures of the molecules are optimized with the software Hyperchem[1]. Prior to the computation of quantum chemical method, all structures of the compounds are submitted to MM+ computation of molecular mechanics for energy optimization. The molecular structures are optimized using the Polak-Ribiere algorithm until the root-mean-square gradient equals to $0.1\ kcal \cdot mol^{-1}$. The structural descriptors are obtained via the computation of semi-empirical method PM3. The computations are carried out at restricted Hartree-Fock level without configuration interaction. At last, the descriptors obtained are as follows: HOMO (highest occupied molecular orbital energy), LUMO (lowest empty molecular orbital energy), H_f (heat of formation), SA (surface area), MV (molecular volume), Dipole (dipole moment) LogP (partition coefficient), MR (molecular refractivity), MW (molecular weight), C_N (charge density of the atom N connecting with R), C_S (charge density of the atom S), etc.

9.5.4　*Selection of descriptors*

A SVC-based feature selection method is applied for descriptor selection. Using floating search technique, the subset of descriptors is selected out via LOO cross-validation. By this way, three descriptors, C_N (charge density of the atom N connecting with R), Dipole (dipole moment) and H_f (heat of formation) are selected for modeling use. Table 9.4 lists the samples with bioactivities MIC and selected descriptors.

---

[1] Release 7.0 for Windows Molecular Modeling System, Hypercube Inc. 2002

Table 9.4   The descriptors of structures and bioactivities of the samples.

No.	R	n	Class	MIC ($\mu g \cdot ml^{-1}$)	Dipole	C_N	H_f
1	H-	0	2	80	3.17	0.0652	38.610
2	CH_3CO-	0	2	80	2.07	0.1037	-4.5822
3	CH_3CH_2CO-	0	2	80	2.15	0.1031	-8.1817
4	CH_3CH_2CO-	2	2	80	5.84	0.1030	-64.756
5	C_6H_5CO-	0	1	40	3.30	0.1032	41.882
6	C_6H_5CO-	2	1	40	6.27	0.1022	-14.749
7	p-BrC_6H_4CO-	0	1	20	2.16	0.1060	51.729
8	p-BrC_6H_4CO-	2	2	80	4.98	0.1040	-4.8797
9	p-ClC_6H_4CO-	0	1	20	2.05	0.1050	36.278
10	p-ClC_6H_4CO-	2	2	80	2.93	0.1040	-19.999
11	p-FC_6H_4CO-	0	2	80	1.11	0.1050	-0.6544
12	p-FC_6H_4CO-	2	2	80	2.84	0.1039	-57.101
13	o-FC_6H_4CO-	0	2	80	4.03	0.0876	0.69755
14	o-FC_6H_4CO-	2	1	40	8.96	0.1013	-54.211
15	p-CH_3O-C_6H_4CO-	0	1	40	4.14	0.1020	5.38131
16	p-CH_3O-C_6H_4CO-	2	2	80	3.68	0.1010	-50.912
17	p-$CH_3C_6H_4CO$-	2	2	80	3.38	0.1020	-23.296
18	p-CH_3O-o-FC_6H_3CO-	0	2	80	4.46	0.0867	-35.760
19	p-CH_3O-o-FC_6H_3CO-	2	1	40	4.89	0.1	-90.781
20	m-$CH_3C_6H_4CO$-	0	1	40	5.19	0.1036	33.346
21	m-$CH_3C_6H_4CO$-	2	2	80	3.68	0.103	-23.118
22	CH_3CO-	2	2	80	2.88	0.106	-61.529
23	p-$CH_3C_6H_4CO$-	0	1	10	3.71	0.1032	33.182

9.5.5   *Selection of the kernel function and parameter C used in SVC model*

Leave-one-out (LOO) cross-validation has been used to find the optimal parameter C and the appropriate kernel function for SVC modeling. The rate of correctness (P_A) of cross-validation test is employed as the criterion. Figure 9.8 illustrates the curve of P_A (with different kernel functions including linear kernel, Gaussian kernel, polynomial kernel of second degree and sigmoid kernel functions) versus the parameter C from 10 to 300. It has been found that the largest P_A can be achieved by using the linear kernel function with the parameter C from 10 to 300. So the best performance of SVC model can be achieved by using linear kernel with C=100.

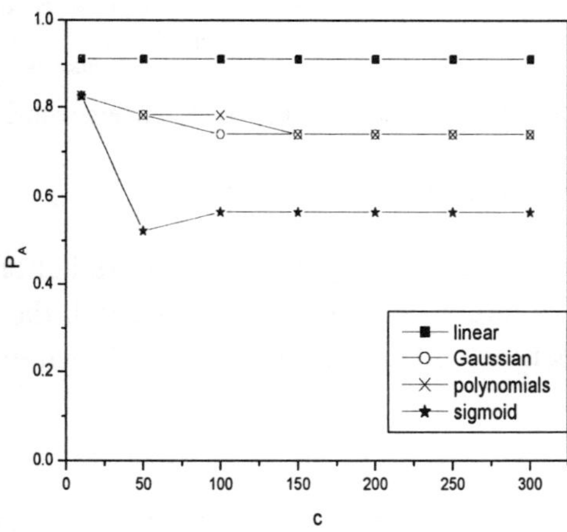

Fig. 9.5 P_A versus the C using LOO cross-validation with different kernel functions.

9.5.6   *Modeling by SVC*

Using the linear kernel function with capacity parameter C=100, the criterion for samples with high activities can be expressed by following inequality:

$$g(X)=4.662558[\text{Dipole}]+3.435568[C_N]+5.164093[H_f]>7.935422 \quad (9.5)$$

According to above mentioned criterion for the classification of samples of triazole-derivatives, the rate of correctness of classification is 91.3%. The training result of SVC is shown in Fig. 9.6. It can be seen that only two samples (compound No.8, No.19) are misclassified.

       Support Vector Machine in Chemistry

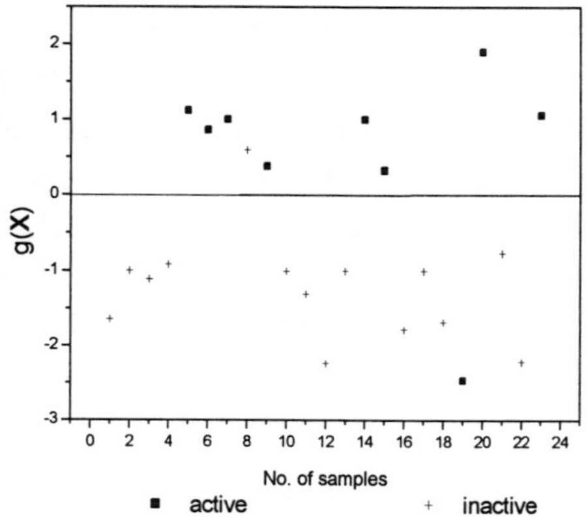

Fig. 9.6   Results by linear kernel function with C=100.

9.5.7   *Results of LOO cross-validation test*

It can be seen from Fig. 9.7 that the result of prediction of SVC in LOO cross-validation is rather good. On the other hand, the prediction results of PCA and Fisher method are not so good, as shown in Fig. 9.8, and Fig. 9.9.

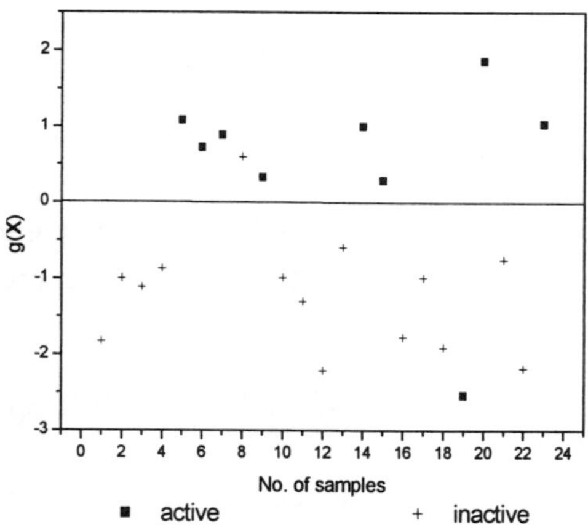

Fig. 9.7   Prediction results by using LOO cross-validation of SVC with C=100.

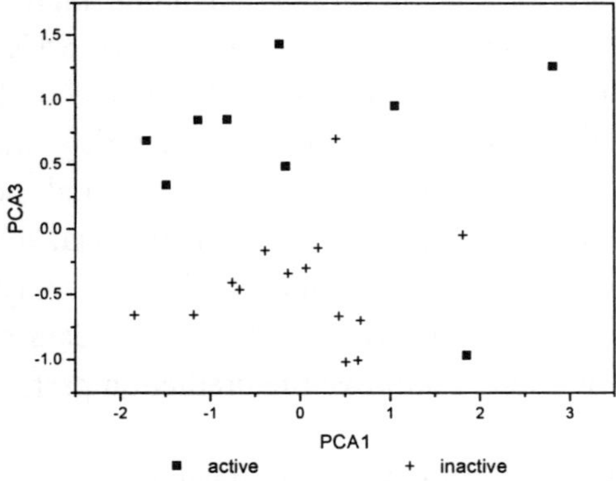

Fig. 9.8   Result of classification by using PCA method.

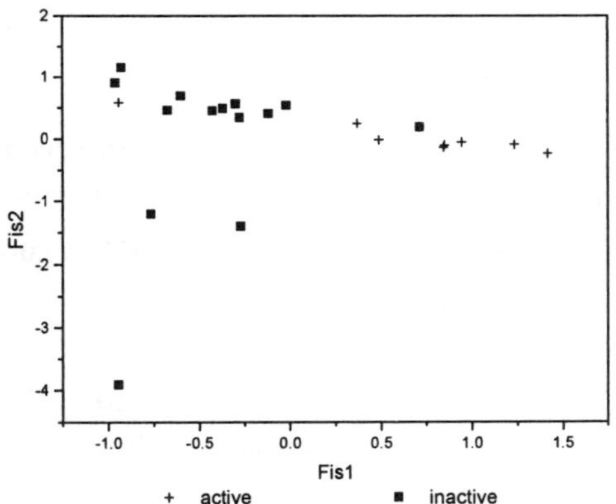

Fig. 9.9   Result of classification by using Fisher method.

9.6   SAR of the 5-hydroxytryptamine Receptor Antagonists

The 5-hydroxytryptamine 5-HT$_3$ receptor antagonists are currently used in the treatment of chemotherapy and radiotherapy induced emesis. The compounds are based on the parent structure shown in Fig. 9.10, the aromatic systems include mono- and bicyclic rings, with and without heteroatoms, and with various substitution patterns. This range of structural variation makes it difficult to analyze SAR of these compounds.

It has been reported that non-linear methods including ANN and non-linear mapping (NLM) could be used to solve SAR problems. For example, SAR of retrosynthetic was investigated by employing self-organizing ANN [13]. And SAR of antagonists was successfully studied by using ReNDeR back propagation neural network, which produced a two-dimensional plot of high dimensional multivariate data set [93]. However, nonlinear treatment of ANN may give rise to overfitting problems in treating finite, multivariate data set. It is a very meaningful research topic to improve the prediction performance of SAR analysis.

Since SVM has been applied to drug design [16], and the prediction of beta-turns and alpha-turn types of proteins [17] was also rather successful. SVM has been used to the SAR study of 5-HT$_3$ antagonists

9.6.1 *Data set*

The calculated molecular descriptors along with observed activity data of 26 5HT$_3$ antagonists are presented in Table 9.5 [93]. The classification of compounds (shown in Table 9.5, 1=inactive; 2=active) are based on their quantitative (ED$_{50}$) data of activities.

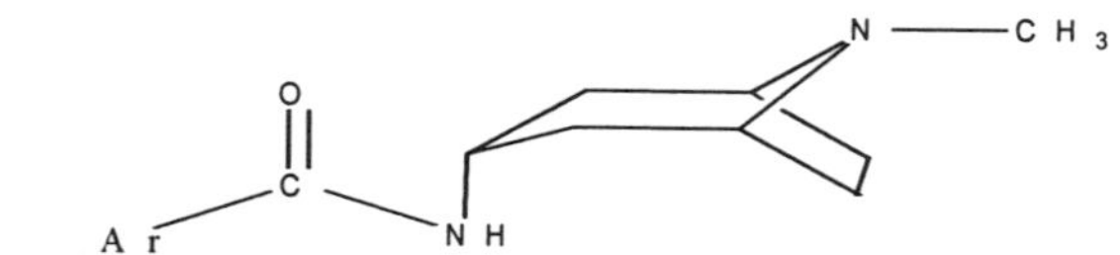

Fig. 9.10   The parent structure of 5-HT$_3$ antagonists.

Table 9.5   The activities and molecular descriptors for 5HT$_3$ antagonists.

No.	Activity	CMR	μZ	HOMO	ALP(3)	FZ(4)	VDWE (4)	FY (6)	FZ (9)	FY (11)
1	1	83.1560	-0.0231	-10.1255	0.1655	0.6910	-0.0182	0.1469	-1.3319	-0.5713
2	1	85.6830	1.1339	-10.3731	0.1642	0.8260	-0.0671	-0.0950	-1.8145	-1.3114
3	1	83.1700	-1.0093	-10.2914	0.1668	0.2000	-0.0075	0.1314	-0.2615	1.6104
4	1	92.4460	-1.1004	-10.2697	0.1661	0.3502	-0.0930	-0.3554	-0.1784	3.9394
5	1	93.9030	-0.8837	-10.7677	0.1652	0.3231	-0.0902	0.2450	-0.4449	-0.1013

6	1	90.3020	1.6816	-9.8969	0.1651	0.2634	-0.0915	-0.2308	-0.6937	-2.0565
7	1	86.6710	1.5092	-10.1418	0.1650	0.0120	-0.0889	-0.4664	-0.2801	-1.6762
8	1	90.3020	0.3749	-9.8533	0.1651	0.0262	-0.1030	-0.4494	-0.0670	-1.3269
9	1	87.7940	-2.0132	-10.1010	0.1665	0.2407	-0.0715	0.2819	-0.0506	1.5652
10	1	92.1900	-1.5911	-10.3540	0.1655	-0.4036	-0.2377	-0.7025	-0.4659	0.1241
11	1	87.8140	1.2661	-10.5717	0.1605	0.1514	-0.0424	-0.1527	-0.5818	0.4746
12	1	78.8890	-0.0176	-11.0561	0.1648	-0.3194	-0.0170	-0.3659	0.7300	0.7013
13	1	85.6830	-2.1269	-10.3731	0.1654	1.1546	-0.0777	0.1066	-1.7731	-0.1158
14	1	85.6830	-1.2604	-10.4574	0.1655	0.9566	-0.0596	0.0896	-1.6307	-0.4895
15	1	86.7500	0.4460	-10.8139	0.1639	0.0515	-0.0137	-0.1020	0.7244	-0.2648
16	1	86.7510	-0.1301	-10.3377	0.1632	0.3429	-0.0547	0.0887	-1.0626	-1.8350
17	1	81.8370	-0.1297	-10.0084	0.1656	0.3040	-0.0827	0.2027	-0.1217	2.2854
18	1	81.8090	-1.4715	-11.0507	0.1648	0.1552	-0.0521	-0.0533	-0.3720	0.2290
19	2	87.7020	-0.5674	-10.7595	0.1633	0.3340	-0.0968	0.0838	-1.0896	-1.5759
20	2	80.5820	-0.5616	-10.4466	0.1648	0.2747	-0.0619	0.0531	-0.5633	-0.2735
21	2	83.0640	-0.9930	-10.8411	0.1636	0.3763	-0.0253	0.0684	-1.0026	-1.5320
22	2	81.8370	0.6651	-10.3840	0.1633	0.2800	-0.0549	-0.0045	-1.0712	-2.1308
23	2	79.9280	-0.6834	-10.1091	0.1663	0.0412	-0.0133	-0.0192	0.0036	2.0010
24	2	89.2330	-0.9552	-10.5527	0.1636	0.6116	-0.2179	0.1166	-1.2019	-1.4577
25	2	85.4960	-0.4648	-10.5037	0.1648	0.1531	-0.0842	-0.0138	-0.4817	-0.3003
26	2	87.4590	-1.8874	-10.8302	0.1631	0.0033	-0.0428	0.0311	-3.2518	-1.0937

The molecular descriptors used are as follows: CMR (Calculated molar refractivity), μZ (Z component of the dipole moment), HOMO (Energy of the highest occupied molecular orbital), FZ (9), FY (6) and FY (11) (Z and Y are the components of the electric field at specified grid points), VDWE (4) (The van der Waal's energy of the interaction of a carbon atom at a specified grid point), ALP (3) (The self atom polarizability of the specified atom).

9.6.2 *Selection of parameters and SVC modeling*

The rate of correctness (P_A) in leave-one-out (LOO) cross-validation is used as the criterion for the choice of parameter. Figure 9.11 shows the curve of P_A versus the capacity parameter C (its value from 1 to 200) with different kernel functions (linear, polynomial, radial basis and sigmoid functions) by using LOO cross-validation of SVC. It can be seen from Fig. 9.11 that the best performance of SVC model can be achieved by using the linear kernel function with the capacity parameter C =10.

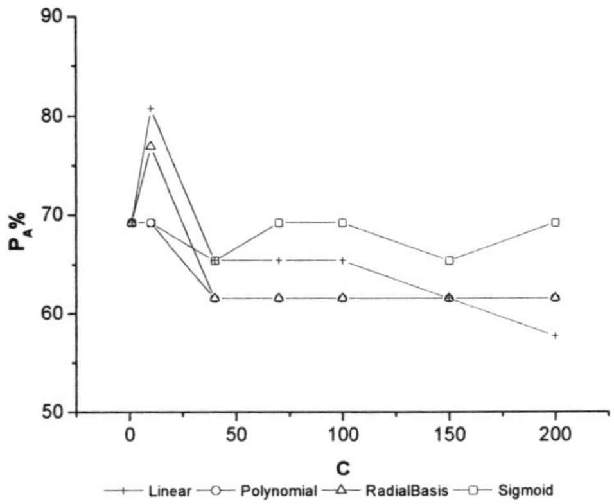

Fig. 9.11 P_A versus C by using LOO cross-validation of SVC.

Based on the above-mentioned results, the criterion for samples with high activity can be expressed as follows:

$$f(x)=0.1301(CMR)+0.2354(\mu Z)-0.2255(HOMO)+230.4224(ALP(3))$$
$$+1.3722(FZ(4))+7.2474(VDWE(4))-2.4128(FY(6))+0.1533(FZ(9)) \quad (9.6)$$
$$+0.5114(FY(11)) \leq 49.5080$$

Table 9.6 lists the rates of correctness (C_A) for the results of training by using SVC, BP-ANN, and Fisher methods respectively.

Table 9.6 C_A obtained by different methods.

Method	SVC	BP-ANN	Fisher
$C_A\%$	88.46	100	88.46

Table 9.7 lists the rates of correctness (P_A) for predicted classification by using LOO cross-validation of SVC, BP ANN, KNN and Fisher methods respectively.

Table 9.7 P_A obtained by using different methods in LOO cross-validation.

Method	SVC	BP-ANN	KNN	Fisher
$P_A\%$	80.77	61.50	61.54	65.39

It can be seen from Table 9.7 that the prediction ability of SVC is better than those of other methods.

9.7   QSAR of *N*-phenylacetamides as Herbicides

9.7.1   *Data set and feature selection*

The objective compounds studied are a series of *N*-phenylacetamides (NPAs) used as herbicides. The molecule structure of *N*-(1-methyl-1-phenylenthyl) phenylacemides is shown in Fig. 9.12, where α represents Me, F, Cl, OMe, CF_3, or Br, etc., and β represents Me, Cl, F, or Br, etc. The molecular structures and the corresponding anti-scipujuncoide activity (PI_{50}, the negative logarithm of the concentration needed for 50% inhibition) data are quoted from literature [77]. The combinations of different α and β can form diversified compounds.

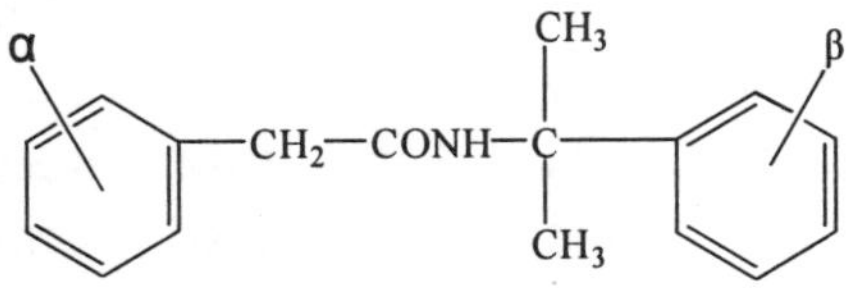

Fig. 9.12   The molecule structure of *N*-phenylacetamides.

Some electronic, geometric, and topological parameters are taken into consideration for the mathematic modeling, they are Log P (hydrophobic constant), square of Log P, σ (Hammett constant), E_s (*Taft* values), MR (molar refractivity), $^1\chi$ and $^2\chi$ (valence molecular connection index).

Table 9.8 Structural parameters and anti-scipusjuncoides activity of
N-phenylacetamides

No.	σ	MR	$(LogP)^2$	Obsd. activity	Pred. activity by ANN	$RE_1\%$	Pred. activity by SVR	$RE_2\%$
1	-0.34	9.24	1.254	5.85	5.953	1.754	5.644	3.519
2	-0.11	4.51	0.49	5.68	6.191	8.988	6.067	6.820
3	0.06	9.62	1.613	5.72	5.817	1.694	5.847	2.220
4	-0.44	11.46	0.292	6.07	5.571	8.214	5.680	6.425
5	-0.24	9.24	1.254	5.89	5.722	2.846	5.790	1.700
6	0.20	9.62	1.613	6.00	5.848	2.530	6.057	0.951
7	0.06	9.62	1.613	5.58	6.125	9.765	5.846	4.759
8	0.29	4.89	0.723	5.44	6.866	26.210	6.400	17.654
9	0.46	10.00	2.016	5.65	6.376	12.853	6.315	11.764
10	-0.04	11.84	0.476	5.46	6.123	12.145	5.575	2.111
11	0.16	9.62	1.613	5.57	5.806	4.237	6.008	7.863
12	0.60	10.00	2.016	5.65	6.368	12.704	6.696	18.522
13	0.10	11.84	0.476	5.42	6.411	18.286	6.390	17.895
14	0.06	9.62	1.613	5.18	5.695	9.938	5.859	13.108
15	0.46	10.00	2.016	5.41	6.453	19.277	6.307	16.585
16	-0.24	9.24	1.254	5.79	5.912	2.107	5.779	0.193
17	-0.01	4.51	0.490	5.92	6.404	8.179	6.252	5.615
18	0.16	9.62	1.613	6.10	6.161	1.002	5.994	1.731
19	-0.34	11.46	0.292	5.79	5.961	2.955	5.638	2.622
20	-0.14	9.24	1.254	5.82	5.764	0.966	5.868	0.825
21	0.30	9.62	1.613	6.22	5.990	3.701	6.205	0.247
22	0.36	8.61	2.074	5.73	6.375	11.258	5.922	3.347
23	0.16	12.47	2.016	4.64	5.885	26.823	5.761	24.159
24	-0.31	16.06	0.884	4.32	4.297	0.537	4.567	5.729
25	0.17	4.51	0.490	6.47	6.180	4.479	6.327	2.213
26	0.57	4.89	0.723	6.57	6.997	6.501	7.027	6.950
27	0.20	9.62	1.613	6.49	5.879	9.408	6.041	6.922
28	0.43	4.89	0.723	7.04	6.382	9.349	6.424	8.752
29	0.60	10.00	2.016	6.71	6.295	6.185	6.595	1.721
30	0.10	11.84	0.476	6.53	5.715	12.482	5.551	14.985

31	0.30	9.62	1.613	6.38	5.834	8.552	6.217	2.555
32	0.74	10.00	2.016	6.95	6.110	12.088	6.883	0.965
33	0.62	12.85	2.465	6.47	5.574	13.853	6.301	2.617
34	-0.34	9.24	0.314	5.82	6.092	4.677	6.217	6.817
35	-0.11	4.51	0.490	6.01	6.192	3.033	6.150	2.332
36	0.06	9.62	1.613	6.22	5.941	4.486	5.737	7.758
37	-0.44	11.46	0.292	5.82	5.703	2.007	5.689	2.258
38	-0.24	9.24	0.314	5.89	6.258	6.250	6.102	3.602
39	0.20	9.62	1.613	5.92	5.891	0.485	6.074	2.602
40	-0.11	4.51	0.490	6.25	6.194	0.890	6.153	1.555
41	0.29	4.89	0.723	7.05	6.292	10.757	6.405	9.153
42	-0.21	6.73	0.014	6.44	6.003	6.783	6.271	2.620
43	-0.01	4.51	0.490	6.50	6.255	3.765	6.240	3.999
44	0.43	4.89	0.723	6.39	6.758	5.757	6.651	4.089
45	0.06	9.62	1.613	6.68	5.902	11.650	5.821	12.859
46	0.29	4.89	0.723	6.63	6.350	4.220	6.385	3.701
47	0.46	10.00	2.016	6.91	5.989	13.336	6.240	9.689
48	0.16	9.62	1.613	6.84	5.941	13.139	5.964	12.811
49	0.29	7.74	1.000	6.60	6.195	6.133	6.440	2.422
50	0.60	12.85	2.465	6.49	5.787	10.827	6.300	2.923
MSE%						8.001		6.305

RE_1: the relative error of ANN model by using LOO cross-validation.
RE_2: the relative error of SVR model by using LOO cross-validation.

Table 9.8 lists the samples with activity and selected descriptors. Three molecular descriptors, including σ (Hammett constant), MR (molecular refractivity) and the square of LogP are used for the modeling by SVR.

9.7.2 *Selection of the kernel function and parameter C*

The performance of SVR is dependent on the combination of several factors, including the parameter C, the kernel type and its corresponding parameters. Leave-one-out (LOO) method is used to find the suitable parameter C and the appropriate kernel function for SVR modeling. In this computation, the mean relative error (*MRE*) is employed as the criterion to obtain the appropriate kernel function and the optimal capacity parameter C. Here *MRE* is defined as follows:

$$MRE = \frac{1}{n}\sum_{i=1}^{n}\left|\frac{p_i - e_i}{e_i}\right| \times 100\% \qquad (9.7)$$

where e_i is the experimental value of sample i, p_i is the predicted value of sample i in LOO cross-validation of SVR, n is the number of the total samples.

Figure 9.13 shows that the polynomial kernel function with degree equal to 2 is the best choice. And the minimum of *MRE* appears with the $\varepsilon = 0.05$.

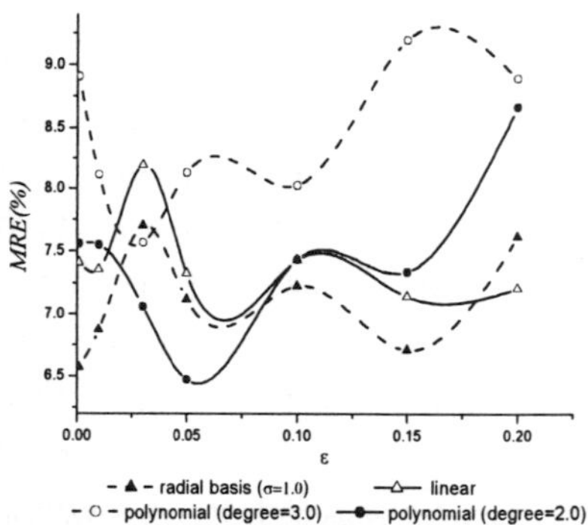

Fig. 9.13   *MRE* versus ε -insensitive loss function with different kernel functions.

By using LOO cross-validation test, it can be found that the best value of C is 80.

9.7.3   *Results of modeling*

Based on the above-mentioned results, the polynomial kernel function (degree=2) with C=80 and ε =0.05 has been used for modeling by SVR. The mathematical model obtained can be expressed as follows:

$$PI_{50} = \sum \beta_i \left[(\mathbf{x}_i \cdot \mathbf{x}) + 1\right]^2 + 0.731 \qquad (9.8)$$

where β_i (listed in Table 9.9) is the Lagrange coefficient corresponding to support vector. $\mathbf{x}$ is a vector (pattern of sample) with unknown activity, $\mathbf{x}_i$ is one of the support vectors. All support vectors obtained are presented in Table 9.9.

Table 9.9   The sample number of support vectors and their corresponding coefficients

Sample	β_i	Sample	β_i	Sample	β_i	Sample	β_i
No.1	32.633	No.14	-80.000	No.27	80.000	No.39	-80.000
No.2	-80.000	No.15	-80.000	No.28	80.000	No.41	80.000
No.4	80.000	No.17	-80.000	No.30	80.000	No.43	80.000
No.7	-80.000	No.18	61.040	No.31	80.000	No.44	-59.123
No.8	-80.000	No.19	29.555	No.33	21.704	No.45	80.000
No.9	-80.000	No.22	-20.402	No.34	-80.000	No.46	80.000
No.10	-80.000	No.23	-80.000	No.35	-11.098	No.47	80.000
No.11	-80.000	No.24	-29.295	No.36	80.000	No.48	80.000
No.12	-80.000	No.25	80.000	No.37	45.987	No.49	44.580
No.13	-35.581	No.26	-80.000	No.38	-80.000	No.50	80.000

9.7.4   *Comparison of prediction ability with other methods*

Table 9.10 lists the MRE by SVR and ANN. It can be seen that the result of LOO cross-validation test of SVR is slightly better than that of ANN.

Table 9.10   Comparison between SVR and BP-ANN in LOO cross-validation.

Methods	MRE
SVR	6.305%.
BP-ANN	8.001%

MRE: error in LOO cross-validation.

It can be seen from the results of the SVM applications mentioned above, the results of prediction of the mathematical models built by SVC or SVR are usually more reliable than that of ANN and some other methods. Therefore, SVM should be considered as a new powerful tool for SAR and drug design.

Chapter 10

SVM Applied to

Data of Trace Element Analysis

10.1　Trace Element Science and Chemical Data Processing

In recent years, the progress of analytical chemistry provides the possibility to detect and determine the contents of trace elements in various materials, and it has been found that these data are rather useful in many scientific and technological fields. One of the most important topics in these fields is to investigate the relationships between some diseases (such as cancer or hypertension) and trace elements [51].

An interesting topic is related to the data of trace element contents in human hair. The results of analysis of trace element contents in human hair are useful for the diagnosis of some diseases. These data can also provide valuable information about the influence of environment on human health. Like the trace element analysis of serum or urine, hair analysis can provide valuable information. Contrary to the results of the analysis of serum or urine, the results of hair analysis can reflect metabolic process in the body over a period of several months. Metal ions, moreover, are permanently built into the α-helical structure of hair keratin, thereby creating chelate links with the acid and basic radicals of amino acids, as a result of which the content of some metallic elements in hair is much higher than that in other tissues. Up to now, 78 elements have been detected in human hair.

Although the data of the content of some single element have also been used in some decision making for practical purposes, many

220

examples of multivariate analysis indicate that the concurrent consideration of many elements together is a more reasonable strategy.

Besides the healthy problems, trace element analysis is also very useful for commodity inspection because trace element content can provide valuable information for the detection of counterfeit products. The products with famous brand are usually the targets of counterfeit activities. Since famous brand foods, wines, cigarettes or tea products are usually produced in certain definite region. So the characteristics of the trace element contents in soil and natural water in this region will give some "finger print" on the products produced in this region. The mode of trace element contents is very difficult to be imitated by counterfeiters.

10.2   SVM Applied to Trace Element Analysis of Human Hair

It has been recognized in medical science that the trace element contents in the hair of hypertension patients are different from those of normal persons. But in most of the published papers the data of hair analysis are treated simply by averaging or by correlation coefficient calculation. To use those simple methods only will give rise to significant loss of meaningful information contained in these data. So it is desirable to use more sophisticated algorithm to do this work. As an example, SVM has been employed to find the criterion of the hair samples of hypertension patients [33].

A total of 53 hair samples, including 26 samples from hypertension patients and 27 samples from normal persons, have been used for trace element analysis. The data of Ca, Mg, Al, Cu and Zn contents are listed in Table 10.1.

Table 10.1   Trace element contents in hair of patients of hypertension disease (μg$\cdot$g^{-1}).

No.	Class*	Al	Mg	Ca	Cu	Zn
1	1	9.7	72.4	407	24.6	37.2
2	1	12.3	87.1	380	19.5	42.7
3	1	12.0	24.6	118	13.6	23.4
4	1	12.3	31.6	178	16.6	30.9
5	1	15.0	64.6	537	17.4	35.5

6	1	12.3	27.5	240	27.5	41.7
7	1	13.7	107	708	15.9	50.1
8	1	10.7	109.8	1148	13.9	55.0
9	1	12.3	47.8	372	14.1	42.7
10	1	13.3	61.7	380	13.5	19.1
11	1	44.0	617	1122	28.2	105
12	1	15.0	132	468	18.6	49.0
13	1	30.0	447	380	18.2	46.7
14	1	26.0	126	398	20.4	56.2
15	1	15.0	794	1479	19.5	85.1
16	1	14.0	52.5	186	14.5	30.9
17	1	24.0	2290	2041	31.6	110
18	1	31.0	276	1698	24.6	83.2
19	1	60.0	41.7	316	53.7	120
20	1	60.0	33.1	285	31.6	95.5
21	1	33.0	363	4677	53.9	141
22	1	14.0	1.91	214	12.0	67.6
23	1	12.3	2.00	204	10.7	72.4
24	1	11.3	67.6	617	11.8	85.1
25	1	13.3	132	1047	20.0	93.3
26	1	16.0	35.5	363	18.6	79.4
27	2	12.3	30.2	389	21.4	41.7
28	2	18.7	70.8	1072	13.5	55.0
29	2	20.3	219	1585	15.1	60.3
30	2	17.3	162	1349	10.5	53.7
31	2	9.0	269	1995	12.6	115
32	2	9.3	79.4	550	10.2	44.7
33	2	8.0	162	955	8.32	41.7
34	2	6.0	20.4	126	7.76	17.8
35	2	9.0	38.9	355	12.0	35.5
36	2	14.0	525	2138	26.9	81.3
37	2	8.7	363	672	15.9	67.6
38	2	8.2	41.7	224	12.9	9.05
39	2	10.0	427	1949	23.4	67.6
40	2	11.7	407	1348	19.1	20.0
41	2	5.7	178	621	8.71	30.9
42	2	16.0	234	1349	9.50	67.6
43	2	7.8	37.2	186	9.55	26.9
44	2	8.7	269	871	11.0	38.9
45	2	8.0	14.1	69.2	9.12	20.0
46	2	16.0	120	1023	17.0	69.2
47	2	8.0	45.7	56.2	8.32	7.08
48	2	15.0	191	1479	17.0	77.6
49	2	20.0	145	1474	24.6	95.5
50	2	9.7	191	1175	20.4	74.1

51	2	8.3	91.2	759	10.2	39.8
52	2	20.0	224	2042	30.9	91.2
53	2	18.0	132	1023	26.9	316

*Class "1" denotes the samples of hypertension patients and class "2" denotes those of normal persons.

It can be seen from Table 10.1 that the relationship between trace element contents and the hypertension seems not very clear if the content of any single trace element is considered, but some regularities can be revealed by multivariate analysis. By Fisher method, the rate of correct separation of the sample points of two classes is 83%, and the rate of correctness of prediction by leave-one-out (LOO) cross-validation method is 81.1%. The rate of correct prediction by KNN method is 77.3%.

By using Gaussian kernel, the data listed in Table 10.1 have been classified by SVM, the rate of correctness of separation is 96.25%, and that of prediction by LOO cross-validation method is 86.7%. It appears that the application of SVM can improve the research work about trace element contents in human hair.

10.3   SVM Applied to Trace Elements Analysis of Cigarettes

China is one of the largest cigarette producers and consumers in the world. Yunnan and Henan provinces are the chief tobacco plantation places in China. There are also several famous brands of cigarettes produced in these two provinces. In recent years, the influence of the trace elements in cigarettes on human health has been widely concerned about. And the possibility of using trace element content data to differentiate the cigarette of famous brand from counterfeit products is also the purpose of the research about the computerized modeling of the trace element contents of different brands of cigarettes. Table 10.2 lists the data of trace element contents in 42 samples of cigarettes produced in Yunnan or Henan provinces. SVC method is used for the processing of these data.

Table 10.2   Trace element contents in cigarette samples ($\mu g \cdot g^{-1}$).

Sample	Place	Mn	Fe	Cu	Zn	Sr	Pb	Cd	Cr
K1a	Yunnan	85.92	227	6.595	25.99	43.73	0.937	1.494	0.767
K1b	Yunnan	86.02	156.9	6.422	24.38	42.26	0.77	1.538	0.762
K1-2a	Yunnan	112.2	234.7	6.84	27.47	45.6	1.577	1.267	0.721
K1-2b	Yunnan	106.4	224.1	6.158	25.99	43.26	1.699	1.191	0.765
K2a	Yunnan	94.05	179.6	6.615	29.09	40.68	0.711	0.905	0.846
K2b	Yunnan	94.61	158.5	7.06	27.56	41.13	0.668	0.81	0.822
K3a	Yunnan	119.2	151.2	5.959	30.51	45.33	0.968	1.02	0.777
K3b	Yunnan	109.9	162.8	5.722	27.67	43.37	1.015	1.134	0.803
K4a	Yunnan	122.9	127.8	8.881	28.98	37.23	1.024	0.914	0.773
K4b	Yunnan	109.2	133.7	4.677	37.94	33.22	1.043	1.083	0.8
K5a	Yunnan	118.7	249.2	8.876	36.04	46.69	2.073	2.13	0.863
K5b	Yunnan	106.2	216.4	7.833	35.3	43.63	2.222	2.038	0.886
K6a	Yunnan	123.8	194.7	8.895	35.61	51.72	0.884	1.187	0.851
K6b	Yunnan	132.2	189.2	8.194	34.47	48.78	0.904	1.28	0.897
K7a	Yunnan	79.54	185.2	11.34	55.82	48.77	1.696	1.13 7	0.855
K7b	Yunnan	70.61	144.2	10.03	49.82	44.51	1.722	1.234	0.853
K8a	Yunnan	126.3	168.3	6.043	41.33	40.25	0.986	1.034	0.881
K8b	Yunnan	144.6	171.5	6.786	33.18	40.75	1.03	1.11	0.892
A1a	Henan	98.5	306.3	10.83	32.69	59.27	0.964	1.004	0.773
A1b	Henan	88.89	244.6	10.86	31.89	56.97	0.972	1.081	0.809
A2a	Henan	79.37	291.7	11.08	26.44	63.33	0.984	1.033	0.836
A2b	Henan	84.04	281.7	11.49	27.02	57.94	0.979	1.019	0.784
A3a	Henan	90.7	232.9	10.79	29.36	61.8	0.838	1.065	0.735
A3b	Henan	96.47	239.4	11.34	32.53	57.33	0.779	0.942	0.887
A4a	Henan	71.87	236.8	9.636	35.74	55.75	0.743	1.226	0.862
A4b	Henan	91.38	310.1	11.27	27.03	63.62	0.717	1.189	0.856
A5a	Henan	153.6	332.1	10.89	27.5	52.42	1.111	0.965	0.867

A5b	Henan	161.9	264.5	12.1	32.5	48.25	1.128	1.064	0.898
A6a	Henan	80.98	292.7	11.85	23.52	65.6	0.792	1.115	0.867
A6b	Henan	91.86	322.6	13.81	30.61	68.54	0.759	0.896	0.842
A7a	Henan	104.5	251.6	11.87	29.81	54.15	1.042	0.678	0.799
A7b	Henan	93.7	216.3	9.51	23.22	54.94	1.084	0.762	0.854
A8a	Henan	99.93	215.1	10.19	25.17	60.38	0.698	0.793	0.885
A8b	Henan	90.53	211	9.396	32.8	71.49	0.668	0.625	0.836
A9a	Henan	88.51	156.3	11.79	29.05	90.92	0.892	1.123	0.938
A9b	Henan	82.64	155.9	12.89	31.2	97.92	0.818	0.905	0.869
A10a	Henan	17.74	52.2	11.97	41.06	65.34	0.875	0.855	0.79
A10b	Henan	26.98	96.9	18.82	54.57	78.6	0.92	0.938	0.826
A11a	Henan	21.82	67	14.76	45.91	65.17	0.702	1.26	0.843
A11b	Henan	7.645	65.4	16.89	34.12	64.41	0.75	1.072	0.75
A12a	Henan	7.137	58	16.73	43.14	53.44	0.833	1.274	0.714
A12b	Henan	12.81	140.1	11.76	20.65	53.52	0.788	1.12	0.728

SVC method has been used to classify the cigarettes from the factories of Yunnan province and those from Henan province. The kernel type has been selected in SVC computation. After several trials, the linear function has been found to be the best one in this case. By feature selection based on the prediction ability of SVC, a data subset including the contents of Cu, Zn, Mn and Cr is used for the mathematical modeling, with 100% rate of correctness of prediction in LOO cross-validation test for this data set. A criterion for the classification of Yunnan cigarettes from Henan ones can be obtained by SVC (linear kernel, C =100) as follows:

$$0.0334[Mn]-0.945[Cu]+0.177[Zn]-10.714[Cr]+9.442>0 \quad (10.1)$$

Although the classification by Fisher method is also very good, but the LOO cross-validation test of Fisher method leads to some misclassification results.

Since Hongtashan is the most famous brand of cigarette in China, it is desirable to find a criterion to detect the cigarette of counterfeit Hongtashan brand. By SVC, the criterion to distinguish samples of Hongtashan brand cigarettes to the samples of other brands of cigarettes can be expressed as follows:

$$44.04-0.0558[Mn]-0.0791[Cu]-0.180[Zn]-40.417[Cr]>0 \quad (10.2)$$

The classification by Fisher method is also clear-cut, but both the result of LOO cross-validation test of Fisher method and the results of K-Nearest Neighbor (KNN) methods (k=1, 3 or 5) cause some misclassification, while the result of the LOO cross-validation test of SVC (linear kernel, C =100) shows 100% rate of correctness.

10.4   SVM Applied to Trace Element Analysis of Tea

China is also one of the largest tea producers and tea consumers in the world. There are many famous brands of green tea, black tea and oolong tea produced in different parts of China. Owing to the different contents of trace elements in the soil and natural water in different regions, the trace element contents of the tea products from different regions are also different. So it is possible to use the trace element contents to carry out data processing for the differentiation of the tea products from different regions [27]. One of the examples of this kind of work will be described as follows:

Twenty one samples of three kinds of tea products (green tea from southeastern China, black tea from south China and oolong tea from Fujian province of China) are collected, and nine trace elements, including K, Ca, Mg, Pb, Fe, Mn, Cu, Al and Si, of these 21 tea samples are measured by the method of atomic absorption spectroscopy. SVC is applied for feature selection. And it has been found that the features including Mn, K, Mg, Fe, Cu, Pb and Ca can be used for the classification of tea products of different regions with good generalization ability. Table 10.3 lists the data of the contents of these seven trace elements in 21 tea samples from different regions of China. And SVC has been used for mathematical modeling based on these data.

Table 10.3 Trace element contents of samples of Chinese tea from different places $(\mu g \cdot g^{-1})$.

No.	Provinces (kinds)[*]	Commercial name	Mn	K	Mg	Fe	Cu	Pb	Ca
1	Fujian(o)	Dahongpao	903	14084	1287	174	11.0	1.04	2200
2	Fujian(o)	Tieguanyin	786	11717	1916	162	8.2	0.845	2520
3	Fujian(o)	Tieguanyin	586	10686	1673	145	7.5	0.858	2490
4	Fujian(o)	Wulong	932	14179	1589	278	8.3	1.06	2155
5	Fujian(o)	Wulong	1006	13726	1616	242	8.4	1.64	2140
6	Yunnan(b)	Dianhong	456	17037	2146	275	19.84	2.68	3450
7	Yunnan(b)	Dianhong	404	17075	2140	280	18.86	2.48	3480
8	Anhui(b)	Chihong	844	13892	1990	249	16.43	1.33	3180
9	Anhui(b)	Chihong	805	13988	1878	354	16.17	1.34	3290
10	Jiangsu(g)	Biluochun	356	14229	1794	166	29.75	10.07	2780
11	Jiangsu(g)	Biluochun	398	14415	1799	175	30.44	9.50	2730
12	Zhejiang(g)	Longjing	861	13128	1695	283	14.52	4.96	2890
13	Zhejiang(g)	Longjing	838	16735	1613	211	14.58	4.65	2840
14	Anhui(g)	Maofeng	351	15974	1709	187	14.56	1.26	2560
15	Anhui(g)	Maofeng	354	15312	1640	163	13.71	1.10	2650
16	Anhui(g)	Chaoqing	439	16274	1793	643	11.74	1.05	2320
17	Anhui(g)	Chaoqing	395	16518	1806	325	13.76	1.02	2280
18	Jiangsu(g)	Yuhuacha	599	14641	1621	153	27.92	2.00	2900
19	Jiangsu(g)	Yuhuacha	576	15134	1640	148	29.53	2.20	2890
20	Jiangxi(g)	Gaoshanyunwu	288	11707	1385	135	31.71	7.98	2740
21	Jiangxi(g)	Gaoshanyunwu	286	12243	1434	192	32.00	7.94	2760

[*]"o" refers to oolong tea, "b" to black tea, "g" to green tea.

The quality of mathematical model made by SVC is dependent on the selection of kernel function and parameter C in computation. In order to find the best choice of kernel function and the value of parameter C, the rate of correctness (P_A) of the prediction in LOO cross-validation test is used as the criterion for this selection work.

Figure 10.1 illustrates the curves of P_A (with different kernel functions, including linear, polynomials, radial basis and sigmoid functions) versus the parameter C from 1 to 100. It has been found that P_A by the linear kernel function is at the highest level among those using different kinds of kernel functions. So SVC model using the linear kernel function with parameter C from 10 to 100 can be used for model building.

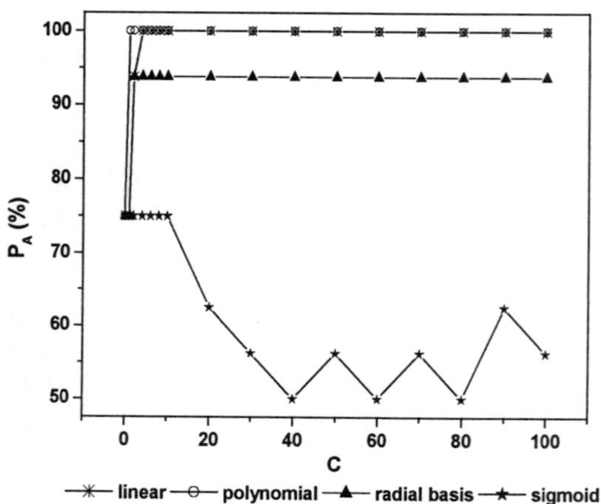

Fig. 10.1   The P_A versus C on LOO cross-validation test.

By using SVC method, it has been found that these three kinds of tea can be classified with linear kernel functions, with the rate of correctness of prediction equal to 100% by using C = 10. By this way, the criterion of samples of oolong tea can be found as follows:

$$1.27[Mn]- 1.93[Ca]-1.00[Pb]- 1.32[Cu]-1.17 [Fe]-1.29[K]+1.00 > 0 \qquad (10.3)$$

And the criterion to differentiate green tea of southeastern China from black tea of south China is as follows:

$$0.607[Mn]+3.08[Ca]-3.05[Pb]- 0.970[Cu]+0.725 [Fe]+0.267[K]-2.32 < 0 \quad (10.4)$$

The classification by Fisher method is also clear-cut, but both the result of LOO cross-validation test of Fisher method and the results of K-Nearest Neighbor (KNN) methods (k=1, 3 or 5) lead to some misclassification, while the result of the LOO cross-validation test of SVC (linear kernel, C =10) gives 100% rate of correctness.

Chapter 11

SVM Applied to

Archeological Chemistry of Ancient Ceramics

11.1 SVM Applied to Archeological Data Processing

Data processing, associated with various methods of chemical analysis, can provide a useful tool for the investigation of samples in archeological research. Since the number of samples in archeological work is usually quite limited, SVM should be a new powerful means in the research work of archeological chemistry. In this chapter, archeological research work related to Chinese ancient ceramics will be used as examples in this field.

China is the fatherland of porcelain technology [85]. The history of ceramics in China started from ten thousand years ago. Since East Han Dynasty Chinese ceramic workers had mastered the technology of porcelain production.

The mathematical modeling of the chemical composition of Chinese ancient ceramics also has practical value, since it can be used for distinguishing true ancient ceramics from the imitation products.

Although the archeological problems discussed in this chapter are limited to the archeological problems of Chinese ancient ceramics, it is reasonable to prospect that these methods and strategies described here can be used for other archeological work based on the data analysis of the composition of samples.

229

11.2   Identification of Jun Wares of Song Dynasty

Jun Kiln was one of the five famous ceramic kilns of Song Dynasty (960–1279 A.D.) in ancient China [63; 90]. Its products were Jun glazing porcelain wares, which were characterized as *integrating of five colors*. The secret recipe of Jun glaze had been lost from Yuan Dynasty (1271–1368 A.D.), and only few Jun porcelain wares were handed down as precious antiques. Since Qing Dynasty (1644–1912 A.D.), people began to imitate Jun glazing porcelain wares and achieved the purpose in some degree. So both the true ancient Jun wares and the modern imitation products appeared on the market. It is necessary to have some scientific methods to differentiate the true ancient Jun wares from the imitation products. Since the raw materials of the glaze of modern imitation products are different from that used in Song Dynasty for Jun ware production, the chemical composition of the glaze of imitation products and that of true Jun wares should have some difference. This difference may be revealed by the data processing of the data obtained by nondestructive analysis of both kinds of samples. By this way, it is possible to distinguish true Jun wares from imitation products. Since the number of examples of Jun wares is rather limited, this problem of data processing is a typical problem of small sample size.

In our previous work [63], PCA, a classical chemometric method, was used to find some mathematical model for the identification of ancient Jun wares. At this time, 19 ancient Jun glazes, 7 glazes of imitation products, and 18 copper red glazes were investigated. The compositions of these examples are listed in Table 11.1.

Table 11.1   Chemical composition of Jun glazes and other glazes(wt%).

Class[*]	SiO_2	Al_2O_3	Fe_2O_3	MgO	CaO	CuO	TiO_2	SnO_2	P_2O_5	K_2O	Na_2O
1	69.74	10.30	1.60	1.50	10.56	0.01	0.24	0.00	0.81	3.48	1.50
1	69.44	10.80	1.66	1.24	9.66	0.01	0.24	0.00	0.79	3.42	1.62
1	70.44	10.70	1.44	1.60	9.45	0.06	0.20	0.00	0.95	3.70	1.40
1	68.89	10.80	2.30	0.85	10.68	0.01	0.31	0.00	0.81	3.86	1.12
1	71.86	9.71	1.95	0.91	9.36	0.01	0.36	0.20	0.68	4.48	0.57
1	70.75	9.60	1.99	0.90	10.04	0.06	0.51	0.49	0.53	3.85	0.55
1	71.89	9.65	2.03	1.10	9.04	0.06	0.36	0.10	0.68	4.75	0.63
1	70.45	9.65	2.53	0.83	9.80	0.45	0.42	0.40	0.58	4.86	0.72

1	71.07	9.75	2.35	0.83	9.46	0.11	0.42	0.15	0.55	4.74	0.65
1	70.45	9.50	2.16	0.95	9.52	0.40	0.37	0.91	0.68	4.28	0.52
1	70.59	9.51	2.16	1.05	9.73	0.13	0.36	0.53	0.63	4.45	0.55
1	70.44	9.50	2.30	1.00	10.55	0.26	0.40	0.49	0.63	3.86	0.67
1	70.50	9.54	2.31	1.10	10.84	0.09	0.31	0.10	0.57	3.64	0.59
1	70.29	9.90	2.72	0.75	10.80	0.19	0.40	0.32	0.58	3.82	0.48
1	69.90	10.45	1.36	0.65	10.91	0.00	0.39	0.00	0.82	4.90	0.78
1	69.76	10.40	2.28	0.50	10.66	0.30	0.42	0.32	0.72	4.10	0.51
1	73.60	9.40	1.09	0.70	8.82	0.01	0.20	0.00	0.74	3.14	1.40
1	69.29	10.20	2.23	0.75	11.81	0.00	0.53	0.00	0.58	4.48	0.86
1	70.29	9.80	1.07	0.90	12.09	0.04	0.23	0.00	0.65	2.35	1.65
2	60.79	10.12	0.95	2.13	14.52	0.80	0.34	0.00	0.37	7.07	2.81
2	59.97	10.61	1.07	1.95	13.35	0.84	0.22	0.00	0.36	6.87	2.83
2	70.52	10.56	0.49	0.93	8.61	0.40	0.22	0.56	0.10	7.38	0.22
2	72.67	11.53	0.83	0.32	10.74	0.22	0.22	0.00	0.06	3.40	0.19
2	62.38	9.11	1.00	2.14	13.85	0.40	0.28	0.00	0.33	6.60	3.78
2	65.07	10.23	0.93	1.60	10.78	0.33	0.27	0.00	0.29	7.52	3.14
2	72.79	9.94	1.58	1.50	8.80	0.30	0.07	0.00	0.54	3.85	0.72
3	57.00	5.66	0.35	0.44	9.10	0.97	0.00	2.49	0.02	3.38	5.18
3	61.29	8.81	0.88	1.61	9.05	0.51	0.08	0.41	0.34	3.46	7.20
3	65.85	9.15	0.52	0.65	10.55	0.77	0.00	0.37	0.01	2.48	4.23
3	63.27	9.53	0.38	1.69	11.85	0.49	0.00	2.60	0.01	2.89	7.22
3	64.60	12.55	0.56	0.86	13.42	0.57	0.10	0.00	0.01	4.64	2.70
3	61.68	13.23	0.71	0.82	15.75	0.43	0.01	0.00	0.01	1.93	4.52
3	63.22	12.87	0.68	0.71	15.11	1.43	0.00	0.00	0.01	1.87	4.89
3	65.07	13.60	0.72	0.63	12.40	0.34	0.00	0.03	0.01	2.01	5.02
3	67.14	15.13	0.74	0.59	9.54	0.26	0.00	0.00	0.01	2.10	4.50
3	64.82	13.54	0.60	1.10	13.46	0.46	0.00	1.02	0.00	2.62	2.26
3	64.87	14.45	0.71	0.73	13.43	0.28	0.01	0.00	0.01	2.14	4.37
3	64.73	14.25	0.71	0.75	12.55	0.51	0.01	0.01	0.00	2.21	4.36
3	64.17	7.92	0.54	0.48	10.24	0.86	0.00	0.00	0.02	3.42	6.38
3	66.28	8.69	0.47	0.36	7.20	0.87	0.00	0.00	0.02	4.45	4.96
3	65.28	7.83	0.48	0.35	4.79	0.74	0.00	0.00	0.02	4.68	9.32
3	56.63	7.63	0.59	1.08	10.01	0.60	0.03	2.27	0.11	4.35	3.30
3	56.30	9.10	0.49	2.60	17.77	3.00	0.00	0.52	0.00	2.44	2.49
3	61.68	13.23	0.71	0.82	15.75	0.53	0.01	0.00	0.01	1.93	4.25

*Class "1" denotes Jun glaze, class "2" denotes glaze of imitation product and class "3" denotes copper red glaze.

Figure 11.1 illustrates the result of projection of PCA method. It can be seen that the glaze samples of the three kinds are distributed in three different regions. And the criterion obtained by PCA method is as follows:

$$Y_1 = 0.393SiO_2 - 0.080Al_2O_3 + 0.407Fe_2O_3 - 0.51MgO - 0.204CaO - 0.328CuO$$
$$- 0.413TiO_2 - 0.137SnO_2 + 0.410P_2O_5 + 0.170K_2O - 0.032Na_2O \qquad (11.1)$$

When $Y_1 < -1.0$ the area is the distribution region of copper red glaze samples;

-1.0 < Y_1 < 1.5 the area is the distribution region of glaze samples of imitation products;

$Y_1 > 1.5$ the area is the distribution region of ancient Jun glaze samples.

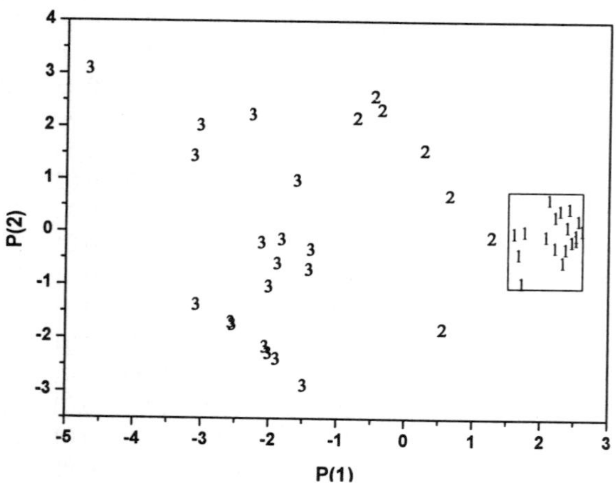

Fig. 11.1   classification of Jun glaze, glaze of imitation products and copper red glaze.

1. Jun glaze, 2. glaze of imitation products, 3. copper red glaze.

Although the separation of ancient Jun wares and imitation products is rather clear-cut in the PCA projection map, the leave-one-out (LOO) cross-validation test indicates that there is wrong prediction result of one sample for this data set. This is evidently due to the overfitting in PCA analysis and the mathematical model mentioned above is not very reliable. So we re-investigate this problem by SVM technique, and it has been found that the rate of correctness of prediction in LOO cross-validation test by SVM is 100%.

Since the data listed in Table 11.1 include some Jun glaze samples produced in Yuan Dynasty. We rebuild the data file shown in Table 11.2.

Here only the Jun wares from Song Dynasty and modern imitation products are included. We use this data set to build mathematical model.

Table 11.2   The composition of Jun glaze of Song dynasty and the glaze of imitation products (wt %).

Class[*]	SiO_2	Al_2O_3	Fe_2O_3	MgO	CaO	CuO	TiO_2	P_2O_5	K_2O	Na_2O
1	69.44	10.80	1.66	1.24	9.66	0.01	0.24	0.79	3.42	1.52
1	69.89	10.80	2.30	0.85	10.66	0.01	0.31	0.81	3.86	1.12
1	71.86	9.71	1.95	0.91	9.36	0.01	0.36	0.68	4.48	0.57
1	70.45	9.65	2.53	0.83	9.60	0.45	0.42	0.58	4.86	0.72
1	71.07	9.75	2.35	0.83	9.46	0.11	0.42	0.55	4.74	0.68
1	70.45	9.50	2.16	0.95	9.52	0.40	0.37	0.68	4.28	0.52
1	70.59	9.51	2.16	1.05	9.73	0.13	0.36	0.63	4.45	0.55
1	70.29	9.90	2.72	0.75	10.80	0.19	0.40	0.58	3.82	0.48
1	71.04	9.60	1.99	0.75	10.63	0.30	0.31	0.46	4.50	0.65
1	69.79	9.52	2.15	1.15	11.20	0.11	0.48	0.53	3.93	0.50
1	73.29	10.73	1.79	1.34	8.74	0.03	0.17	0.64	3.16	1.02
1	70.86	9.52	1.76	0.80	11.32	0.02	0.36	0.41	3.39	0.57
1	72.54	9.80	1.79	0.83	8.29	0.012	0.21	0.66	5.23	0.47
1	71.10	10.28	1.72	0.79	10.25	0.02	0.34	0.40	3.98	0.53
1	71.87	10.84	2.16	0.85	8.28	0.03	0.34	0.46	3.40	0.57
2	60.79	10.12	0.95	2.13	14.52	0.80	0.34	0.37	7.07	2.81
2	59.67	10.61	1.07	1.95	13.35	0.84	0.22	0.36	6.87	2.83
2	70.52	10.56	0.49	0.93	8.61	0.40	0.22	0.10	7.38	0.22
2	72.67	11.53	0.83	0.32	10.74	0.22	0.22	0.06	3.40	0.19
2	62.38	9.11	1.00	2.14	13.85	0.40	0.28	0.33	6.60	3.87
2	85.07	10.23	0.93	1.60	10.78	0.33	0.27	0.29	7.52	3.14
2	72.79	9.94	1.58	1.50	8.80	0.30	0.07	0.54	3.85	0.72

[*]Class "1" denotes ancient Jun glaze and class "2" denotes glaze of imitation products.

Using the feature selection algorithm based on SVM, it has been found that a data set having three features, MgO, TiO_2 and P_2O_5 contents, can get 100% rate of correct prediction in LOO cross-validation test. And the criterion for ancient Jun glaze is as follows:

$$3.81(TiO_2)-3.50(MgO)+5.72(P_2O_5)-2.40 > 0 \qquad (11.2)$$

In order to test the reliability of this criterion, we try to classify other seven samples not listed in the Table 11.2. The prediction results of all 7 samples are classified correctly. It implies that the above-mentioned criterion has good generalization ability. From the above-mentioned criterion it can be seen that the glaze of imitation products contains more

MgO and less TiO_2 and P_2O_5 as compared with ancient Jun glazes. Besides, our other computation results indicate that there are also other differences between the ancient Jun glaze and the glaze of imitation products. For instance, after the data of P_2O_5, TiO_2 and P_2O_5 are removed from the data set, it is still possible to find another effective criterion with good generalization ability to differentiate the ancient Jun glaze from the glaze of imitation products. For example, the data of Fe_2O_3, Na_2O and SiO_2 contents can be used to build an effective mathematical criterion for ancient Jun glaze identification.

Another method of identification of Jun wares can be obtained by data processing with SVR. It can be found that the relationship between TiO_2 content and the contents of the other components in ancient Jun glaze samples can be expressed by the following equation:

$$(TiO_2) = 0.581(Fe_2O_3) - 0.471(Al_2O_3) - 0.335(SiO_2) + 0.215(MgO)$$
$$-0.010(CaO) - 0.198(CuO) + 0.507(P_2O_5) + 0.677(Al_2O_3/P_2O_5) - 0.225 \tag{11.3}$$

While the glaze of imitation products does not obey this equation, and it is possible to differentiate them by a figure shown in Fig. 11.2.

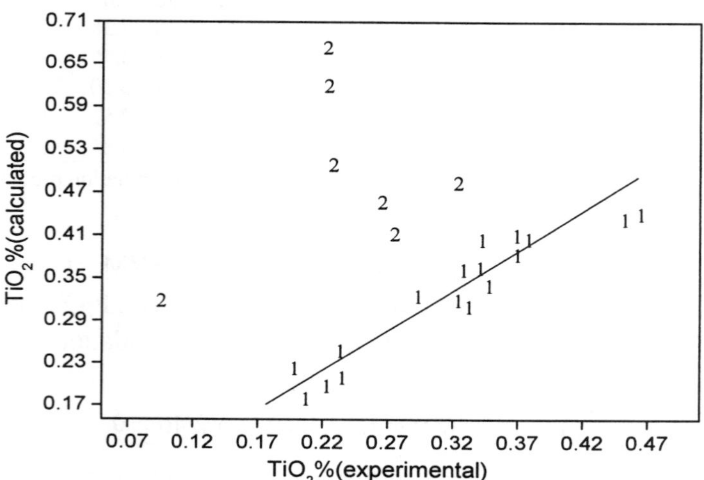

Fig. 11.2   Relationship between TiO_2 content and other components in Jun glazes produced in Song dynasty.

1. sample points of glaze of ancient Jun wares   2. sample points of glaze of imitation products.

11.3   Modeling of Official Ru Wares

Ru wares are the porcelain produced in Ruzhou (Henan province). From the early years in Song Dynasty (960 to 1022A.D.), there were many folk porcelain kilns. Since the beautiful color was appreciated by king palace, an official kiln was built to produce high quality porcelain for palace use in the late period of North Song Dynasty (1086 to 1106 A.D.). Owing to its high quality, the official porcelain Ru kiln was one of the five famous porcelain kilns in Song Dynasty. Since the secret of technology of this kiln had been lost in the period of South Song Dynasty, the production period of Ru kiln was only about 20 years. Only a few products have been handed down. On the other hand, folk porcelain kilns still operated after North Song Dynasty for a long time. So how to differentiate the porcelain produced by official Ru kiln and the porcelain made by folk kilns in Ruzhou region is an identification task for the investigation of Ru porcelains.

In 1986, the site of ancient official Ru kiln has been found in Henan province, and a series of porcelain chips has been found there. Cheng Huansheng has used PIXE technique to make trace element analysis [35]. The results are compared with the products of folk Ru kilns. The results are listed in Table 11.3 and Table 11.4.

Table 11.3   The trace element contents of Ru-kiln glazes ($\mu g \cdot g^{-1}$).

Class[*]	Ni	Cu	Zn	Ga	Rb	Sr	Zr	Pb
1	14	80	88	36	98	249	92	15
1	3	56	64	35	114	257	47	20
1	6	63	68	35	134	259	113	10
1	13	152	45	27	279	424	229	105
1	20	95	126	35	203	312	55	38
1	17	71	113	34	139	249	78	39
1	14	126	73	47	165	244	102	28
1	20	148	67	40	165	227	76	35
1	25	99	75	22	140	257	54	38
1	20	73	67	36	137	208	57	31
1	8	52	72	35	160	236	80	17
1	9	89	77	32	147	288	120	37
1	51	85	72	34	126	252	109	30
1	10	63	54	26	115	176	94	39
1	21	122	92	39	162	278	102	35

1	13	47	81	30	148	279	74	62
1	4	93	53	20	144	241	68	27
1	18	117	76	25	178	285	89	30
1	10	80	44	15	132	206	74	12
2	71	47	25	85	90	177	77	20
2	146	79	20	26	135	327	60	31
2	45	97	43	34	99	309	63	41
2	136	66	16	28	123	247	64	19
2	55	37	62	45	105	223	68	27
2	233	90	37	46	104	276	87	22
2	64	101	32	14	83	359	65	22
2	71	72	9	21	106	257	178	7

*Class "1" denotes the product of official Ru kiln and class "2" denotes product of folk Ru kiln.

Table 11.4   The trace element contents of Ru kiln porcelain body (μg $\cdot$ g^{-1}).

Class*	Ni	Cu	Zn	Ga	Rb	Sr	Zr	Pb
1	37	20	47	37	80	83	201	64
1	26	35	83	19	78	59	138	70
1	32	65	85	19	68	108	128	77
1	24	71	132	20	61	74	143	106
1	40	35	143	38	86	55	224	58
1	60	75	147	24	76	57	215	76
1	17	19	59	22	96	90	230	117
1	20	44	266	13	65	54	128	67
1	23	4	38	33	114	63	192	64
1	21	13	37	19	57	65	138	68
2	23	3	77	39	61	111	344	15
2	13	5	55	33	111	212	207	102
2	27	22	66	32	136	178	200	54
2	21	7	75	36	76	81	174	63
2	25	0	73	39	82	94	216	50
2	23	32	94	40	86	86	197	38

*Class "1" denotes the product of official Ru kiln and class "2" denotes the product of folk Ru kiln.

The data set listed in Table 11.3 and Table 11.4 have been treated by cluster analysis, but it was found that the separation of the products of official Ru kiln and those of folk Ru kilns was not clear-cut. Since SVM is usually more powerful for data processing, we have used it to treat these data again [136].

(1) Classification of the glaze samples of official Ru wares and folk Ru wares:

Using SVM-based feature selection methods, it can be found that a three dimensional space spanned by (Ni), (Zn) and (Rb/Ga) can be used for the classification. Within this space, the criterion of official Ru wares can be expressed by the following inequality:

$$7.38(Zn)-4.47(Ni)-6.51(Rb/Ga)-0.711 > 0 \qquad (11.4)$$

Using the same method, the correctness rate of prediction in LOO cross-validation test is 100%.

Using SVR, the relationship between the Sr content and the contents of other trace elements in the glaze samples of official Ru wares can be roughly expressed by the following equation:

$$(Sr) = 0.259(Zn)-0.093(Ni)+0.006(Cu)-0.714(Ga)$$
$$+1.41(Rb)+0.158(Zr)-0.011(Pb)+0.886\left(Rb/Ga\right)-0.204 \qquad (11.5)$$

while some imitation products do not obey this equation (see Fig. 11.3). Perhaps this relationship can also be used for the identification of official Ru wares.

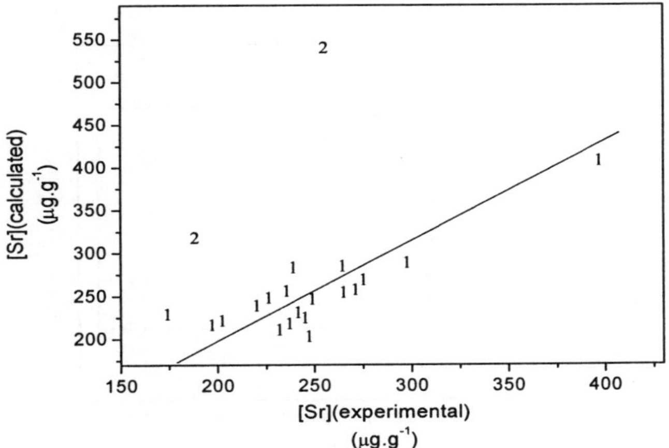

Fig. 11.3   A relationship useful for identification of official Ru wares.
1. official Ru wares   2. folk Ru wares.

(2) Classification of the official Ru wares and folk Ru wares based on the composition of porcelain body:

It has been reported that the cluster analysis cannot completely separate official Ru wares from folk Ru wares based on the data of trace element contents, and according to our result of computation, KNN method also cannot separate these data very clearly. This may be due to the fact that the sample points distribution in the feature space are far from uniform, so that the methods depending on local structure are not very suitable for this purpose.

By SVM technique, it can be found that the space spanned by (Ni), (Sr) and (Ga) can be used to classify the porcelain body of official Ru wares and folk Ru wares, by the following inequality:

$$6.96(Ni)-7.81(Sr)-7.32(Ga) +5.39 > 0 \qquad (11.6)$$

The correctness of prediction in LOO cross-validation test in this space is also 100%.

By some classical chemometric methods such as Fisher method, the samples of official Ru wares can be also clearly separated from those of folk Ru wares based on the trace element contents of porcelain body. Their projection map is shown in Fig. 11.4. But by cross validation there are some wrong prediction results. It implies that there are some overfitting effects.

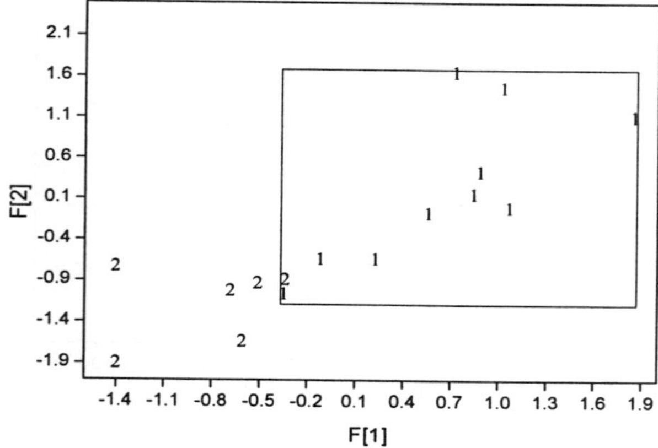

Fig. 11.4 Classification of official Ru wares from folk Ru wares by Fisher method.
1. official Ru wares 2. folk Ru wares

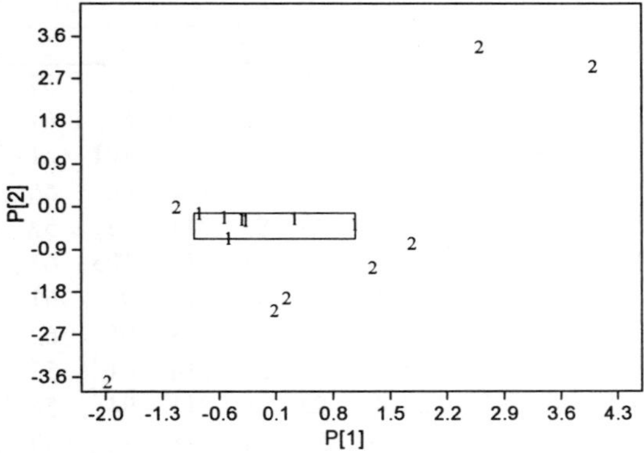

Fig. 11.5   Classification of official Ru wares and folk Ru wares by PCA projection.
1. official Ru wares   2. folk Ru wares

It is interesting to note that even by the contents of major elements the official Ru wares and folk Ru wares can be also clearly classified by PCA method, as shown in Fig. 11.5. From this figure we can also see that the composition of official Ru wares is kept within a narrow range as compared with that of folk Ru wares. It should be the result of strict control of the raw materials for official Ru ware production.

11.4   Modeling of Composition of Yue Wares

Yue ware is widely recognized as the earliest form of Chinese porcelain and even the earliest within worldwide range. It was first produced during the Eastern Han Dynasty. The kiln sites mainly were spreaded through Cixi, Ningbo, Shaoxin, Shangyu cities in Zhejiang province. Yue ware has remained one of the most important representatives of ancient Chinese ceramics and has excerted a powerful influence on the production and development of Chinese porcelain. Wu and Li have collected 47 specimens of Yue ware, excavated from Cixi, Ningbo, Shaoxin, Shangyu kiln sites and Hangzhou palace. These samples are used to make trace element analysis by energy-dispersive X-ray fluorescence (EDXRF) analysis [139]. The results of analysis are listed in Table 11.5.

Table 11.5   The contents of trace elements of Yue wares($\mu g \cdot g^{-1}$).

Kiln place	Ba	Cr	Mn	Ni	Cu	Zn	Pb	Rb	Sr	Y	Zr
Ningbo	947	366	542	120	14	39	81	217	81	57	524
Ningbo	875	314	539	130	21	43	56	219	87	64	519
Ningbo	708	313	540	133	37	50	24	217	101	61	505
Ningbo	1056	385	505	123	41	31	17	211	96	56	527
Ningbo	1186	391	539	154	21	33	8	211	93	52	495
Ningbo	934	377	520	135	38	26	17	215	105	61	480
Ningbo	820	390	507	143	40	41	34	215	97	61	617
Ningbo	851	304	560	147	32	45	12	213	96	62	474
Ningbo	731	371	513	122	37	23	25	214	91	53	511
Ningbo	881	370	507	128	44	52	25	211	94	56	535
Ningbo	1331	339	521	122	40	53	44	223	82	60	516
Shaoxin	1618	311	488	144	40	106	4	200	91	52	501
Shaoxin	1312	306	472	110	41	64	7	189	84	52	522
Shaoxin	1025	293	564	120	33	65	12	220	96	54	413
Shaoxin	1021	291	525	125	51	72	1	198	93	46	432
Shaoxin	1713	306	538	104	50	125	2	179	108	36	388
Shaoxin	973	298	557	111	44	74	1	191	122	38	426
Shaoxin	1296	203	556	100	21	104	1	188	109	49	462
Shaoxin	1397	259	482	127	19	79	2	185	92	60	354
Shaoxin	750	267	440	115	18	77	2	220	102	41	351
Shaoxin	654	303	452	147	36	99	3	192	106	47	448
Shaoxin	1006	256	576	132	16	116	1	195	140	39	355
Shaoxin	875	342	537	142	24	67	3	185	83	63	496
Shangyu	557	333	515	139	44	49	2	273	73	67	449
Shangyu	551	291	544	133	33	13	2	274	83	62	410
Shangyu	413	295	429	108	23	62	13	249	87	54	527
Shangyu	445	297	539	133	23	61	4	235	104	46	444
Shangyu	507	272	430	123	17	62	21	236	82	59	522
Shangyu	393	322	500	97	44	48	3	255	112	39	512
Cixi	927	308	484	124	33	39	12	236	109	42	337
Cixi	1309	264	482	123	22	50	26	235	94	53	313
Cixi	1089	271	511	118	34	67	21	227	107	47	387
Cixi	1218	303	641	104	28	47	36	233	92	47	399
Cixi	1324	297	534	86	37	61	9	231	109	48	396
Cixi	1264	264	507	125	27	50	21	235	109	47	389
Cixi	1326	241	539	140	9	35	10	222	100	49	314
Hangzhou	1038	317	485	130	26	57	24	223	109	50	380
Hangzhou	879	364	543	142	55	76	38	230	112	45	371
Hangzhou	1194	300	679	136	34	74	7	238	109	51	341
Hangzhou	1005	331	567	112	27	56	6	233	102	41	363
Hangzhou	1074	330	671	84	25	78	14	225	145	41	378
Hangzhou	1132	307	648	102	33	55	37	218	98	48	391

Hangzhou 933	334	525	122	26	52	29	231	92	68	382
Hangzhou 995	296	481	128	14	51	30	223	104	44	387
Hangzhou 1084	293	696	115	16	72	24	227	121	36	375
Hangzhou 1021	314	542	131	30	64	9	232	113	56	421

By support vector classification using linear kernel, the porcelain of Hangzhou palace can be separated clearly from the porcelain samples from Ningbo, Shangyu and Shaoxin, but cannot separated from that of Cixi, so it appears that the porcelain samples found in Hangzhou palace were produced at Cixi.

11.5 Modeling of Composition of Blue and White Porcelain Samples

Ming Dynasty was a very important period for the development of Chinese porcelain. Chinese porcelain, especially blue and white porcelains were exported in large quantities at that time. They were distributed in many regions of Asia, Africa and Europe, and had great influences upon the cultural exchange between the East and the West. Besides the chief porcelain production center, Jingdezhen, several places in Fujian, Guangdong and Jiangxi provinces of China also produced and exported porcelain on a large scale. How to find the exact production places of these exported blue and white porcelains? This question is of interest to art historians and ceramic scientists. In order to use data analysis to solve this problem, Li Jiazhi and his coworkers have used X-ray fluorescence technique to determine the contents of trace elements of 42 typical porcelain body samples produced at different places in Ming Dynasty [140]. The data are listed in Table 11.6.

Table 11.6   The contents of trace elements in blue and white porcelains produced in Ming Dynasty ($\mu g \cdot g^{-1}$).

Production place	Zn	As	Pb	Rb	Sr	Y	Zr	Ba	P
Zhangzhou	78	0	0	252	121	129	202	636	304
Zhangzhou	68	0	0	238	90	54	201	620	195
Zhangzhou	66	0	0	219	70	162	194	550	254

Zhangzhou	29	0	0	216	51	130	227	597	176
Zhangzhou	60	0	0	231	83	119	206	601	232
Zhangzhou	68	0	0	238	94	113	202	610	248
Leping	50	0	12	232	75	47	115	377	89
Leping	8	8	1	188	214	22	329	146	235
Leping	33	10	1	402	77	34	172	122	96
Leping	13	0	0	238	190	32	184	284	53
Leping	12	0	0	223	212	25	169	331	2
Leping	56	4	23	285	121	29	182	243	71
Dehua	28	0	26	227	62	58	120	300	0
Dehua	120	0	107	239	47	74	81	367	0
Dehua	76	0	70	253	73	86	89	454	0
Dehua	89	0	66	237	32	85	87	223	0
Dehua	37	0	0	309	35	203	114	184	0
Dehua	117	27	60	243	91	129	174	612	0
Dehua	98	34	81	248	21	118	176	109	0
Dehua	31	17	0	300	81	186	134	129	0
Dehua	48	0	22	362	35	58	61	34	0
Dehua	44	0	0	173	60	87	152	512	0
Dehua	41	0	3	229	39	72	87	286	0
Jingdezhen (f)[*]	31	210	0	405	51	55	92	309	1700
Jingdezhen (f)[*]	0	189	0	458	47	58	81	313	811
Jingdezhen (f)[*]	15	199	0	432	49	57	91	311	1255
Jingdezhen (f)[*]	20	203	0	423	49	56	91	310	1404
Linjiang	86	4	0	459	84	27	92	201	464
Linjiang	71	7	0	421	92	25	86	187	551
Linjiang	73	6	0	440	88	26	89	194	508
Linjiang	81	5	0	447	87	27	90	196	493
Jingdezhen (o)[*]	81	109	81	312	32	55	16	260	0
Jingdezhen (o)[*]	58	144	122	278	0	59	24	184	0
Jingdezhen (o)[*]	67	23	73	213	0	50	55	183	186
Jingdezhen (o)[*]	70	34	65	333	0	51	70	208	314
Jingdezhen (o)[*]	59	42	78	312	0	45	61	141	53
Jingdezhen (o)[*]	64	25	56	330	0	54	82	217	0
Jingdezhen (o)[*]	69	37	95	352	17	58	44	183	103
Jingdezhen (o)[*]	46	25	47	294	19	57	53	293	226
Jingdezhen (o)[*]	66	38	83	418	9	68	34	253	401
Jingdezhen (o)[*]	65	87	46	429	10	73	52	207	445
Jingdezhen (o)[*]	63	40	62	387	23	49	40	290	325
Jingdezhen (o)[*]	53	25	52	341	13	70	58	166	367

[*]Jingdezhen (f) denotes folk kiln of Jingdezhen and Jingdezhen (o) denotes official kiln of Jingdezhen.

(1) SVM criterion to differentiate Jingdezhen porcelain from Dehua porcelain: Using the feature selection algorithm based on SVM techniques, it has been found that the criterion of products of Jingdezhen official kiln can be expressed as follows:

$$1.47(Ba)-6.54(Sr)-4.29(Zr)+2.72 > 0 \qquad (11.7)$$

It means that, as compared with the products of Jingdezhen official kiln, Dehua kiln products contain more Sr and Zr, and less Ba. These differences should be the results of different trace element contents in the raw materials used in the two places. Using the data of these three elements, the criteria of LOO cross-validation test can give 100% correctness of prediction by SVM technique.

In order to make comparison, we also try to use some classical pattern recognition methods to treat the same data set. Figure 11.6 illustrates the map of projection by PCA. It can be seen that the classification is rather good. The result of LOO cross-validation test, however, has one sample misclassified. At the same time, the rate of KNN method can only give the rate of correct classification equal to 82.6%.

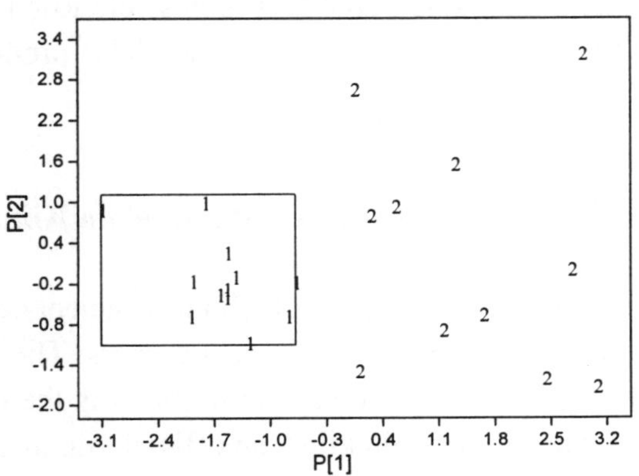

Fig. 11.6   Classification of porcelain from Jingdezhen and Dehua.
1. Jingdezhen products   2. Dehua products

(2) SVM criterion to differentiate official Jingdezhen porcelain from folk Jingdezhen porcelain:

By SVM computation, it can be found that the products of Jingdezhen official kiln can be differentiated from those of Jingdezhen folk kiln by the following criterion:

$$3.85-2.56(Zr)-1.42(Ba)-2.59(P) > 0 \tag{11.8}$$

The result of LOO cross-validation test given by SVM can achieve 100% correctness of prediction.

(3) SVM models to differentiate official Jingdezhen porcelain from Zhangzhou or Linjiang kiln products:

The criterion to differentiate the products of Jingdezhen official kiln from those of Zhangzhou porcelain kiln can be expressed as follows:

$$2.08-3.46(Zr)-0.28(P) >0 \tag{11.9}$$

The criterion for differentiation of Jingdezhen official kiln products from those of Linjiang porcelain kiln can be expressed as follows:

$$0.447(As) +1.15(Pb)-2.11(Sr) +0.925> 0 \tag{11.10}$$

Both of these criteria have 100% correctness of classification. The results of LOO cross-validation are also satisfactory.

Besides, it can be shown that the differentiation results of the products of other pairs of production places by SVM are also very good. Based on the criteria mentioned above, it is possible to build a decision tree for the classification of blue and white porcelains produced in Ming Dynasty.

11.6 Archeological Research of Ancient Porcelain Kilns

In the field of archeological chemistry of Chinese ancient ceramics, there are two difficult problems not solved for many years. The first problem is the location of Ge kiln. And the second problem is the existence and location of the "Xiuneisi porcelain kiln", an official kiln in Song Dynasty near Hangzhou city.

Ge kiln was one of the five famous porcelain kilns in Song Dynasty. Although a few porcelain products of Ge kiln have been handed down, with some description records in ancient books, the exact location of Ge kiln is still an open question. Some authors believe that Ge kiln was located near Longquan kiln near the south border of Zhejiang province, and others believe that Ge kiln was located near Hangzhou city. This is a controversy for many years.

The existence of Xiuneisi kiln is another problem having in controversy for many years. According to a book written in Song Dynasty, after the government of Song Dynasty moved to Hangzhou, "a porcelain kiln was built at Xiuneisi, produced porcelain, ...the products were very fine, ... Later, another new kiln was built at Jiaotanxia ...". According to this record, in Southern Song Dynasty there should be two official porcelain kilns, Xiuneisi kiln and Jiaotanxia kiln, in Hangzhou region. The site of Jiaotanxia kiln was discovered by archeologists many years ago, but the Xiuneisi kiln had not been found even after more than 50 years exploration. So some archeologist suspected the existence of Xiuneisi kiln.

Recently, the archeologists in Zhejiang province have found a site of ancient porcelain kiln at Laohudong region of Hangzhou city. There were a layer of Song Dynasty and a layer of Yuan Dynasty. And many porcelain chips have been found there. Besides, some porcelain chips have been found near this kiln. The chemical composition of these porcelain chips [87; 86] is shown in Table 11.7.

Table 11.7   Composition of porcelain body of products of two official kilns in Hangzhou (wt %).

Class[*]	SiO_2	Al_2O_3	Fe_2O_3	TiO_2	CaO	MgO	K_2O	Na_2O
1	70.11	22.86	2.55	1.19	0.14	0.25	3.11	0.18
1	69.42	23.86	2.41	1.28	0.13	0.24	1.93	0.23
1	68.72	24.41	2.33	1.29	0.12	0.25	2.50	0.22
1	65.51	25.09	2.75	1.28	0.15	0.21	2.11	0.30
1	66.15	25.81	2.92	1.30	0.10	0.19	2.14	0.39
1	66.30	25.59	2.54	1.10	0.14	0.36	2.86	0.45
2	67.01	25.51	2.91	1.32	0.16	0.21	2.29	0.30
2	64.10	27.16	2.72	1.32	0.25	0.49	3.73	0.62
2	66.21	25.56	2.48	1.26	0.16	0.22	2.61	0.50
2	68.41	24.33	2.28	1.30	0.13	0.24	2.27	0.30
2	65.28	24.74	3.28	1.24	0.20	0.41	3.22	0.33
3	61.27	28.81	4.12	0.67	0.21	0.62	4.16	0.19
3	64.53	26.45	2.75	1.30	0.17	0.25	3.68	0.73
3	66.56	24.24	2.63	1.08	0.32	0.36	3.71	0.28
3	68.72	23.59	2.07	1.10	0.60	0.34	3.12	0.30
3	69.79	20.59	3.09	0.73	0.32	0.70	3.75	0.39

3	65.05	26.75	2.81	1.47	0.33	0.14	3.04	0.24
3	65.29	23.56	4.22	2.11	0.20	0.35	4.22	0.24
3	67.04	23.43	1.92	2.17	0.15	0.36	3.63	0.33
3	67.45	23.42	3.00	1.25	0.32	0.17	3.69	0.26
3	66.69	22.68	3.86	1.33	0.08	0.76	4.12	0.18

[*]Class "1" denotes the porcelain body of chips in Song Dynasty layer in Laohudong kiln site, class "2" denotes the porcelain body of chips in Yuan Dynasty layer in Laohudong kiln site and class "3" denotes the porcelain body of chips of Jiaotanxia kiln site.

By SVM method, a criterion can be found to differentiate the porcelain body of the products of Laohudong official kiln from that of Jiaotanxia kiln at Hangzhou in Song Dynasty:

$$5.77 - 3.30(TiO_2) - 4.29(CaO) - 6.07(K_2O) > 0 \qquad (11.11)$$

By the same method, the rate of correctness of prediction in LOO cross-validation test is 100%. It means that the raw materials of these two kilns were not the same. So it appears that the new kiln site found in Laohudong region should just be another official kiln, i.e., Xiuneisi kiln in Hangzhou region.

Besides, by SVM computation, it has been found that the composition of porcelain chips found in the Yuan Dynasty layer are rather close to the composition of handed down Ge ware sample stored by Palace Museum in Beijing. It suggests that probably the Ge ware is just the product of Xiuneisi kiln in Yuan Dynasty.

11.7   Period Discrimination of Ancient Samples

Since the production technology and the composition of raw materials were usually different in different ancient periods, data processing by SVM can be used for the differentiation of the periods of production of porcelain samples. Table 11.8 lists the glaze composition of white wares produced at Jingdezhen. These samples were produced in Song dynasty or Yuan dynasty.

Based on SVM computation, it is found that the chief elements for differentiation of the samples produced in Song dynasty from that of Yuan dynasty are K and Na. Using a data file with the contents of these elements, a criterion for the porcelain produced in Song dynasty can be expressed as follows:

$$8.23 - 12.0(K_2O) - 7.11(Na_2O) > 0 \qquad (11.12)$$

Table 11.8   Composition of white wares produced in Song and Yuan dynasty (wt%).

Period	SiO$_2$	Al$_2$O$_3$	Fe$_2$O$_3$	TiO$_2$	CaO	MgO	K$_2$O	Na$_2$O	MnO
1	66.68	14.30	0.99	0.00	14.87	0.26	2.06	1.22	0.10
1	67.26	17.08	0.93	0.12	10.05	1.90	2.27	0.31	0.15
1	66.40	14.39	1.16	0.00	14.08	0.56	1.46	1.64	0.00
1	66.69	15.17	1.11	0.07	13.94	0.44	1.47	0.64	0.06
1	65.40	13.99	1.06	0.05	15.43	0.60	2.04	1.01	0.09
1	65.85	13.85	0.83	0.06	14.15	0.64	1.55	2.74	0.09
1	65.84	14.08	0.70	0.06	16.01	0.72	1.58	0.55	0.05
1	65.45	15.94	1.33	0.10	11.99	0.53	2.00	2.16	0.09
1	65.99	14.44	1.11	0.07	14.00	0.62	1.58	1.93	0.07
1	68.68	14.43	0.93	0.09	10.01	0.75	1.85	2.69	0.07
2	66.48	12.96	0.90	0.12	12.85	0.18	2.24	4.00	0.10
2	67.56	14.07	0.86	0.08	11.98	0.45	2.07	2.92	0.09
2	73.36	14.61	0.78	0.00	5.33	0.16	2.89	3.31	0.08
2	72.70	15.23	0.78	0.00	4.81	0.18	2.99	3.72	0.10
2	71.98	15.58	0.85	0.00	5.58	0.20	3.06	3.47	0.10
2	73.41	15.63	0.95	0.00	4.03	0.24	3.22	3.34	0.10
2	72.15	15.17	1.01	0.00	6.06	0.26	2.88	2.27	0.11
2	71.78	13.68	0.83	0.22	5.59	0.19	3.17	3.60	0.09
2	70.09	15.24	0.83	0.16	6.40	0.18	3.22	3.13	0.09

*Period "1" denotes the porcelain produced in Song dynasty and period "2" denotes that of Yuan dynasty.

The rate of correctness of classification is 100%, and the rate of correctness of prediction in LOO cross-validation test is 94.7%. This computation result means that the contents of alkali metal oxides in glaze had been significantly increased after Song Dynasty. The original CaO containing flux changed to alkali oxide-calcium oxide flux. Actually this was just a technical achievement of Jingdezhen porcelain production in ancient time. The increase of contents of alkali metal oxides significantly improved the quality of porcelain products.

Chapter 12

SVM Applied to Cancer Research

12.1   SVM Applied to Cancer Epidemiology

12.1.1   *Relationships of trace elements and carcinogenesis*

Although the exact mechanism of carcinogenesis and the exact relationships between the environmental factors and the cancer mortality are not clearly understood yet, it is already widely recognized that many trace elements are involved in the metabolism of carcinogenesis, because there are plenty of evidences to support this point of view: In clinical study it has been confirmed that the contents of trace elements (especially the contents of Se, Zn, Cu, Co, Mo) of cancer tissue and that of normal tissue around are always different [51; 66; 128]. The content of Se and Zn/Cu ratio in the serum of cancer patients are also abnormal compared with that of normal persons. And there are data that the intake quantity of trace elements and the breast cancer mortality in different regions indeed exhibit obvious correlation [141], as illustrated by Table 12.1.

Based on the data listed in Table 12.1, the relationship between trace element intake and breast cancer mortality found by support vector regression can be expressed as follows:

$$\text{Mortality} = 0.7867\text{-}0.645(Se)\text{-}0.335(Zn)\text{-}0.245(Cu)\text{-}0.374(Cu/Zn)$$
$$-0.197(As)+0.664(Cd)+0.048(Mn)+0.234(Cr) \tag{12.1}$$

The multiple correlation coefficient is as high as 0.901. Figure 12.1 illustrates the comparison of the real and predicted mortality by leave-one-out (LOO) cross-validation test. It can be seen that the regularity is rather good and believable.

Table 12.1   Results of statistics of food intake of trace elements and breast cancer mortalities in 27 different regions
(food intake: mg/year. person; mortality: death number in 100000 persons).

No. of regions	Se	Cu	Zn	Cd	Cr	Mn	As	Mortality
1	75.9	1125	6948	123.9	21.4	722	152.4	19
2	71.6	824	4272	77.0	22.0	858	102.4	17
3	70.6	741	4425	80.5	21.1	803	158.2	21
4	61.8	874	5313	97.2	21.9	711	139.1	23.5
5	71.1	725	4387	76.9	18.6	677	166.0	24
6	64.3	784	4473	74.4	20.6	751	136.3	17.5
7	75.1	836	3712	99.7	16.4	889	114.4	21.5
8	77.3	822	4449	77.1	23.2	923	136.9	21
9	57.5	693	3741	70.8	20.3	701	108.5	26
10	82.2	608	4126	87.5	17.0	672	269.1	17.5
11	65.8	703	3783	74.8	18.3	646	167.1	18.5
12	65.6	850	4169	83.6	23.2	819	109.6	21.5
13	61.7	729	4502	85.4	17.7	717	132.2	25
14	61.0	849	5108	87.3	25.4	652	132.8	21.5
15	107.6	861	3924	79.5	16.7	1129	102.1	9.0
16	85.1	838	4712	82.6	15.3	1029	120.2	15.5
17	67.1	690	3623	88.2	13.7	716	132.1	13
18	76.4	881	5339	91.5	21.2	965	173.8	16
19	91.9	911	4452	92.3	17.7	1154	185.4	8.5
20	91.2	639	2959	54.6	19.6	463	273.4	10
21	86.5	768	3958	72.4	14.9	1075	91.3	14.5
22	82.5	855	4304	83.8	16.4	1161	134.4	16.5
23	85.8	643	2420	43.4	17.4	674	233.4	3.5
24	93.8	634	5231	80.3	15.9	876	138.3	11
25	87.2	714	4347	74.6	15.8	1008	268.5	12.5
26	84.1	592	1674	83.3	12.8	522	184.3	4.0
27	98.6	722	3288	72.9	11.7	1169	82.1	8.0

It is well known that the cancer mortality of different regions is quite different. And there are some local regions having very high cancer mortality as compared with other regions. How to depress the abnormal high mortality in these regions is a very interesting problem for epidemiologists.

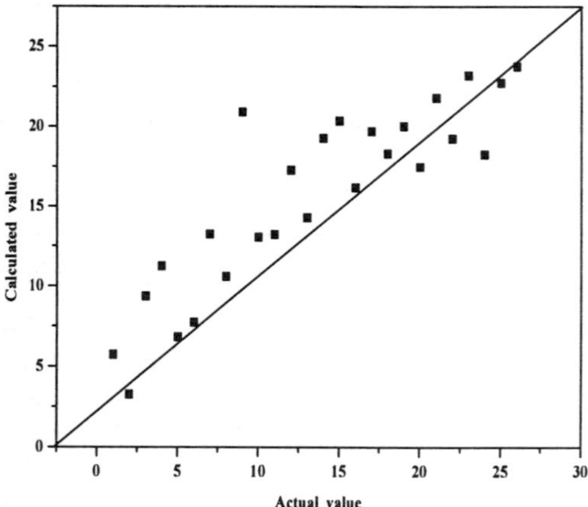

Fig. 12.1   Comparison between actual mortality and the mortality predicted by SVR.

There are many evidences that the geographical abnormal distribution of trace elements in soil or natural water has obvious correlation with the high mortality regions. It means that the shortage or surplus of some trace elements in environment may be an important factor affecting cancer mortality. Research work in this field may lead to useful results, because perhaps we can find some effective ways to control the cancer mortality by adjusting the intake of trace elements of people in some local regions. This task is, however, a very complicated and very difficult one, due to the following reasons: (1) The content of trace elements is only one of many factors for carcinogenesis and cancer mortality. Other factors, such as the organic carcinogens (such as polycyclic aromatic hydrocarbons, nitroso-amines, organic carcinogens related to the habit of life of local people, etc.) are also very influential. So it is difficult to isolate the influence of trace elements from that of other factors. (2) There are many trace elements involved in the metabolism of carcinogenesis. The relationships between the different elements and different amount of elements are rather complicated. For example, it is already known that large quantity of arsenic can induce cancer, but it is already realized that small amount of arsenic oxide can be used to cure some kinds of cancer. Another example is that selenium is useful for prevention of cancer, while it is already known that many other elements such as mercury can depress the effect of selenium. (3) The process of carcinogenesis in human body is a rather slow process. And many people can move from one place to the other. So it is difficult to find an exact correlation between the cancer mortality and the trace element contents in soil and natural water of this region. In other words, data processing in this field suffers from the difficulties induced by high noise/signal ratio problems. The application of SVM may be helpful for overcoming these difficulties.

12.1.2 *SVR applied to find multiple correlation coefficients between the cancer mortality and trace element content in soil of different provinces and regions of China*

For many years, it has been widely recognized that the trace element content in environment is one of the important factors influencing the averaged cancer mortality of local regions. Soil is the chief source of trace elements in environment. Therefore there are already many research projects dealing with the possible correlation between the trace element content in the soil of local regions and the averaged cancer mortality in these regions. China is a country having vast territories and regions with very different trace element contents. China has statistical data records of mortality of different kinds of cancer in different regions for many years. At the same time, China has published a handbook "The background values of soil in China" [36] having complete data of trace elements of soil of different regions based on the results of a large scale investigation project. Besides, until twenty years ago, most of Chinese people usually did not move and lived in a fixed region due to their social situation and old tradition of their life. This is very beneficial for the statistical work to find the correlation between the data of cancer mortality [151] and trace element content in local soil. In recent years, some Chinese geologists were dealing with a research project to study the relationship between cancer mortality and trace element content in local soil. Since the number of trace elements involved is several tens, and the number of provinces and regions are limited, this is a typical problem with small sample size. SVM should be useful to solve this problem.

(1) Multiple correlation between the mortality of leucocythemia and trace element content of soil in 29 provinces or regions in China.

The detailed metabolism of leucocythemia is not very clear yet, but it has been known that Li_2CO_3 or As_2O_3 can be used to treat the acute leucocythemia. And it is also found that the trace element contents in the hair or blood of leucocythemia patients are abnormal compared with that of the ordinary persons. So it is interesting to investigate if there are some correlations between the amount of trace elements in soil and the mortality of leucocythemia of different local regions.

Table 12.2 lists the trace element contents and the mortalities of leucocythemia of 29 regions in China.

Table 12.2   Trace element contents in soil and leucocythemia mortalities in 29 regions of China.

Regions	As	Cr	Mn	Ni	Zn	Cd	Cu	Ca[*]	Hg	Mortality
Liaoning	8.8	57.9	564	25.6	63.5	0.108	19.8	0.86	0.030	2.88
Hebei	13.6	68.3	608	30.8	78.4	0.094	21.8	2.18	0.036	3.02
Shandong	9.3	66.0	644	25.8	63.5	0.084	24.0	1.67	0.019	2.05
Jiangsu	10.0	77.8	585	26.7	62.6	0.126	22.3	1.82	0.289	3.75
Zhejiang	9.2	52.9	448	24.6	70.6	0.070	17.6	0.12	0.086	3.98
Fujian	6.3	44.0	391	18.2	86.1	0.074	22.8	0.05	0.093	3.42
Guangdong	8.9	50.5	279	14.4	47.3	0.056	17.0	0.06	0.078	2.33
Guangxi	20.5	82.1	446	26.6	75.6	0.267	27.8	0.13	0.152	1.03
Heilongjiang	7.3	58.6	1065	22.8	70.7	0.086	20.0	0.94	0.037	2.15
Jilin	8.0	46.7	636	21.4	80.4	0.099	17.1	1.26	0.037	2.96
Neimong	7.5	41.4	520	19.5	59.1	0.053	14.4	1.80	0.040	2.53
Shanxi	9.8	61.8	554	32.0	75.5	0.128	26.9	4.15	0.027	2.65
Henan	11.4	63.8	579	26.7	60.1	0.074	19.7	2.43	0.034	2.59
Anhui	9.0	66.5	530	29.8	62.0	0.097	20.4	1.20	0.033	3.08
Jiangxi	14.9	45.9	328	18.0	69.4	0.108	20.3	0.09	0.084	2.67
Hubei	12.3	86.0	712	37.2	83.6	0.172	30.7	0.52	0.080	2.32
Hunan	15.7	71.4	459	31.9	94.4	0.126	27.3	0.13	0.116	2.85
Shanxi	11.1	62.5	557	28.8	69.4	0.094	21.4	2.95	0.030	2.20
Sichuan	10.4	79.0	657	32.6	86.5	0.079	31.1	1.13	0.061	1.90
Guizhou	20.0	95.9	794	39.1	99.5	0.659	32.0	0.62	0.110	1.27
Yunnan	18.4	65.2	626	42.5	89.7	0.218	46.3	0.16	0.058	1.26
Ningxia	11.9	60.0	524	36.5	58.8	0.112	22.1	4.18	0.021	2.47
Gansu	12.6	70.2	653	35.2	68.5	0.116	24.1	4.41	0.020	0.84
Qinghai	14.0	70.1	580	29.6	80.3	0.137	22.2	3.79	0.020	1.29

Xinjiang	11.2	49.3	688	26.6	68.8	0.120	26.7	4.99	0.017	1.90
Xizang	19.7	76.6	625	32.1	74.0	0.081	21.9	1.22	0.024	0.87
Beijing	9.7	68.1	705	29.0	102.6	0.074	23.6	1.52	0.069	3.06
Tianjin	9.6	84.2	686	33.3	79.3	0.090	28.8	2.34	0.084	2.79
Shanghai	9.1	70.2	548	29.9	81.3	0.138	27.2	1.26	0.095	3.37

*The content of Ca is weight percent, and all other contents of elements are expressed in $mg \cdot kg^{-1}$. The unit of mortality is the death number in 100000 persons.

By SVR, it has been found that the value of multiple correlation coefficient between the leucocythemia mortalities and the linear combination of the data of trace element contents is as high as 0.896:

$$\text{Mortality} = 0.745 - 0.693(As) - 0.161(Cr) - 0.242(Mn) + 0.315(Ni) + 0.162(Zn) \\ -0.162(Cd) - 0.235(Cu) - 0.176(Ca) + 0.471(Hg) \tag{12.2}$$

If we define the regions with mortality lower than 2×10^{-5} as the samples of class "1", and that higher than 2×10^{-5} as samples of class "2', and make classification by SVM. The criterion of class "1" obtained can be expressed as follows:

$$4.42(As) + 0.835(Cr) + 2.549(Mn) - 2.601(Ni) + 1.204(Cd) \\ +2.250(Cu) + 2.603(Ca) - 1.057(Hg) > 4.772 \tag{12.3}$$

The rate of correctness of classification is 93.1%, and the rate of prediction correctness by leaving-one out (LOO) cross validation test is 89.6%.

Using the same data, the rate of correctness of classification by KNN method is only 72.4%, much lower than that of SVM.

The above-mentioned result indicates that the correlation between the trace element content in soil and the mortality of leucocythemia appears rather obvious. And it is especially interesting to note that the contents of arsenic and mercury in soil exhibit evident relationships with leucocythemia mortality in different regions. The negative coefficient of arsenic content implies that the soil with relatively higher arsenic content corresponds to lower mortality of leucocythemia. According to some literatures, although very high amount of arsenic can induce cancer (this has been confirmed in animal experiments), but it has been proved that small amount of arsenic oxide can induce the apoptotic cell death, and it has been recognized that the process of apoptotic cell death may be a new way to cure cancer. So some biochemists have claimed that arsenic may be a beneficial element for the human health at very low concentration. On the other hand, the large positive coefficient of mercury content indicates that mercury is a harmful element. This can be explained by the antagonistic mechanism of mercury to the beneficial action of selenium reported in literature.

(2) Multiple correlation between the breast cancer mortality and the trace elements in soil of 29 provinces or regions in China:

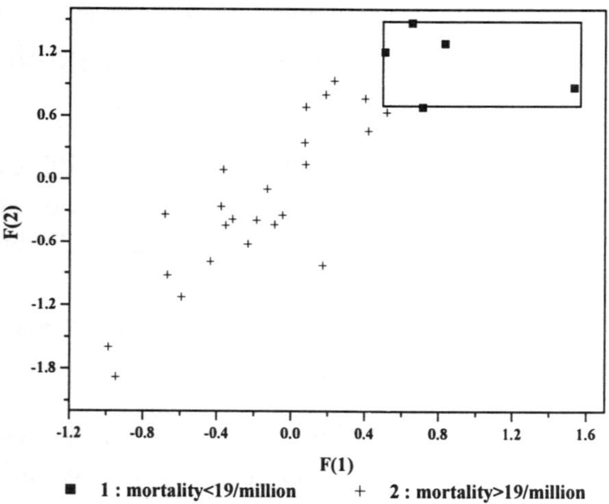

Fig. 12.2 Distribution of high mortality and low mortality regions of breast cancer in China (Fisher method).

According to the data of trace element contents in the cancerous tissues of breast cancer, the contents of 11 elements (As, Co, Cr, Cu, Hg, K, Mg, Mn, Na, Sr and Zn) are abnormal as compared with those of normal tissues. So that it is reasonable to expect that these 11 elements are relevant to the metabolism of breast cancer. Figure 12.2 illustrates the distribution of the sample points of high mortality regions (mortality of breast cancer higher than 19/million) and low mortality regions of breast cancer in China by Fisher method. It can be seen that the regularity is rather clear.

By SVR, the multiple correlation coefficient found between the mortality of breast cancer and the contents of the following 7 elements in soil is as high as 0.86:

$$\text{Mortality} = 0.660(\text{Hg}) + 0.078(\text{Mg}) + 0.208(\text{K}) + 0.190(\text{Na})$$
$$-0.472(\text{As}) - 0.305(\text{Sr}) - 0.006(\text{Mn}) + 0.301$$

(12.4)

12.2 Carcinogenic and Environmental Behaviors of Polycyclic Aromatic Hydrocarbons

Polycyclic aromatic hydrocarbons are the most notorious and widely distributed carcinogens. These compounds are produced in many burning processes in nature and human life. Many chemical industrial processes, especially coke production, are also the sources of polycyclic aromatic hydrocarbons. The chemical behaviors of polycyclic aromatic hydrocarbons are relatively inert, so they are relatively stable and decompose rather slowly. Therefore the accumulation of these compounds in environment has been one of the most serious problems to human health.

12.2.1 *Molecular structure and carcinogenic activity of polycyclic aromatic hydrocarbons*

It is well-known that some polycyclic aromatic hydrocarbons, such as benzo(a)pyrene, are strong carcinogens, while some other polycyclic aromatic hydrocarbons with quite similar molecular structure, such as benzo(a)naphthacene, having no carcinogenic activity. In order to find the regularities of the carcinogenic activity of polycyclic aromatic hydrocarbons, quantum chemical parameters and molecular descriptors have been used for SVM computation.

According to the two-region theory [44] of carcinogenesis of polycyclic aromatic hydrocarbons, following four parameters are used for the correlation between the molecular structure of polycyclic aromatic hydrocarbons and their carcinogenic activities:

Highest energy of delocalization of bay region ΔE

The delocalization energy of two active regions ΔE_1 and ΔE_2

The number of detoxification region n

Since the degree of carcinogenic activity can only be expressed semi-quantitatively, SVR with ε-insensitive loss function is especially suitable to investigate this problem. Figure 12.3 illustrates the comparison of the actual and calculated values of degree of activity of carcinogenesis of 43 polycyclic aromatic hydrocarbons.

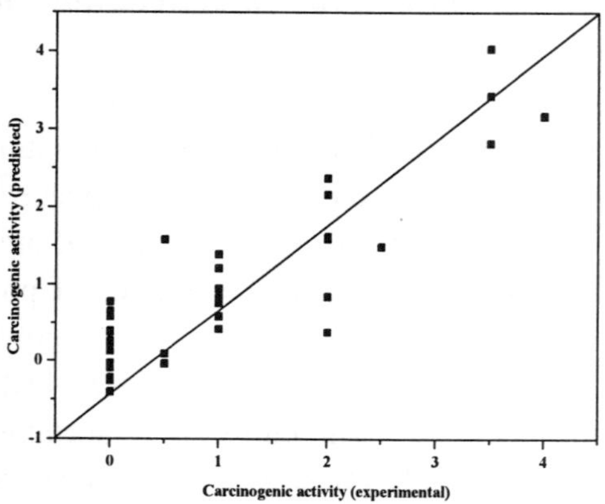

Fig. 12.3   Semi-quantitative relationships between molecular parameters and carcinogenic activity of polycyclic aromatic hydrocarbons.

12.2.2 *QSPR of polycyclic aromatic hydrocarbons*

Since many polycyclic aromatic hydrocarbons are strong carcinogens, it is necessary to investigate their distribution or degradation related properties, including the distribution coefficients between air and octanol (K_{OA}), the absorption parameters in soil (K_{OC}), and the bioconcentration factor (BCF), etc. These data of most polycyclic aromatic hydrocarbons, however, have not been measured yet owing to the experimental difficulties. So it is desirable to make mathematical modeling based on the known data and then make computerized prediction for unknown data. In our previous work, SVR has been used for this purpose. And it has been found that the results of prediction are usually better than those of some other methods. Table 12.3 lists the structure descriptors and some known properties of some polycyclic aromatic hydrocarbons [52].

Table 12.3   Molecular descriptors and Properties of polycyclic Aromatic hydrocarbons

Compound	N^*	B^*	L^*	V^*	X_v^*	X_e^*	K_{OA}^{**}	K_{oc}^{**}	BCF^{**}
Benzene	1	6.740	7.406	193.8	3	3	-	1.57	-
Naphthalene	2	7.428	9.913	286.0	4.966	5.454	5.13	3.04	2.12
Anthracene	3	7.439	11.651	336.4	6.933	7.942	7.34	4.37	2.95
Phenanthrene	3	8.031	11.752	366.9	6.950	7.926	7.45	4.36	2.51
Naphthacene	4	7.446	14.116	408.3	8.899	10.43	-	5.81	-
Pyrene	4	9.279	11.662	420.7	7.933	9.409	8.43	5.03	3.43
Chrysene	4	8.039	13.939	447.3	8.933	10.93	10.44	-	3.785
Perylene	5	9.247	11.809	553.6	9.933	11.89	11.70	-	3.86
Benz(a)anthracene	4	8.717	13.942	472.3	8.916	10.41	10.80	6.30	4.00
Triphenylene	4	10.44	11.682	411.5	8.950	10.41	-	-	3.96
Benzo(c)phenanthrene	4	9.323	11.909	553.6	8.933	10.40	-	-	-
Benzo(a)pyrene	5	9.297	13.882	502.1	9.916	11.89	10.71	6.46	3.82
Benzo(e)pyrene	5	10.52	11.765	481.0	9.933	11.89	-	-	
Dibenz(a,h)anthracene	5	8.726	15.898	539.6	10.89	12.88	13.91	-	4.00
Benzo(ghi)perylene	6	10.48	11.779	480.0	10.91	13.38	-	6.80	4.45
coronene	7	11.70	11.722	533.1	11.89	14.86	-	-	-

*N denotes the number of benzene rings; B denotes the width of molecule (in A), L denotes the length of molecule (in A); V denotes the volume of molecule (in A^3); X_v denotes the vertex connectivity of molecule; X_e denotes the edge connectivity of molecule.

$^{**}K_{OA}$ is the distribution coefficients between octanol and air; K_{OC} is the parameter describing sorption by soil, and BCF is the distribution ratio of aquatic organism to ambient environment.

Some of the mathematical models obtained in our computation are as follows [28]:

(1) Modeling of K_{OA}:

Since the polarity of octanol is similar to the cell membrane of human body or other animals, the distribution coefficient of polycyclic aromatic hydrocarbons between air and octanol (K_{OA}) can be considered as a key descriptor for the partitioning of polycyclic aromatic hydrocarbons between the atmosphere and terrestrial animals. Using three molecular descriptors: V (molecular volume), L/B (ratio of the length to width of molecule) and W (molecular weight) as independent variables, a mathematical model can be obtained by support vector regression with ε = 0.05:

$$\text{Log } K_{OA} = 0.00409(W) + 1.911(L/B) + 0.0159(V) - 2.239 \quad (12.5)$$

By LOO cross-validation method, the predicted values and experimental values are compared in Table 12.4.

Table 12.4   Log K_{OA} of polycyclic aromatic hydrocarbons

Polycyclic aromatic hydrocarbons	Log K_{OA}(predicted))	Log K_{OA}(experimental)
Naphthalene	5.165	5.13
Anthracene	7.257	7.34
Phenanthrene	7.570	7.45
Benzo(a)pyrene	11.470	10.71
Dibenz(A,B)anthracene	13.475	13.91
Chrysene	10.465	10.44
Benz(a)anthracene	10.601	10.80
Pyrene	8.311	8.43
perylene	11.723	11.70

The averaged absolute error of the data listed in Table 12.4 is slightly lower than those predicted by PLS.

(2) Modeling of K_{OC}

K_{OC} represents the degree of sorption of polycyclic aromatic hydrocarbon in soil. Using three molecular parameters: the length of molecule (L), the vertex connectivity index (X_v) and the edge connectivity index (X_e) and support vector regression with ε = 0.05, following expression is obtained for the prediction of K_{OC}:

$$\text{Log } K_{OC} = 0.2866(X_v) + 0.2198(X_e) + 0.1450(L) - 1.0094 \quad (12.6)$$

By LOO cross validation method, the predicted values obtained are compared with the experimental values in Table 12.5.

The predicted values by SVR are slightly better than those by PLS method.

Table 12.5 K_{OC} of polycyclic aromatic hydrocarbons.

Polycyclic aromatic hydrocarbon	Log K_{OC}(predicted)	Log K_{OC}(experimental)
Benzene	6.568	6.57
Naphthalene	3.033	3.04

Phenanthrene	4.428	4.36
Anthracene	4.393	4.37
Pyrene	5.004	5.03
Naphthacene	5.934	5.81
Benz(a)anthracene	5.812	6.30
Benzo(a)pyrene	6.441	6.46
Benzo(ghi)perylene	6.659	6.80

(3) Modeling of BCF factor

BCF denotes the distribution ratio of a compound between an aquatic organism and the ambient environment:

$$BCF = (C_{org} / C_{water})$$

By using the molecular parameter X_e (edge connectivity), B (width of molecule), and SVR with $\varepsilon = 0.05$, the following equation for BCF can be obtained:

$$Log\ (BCF) = 0.2674(X_e) + 0.2216(B) - 0.8256 \qquad (12.7)$$

The predicted values of BCF are compared with the experimental values in Table 12.6.

Table 12.6   BCF of polycyclic aromatic hydrocarbons.

Polycyclic aromatic hydrocarbons	Log(BCF) (predicted)	Log(BCF) (experimental)
Naphthalene	2.488	2.21
Anthracene	2.663	2.95
Phenanthracene	2.980	2.51
Benz(a)anthracene	3.494	4.00
Chrysene	3.321	3.785
Triphenylene	3.910	3.96
Pyrene	3.38	3.43
perylene	3.903	3.86
benzo(a)pyrene	3.911	3.82
Dibenz(a,h)anthracene	4.096	4.00
Benzo(ghi)perylene	4.442	4.45

The prediction results of SVR are slightly better than that of PLS method.

12.3   SVM Applied to Cancer Diagnosis

SVM is also useful for solving the multivariate problems in cancer diagnosis work. H.X. Liu and his coworkers have used nine features in the diagnosis of breast cancer: clump thickness, uniformity of cell size, uniformity of cell shape, marginal adhesion, single epithelial cell size, bare nuclei, bland chromatin, normal nucleoli and mitosis.

According to these parameters, the breast tumor samples can be classified into two classes: benign breast tumor and breast cancer. It has been found that among the features the parameter 6 (bare nuclei), parameter 2 (uniformity of cell size), parameter 8 (normal nucleoli), parameter 4 (marginal adhesion) and parameter 1 (clump thickness) are the chief factors influencing the classification results. The Gaussian kernel function and polynomial kernel functions have been used for the classification. The results of classification of the data by SVM and ANN are compared: Although the data fitting by ANN is better than that of SVM, the number of mistaken samples in LOO cross-validation test of SVM is smaller than that of ANN [89]. It means that SVM has better prediction reliability in breast cancer diagnosis.

Chapter 13

SVM Applied to

Some Topics of Chemical Analysis

13.1　Multivariate Calibration in Chemical Analysis

In chemometrics, multivariate calibration methods provide a convenient way to determine several components in a mixture within one experimental step, without the tedious operation of separation of these components. The method of calculation usually used is PLS method. Artificial neural network is also often used especially when the data set exhibits obvious nonlinearity. But it is prone to overfitting. Therefore, several types of techniques have been developed to prevent overfitting. At the same time, support vector regression, as a method suitable to treat nonlinear data without serious overfitting, can be used as a new method of computation in multivariate calibration. An example of using SVR in multivariate calibration will be described as follows.

Amino acids are the structural units of protein, and some of them have been used as drugs or food additives, so the determination of amino acids is useful for biochemical research and for commercial product analysis. Among essential amino acids, there are three aromatic amino acids, phenylalanine, tyrosine and tryptophan, which exhibit fluorescence when they are excited by ultraviolet rays. So it is possible to determine them by the fluorescence spectroscopic method. The λ_{max} of phenylalanine, tyrosine and tryptophan are 282 nm, 303 nm and 348 nm respectively, but their fluorescence spectra are partially overlapped. Since the separation operation of these three amino acids is tedious and

261

troublesome, it is desirable to use multivariate calibration method to determine them in their mixture by fluorescence spectroscopy. Table 13.1 lists the contents of 23 samples of mixtures of the three aromatic amino acids, and Fig. 13.1 illustrates the fluorescence spectra of these 23 samples.

Table 13.1   The contents of three aromatic amino acids in training samples ($\mu g \cdot ml^{-1}$).

No. of samples	Tyrosine	Tryptophan	Phenylalanine
1	2.004	0.0512	4.048
2	1.503	0.1024	3.542
3	1.002	0.256	3.036
4	0.501	0.512	2.53
5	0.2505	1.024	2.024
6	0.1002	1.536	1.518
7	0.0501	2.048	1.012
8	0.1002	1.536	4.048
9	0.2004	2.048	0.506
10	2.004	0.0205	0.506
11	0.4008	2.048	6.072
12	1.002	1.024	5.06
13	4.008	0.1536	2.024
14	3.006	0.2048	2.024
15	0.1002	2.048	4.048
16	0.3006	2.048	4.048
17	0.2004	1.536	3.036
18	0.501	1.048	2.53
19	1.002	0.512	2.024
20	1.503	0.3072	4.554
21	2.004	0.2048	3.036
22	2.505	0.1024	1.012
23	2.004	0.1048	2.024

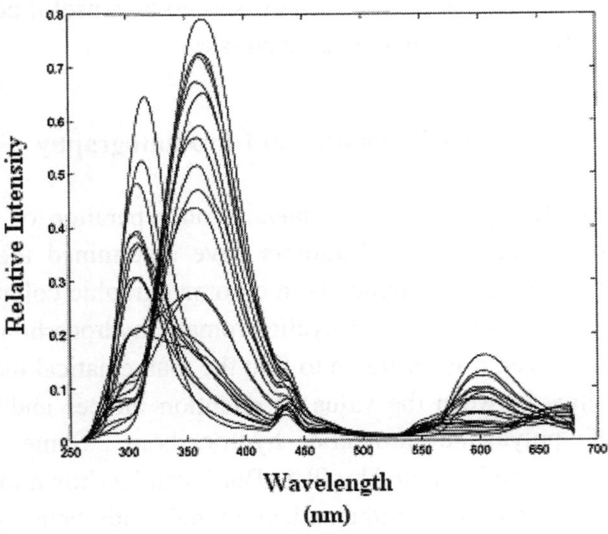

Fig. 13.1   The fluorescence spectra of 23 samples.

As the methods of computation, the data of 23 samples are treated by support vector regression, artificial neural networks and PLS concurrently. The errors of leave-one-out (LOO) cross-validation test are listed in Table 13.2.

Table 13.2   Comparison of errors of cross validation by different algorithms.

Algorithm	Root mean square error	Mean absolute error
SVR (linear kernel)	0.2347	0.1427
SVR (Gaussian kernel)	0.1884	0.1258
ANN	0.2436	0.1753
Weight decay ANN	0.2364	0.1677
Early stopping ANN	0.3488	0.2182
PLS	0.3011	0.2076

From Table 13.2 it can be seen that the best result is obtained by SVR with Gaussian kernel.

SVR has been also applied to simultaneous determination of Pb, Cd, Zn in aqueous solutions and simultaneous determination of NO_3^- and NO_2^- in aqueous solutions. Both of these results are also rather satisfactory [13; 48]. It appears that SVR should be a useful new tool for multivariate calibration in analytical chemistry.

13.2   Retention Indices Estimation in Chromatography

Motivated by the requirement of analysis and separation of polycyclic aromatic hydrocarbons, several authors have determined the retention indices of some of these compounds in chromatographic columns. Since the retention indices of many polycyclic aromatic hydrocarbons have not been determined yet, it is desirable to find the mathematical model about the relationships between the value of retention indices and molecular structure of polycyclic aromatic hydrocarbons. Some of these relationships have been studied by PLS. But it can be shown that support vector machine can give mathematical model with better prediction ability. Table 13.3 illustrates the experimental values of 33 polycyclic aromatic hydrocarbons and their molecular parameters [52; 5]. Support vector regression has been used for this modeling work.

The data listed in Table 13.3 are used for model-building by following steps: at first KL transformation is carried out, and then SVR-based algorithm is used to make feature selection, then the data file after feature selection is used for mathematical modeling. The same data file is also treated by PLS method. The errors of prediction in LOO cross-validation test are listed in Table 13.4.

It can be seen that the best result is obtained by SVR computation of the data after KL transformation and feature selection.

Similar calculation has been also carried out for 32 samples of alkyl-substituted polycyclic aromatic hydrocarbons. Table 13.5 lists the values of errors obtained by different methods. It can be seen that the result of SVR with a data file obtained by SVR-based feature selection is also the best one.

Table 13.3   Retention indices and molecular parameters of 33 polycyclic aromatic hydrocarbons.

PAH	RI[*]	IP[*]	EA[*]	GAP[*]	W[*]	L/B[*]	X_v[*]	X_e[*]	Width[*]	Length[*]	Volume[*]	Sarea[*]
		eV	eV	eV					A	A	A^3	A^2
Benzene	1	9.391	-0.368	9.759	27	1.099	3	3	6.74	7.406	193.8	104.8
Naphthalene	2	8.5748	0.332	8.242	109	1.238	4.966	5.454	7.428	9.915	286.1	141
Anthracene	3.2	8.049	0.842	7.207	279	1.566	6.933	7.942	7.439	11.65	336.5	160.8
Phenanthrene	3	8.478	0.481	7.997	271	1.463	6.95	7.926	8.031	11.75	366.9	171.3
naphthacene	4.51	7.721	1.185	6.536	569	1.806	8.899	10.43	7.446	14.12	408.3	188.9
benz(a)anthrcene	4	8.114	0.83	7.284	557	1.599	8.916	10.41	8.717	13.94	472.4	209.6
chrysene	4.1	8.261	0.719	7.542	545	1.734	8.916	10.4	8.039	13.94	447.3	199.8
triphenylene	3.7	8.505	0.557	7.947	513	1.119	8.933	10.41	10.44	11.68	411.5	196.6
pyrene	3.58	8.039	0.902	7.137	362	1.257	7.933	9.409	9.279	11.66	420.7	189.6
benzo©phenanthrene	3.64	8.243	0.692	7.591	529	1.277	8.933	10.4	9.323	11.91	553.7	216.9
pyrylene	4.33	7.815	1.119	6.695	654	1.276	9.933	11.9	9.247	11.8	427	191.5
benzo(a)pyrene	4.53	7.861	1.088	6.773	680	1.493	9.916	11.9	9.297	13.88	502.2	219.3
benzo(e)pyrene	4.28	8.108	0.889	7.219	652	1.118	9.933	11.9	10.52	11.77	481.1	210.4
picene	5.18	8.237	0.751	7.486	963	2.005	10.916	12.868	8.037	16.112	504.6	223.6
pentaphene	4.67	8.113	0.844	7.268	993	1.748	10.882	12.902	9.207	16.093	576.8	246.7
benzo(b)chrysene	5	7.972	0.995	6.997	975	1.858	10.899	12.885	8.775	16.264	553.9	239.7
dibenz(a,h)anthracene	4.73	8.149	0.829	7.32	975	1.822	10.899	12.885	8.726	15.898	539.6	234.5

dibenz(a,j)anthracene	4.56	8.174	0.829	7.32	973	1.53	10.899	12.885	9.502	14.542	517.1	231.6
dibenz(a,c)anthracene	4.4	8.174	0.823	7.351	911	1.238	10.916	12.902	11.247	13.922	608.9	254.5
benzo©chrysene	4.45	8.269	0.747	7.522	931	1.519	10.916	12.874	9.342	14.193	715.3	259.6
dibenzo(c,g)phenanthrene	4.07	8.217	0.822	7.352	899	1.165	10.916	12.88	10.146	11.815	748	353.9
benzo(a)naphthacene	4.99	7.797	1.145	6.652	987	1.801	10.882	12.902	9.004	16.217	567.6	244
dibenzo(a,h)pyrene	6	7.646	1.318	6.327	1142	1.732	11.899	14.385	9.287	16.084	581.8	248.2
anthanthrene	5.08	7.618	1.33	6.287	839	1.345	10.899	13.397	9.588	12.898	480.3	211
dibenzo(de,qr)naphthacene	4.92	8.166	0.866	7.301	1113	1.288	11.974	14.385	11.161	14.375	707.7	273.1
benzo(ghi)perylene	4.76	7.94	1.071	6.869	815	1.124	10.916	13.38	10.484	11.779	480	210
benzo(b)perylene	5.04	7.819	1.144	6.675	1088	1.396	11.916	14.385	10.301	14.375	717.3	267.6
dibenzo(a,e)pyree	4.97	7.947	1.047	6.9	1082	1.241	11.916	14.385	11.207	13.904	653.5	261.1
dibenzo(a,i)pyrene	4.89	7.884	1.09	6.794	1066	1.171	11.916	14.39	11.696	13.693	827.5	291.3
dibenzo(a,i)pyrene	5.73	7.816	1.142	6.673	1142	1.732	11.899	14.385	9.291	16.089	581.9	248.3
benzo(g)chrysene	4.27	8.157	0.815	7.341	899	1.314	10.993	12.891	10.488	13.78	769.3	273.7
dibenzo(b,g)phenanthrene	4.33	7.972	0.967	7.005	942	1.373	10.899	12.891	10.13	1.909	736.8	266.6
naphtha(2,1,8-qra)naphthacene	5.87	7.692	1.264	6.429	1165	1.693	11.882	14.052	9.587	16.23	605.3	256

*In this table, RI denotes the retention indices of reverse phase liquid chromatography, determined by Sander and Wise, IP denotes the ionization potential of molecules, EA denotes the electron affinity of molecules, GAP denotes the difference of LUMO and HOMO of molecules, W denotes Wiener topological indices, L/B denotes the length/breath ratio of molecules, X_v and X_e denote edge connectivity and vertex connectivity respectively, "Width", "Length", "Volume" denote the width, length, and volume of molecules respectively, and Sarea denotes the surface area of molecules.

Table 13.4　Errors of prediction of mathematical models obtained by different algorithms

Method of computation	Mean value of absolute error
SVR for feature selection after KL transformation	0.0995
SVR without feature selection	0.112
Linear regression	0.133

Table 13.5　Errors of prediction of retention indices of 32 Alkyl-substituted polycyclic aromatic hydrocarbons

Method of computation	Mean value of absolute error
Linear regression	0.107
SVR after SVR-based feature selection	0.097
SVR after SVR-based feature selection with data file after KL transformation	0.088

13.3　Detection of Hidden Explosives

Since the large number of terrorist bomb attacks has been happened around the world in the past several years, the security check has become a very important task in the aviation baggage control and the protection of some places that are likely targets for bomb attacks. Since bomb or explosives can be easily concealed in some harmless objects. To detect the hidden explosives has become a great challenge to analytical chemists. It is very meaningful to bring forward some new methods to differentiate hidden explosives from ordinary materials quickly and accurately. It was reported that the information of element contents of N, O, C and density of materials would be useful to detect hidden explosives [70; 81; 58]. Moreover, the equipment of γ-ray resonance has been developed to determine the element contents of N, O, H and C of object in baggage immediately. Since both explosives and many harmless substances of daily life such as wool, protein-containing food and some plastics are composed of nitrogen, oxygen, carbon and hydrogen, it is necessary to find some mathematical model to differentiate commonly used explosives from harmless objects based on the contents of these four elements. Table 13.6 lists the ratio of the contents of N, O, C and H in 49 different substances, including 34 explosives (Class 1) and 15 everyday life harmless substances (Class 2). Different data processing methods, including Fisher method, KNN and support vector classification method, are used to make data fitting and prediction by LOO cross-validation method. The results of classification by different methods are also listed in Table 13.6 [94].

Table 13.6	Data and predicted results using SVC, Fisher and KNN methods for the differentiation of explosives and harmless substances*.

No.	Objects	H/N	C/N	O/N	Actual Class	T_{SVC} Class	P_{SVC} Class	T_F Class	P_F Class	KNN Class
1	2, 4, 6-Trinitroaniline	1	1.5	1.5	1	1	1	1	1	1
2	Ammonium hexanitrodiphenylamide	1	1.5	1.5	1	1	1	1	1	1
3	Hexanitrodiphenylamine	0.71	1.71	1.71	1	1	1	1	1	1
4	2, 3, 4, 6-Tetranitroaniline	0.6	1.2	1.6	1	1	1	1	1	1
5	Ammonium picrate	1.5	1.5	1.75	1	1	1	1	1	1
6	2,3,4,6-Tetranitrophenylamine	0.6	1.2	1.6	1	1	1	1	1	1
7	Cyclotrimethylenetriinitramine	1	0.5	1	1	1	1	1	1	1
8	Diazodinitrophenol	0.5	1.5	1.25	1	1	1	1	2	1
9	Hydrazine nitrate	1.6	0	1	1	1	1	1	1	1
10	Guanidine nitrate	1.5	0.25	0.75	1	1	1	1	1	1
11	Nitroguanidine	1	0.25	0.5	1	1	1	1	1	1
12	Nitrourea	1	0.33	1	1	1	1	1	1	1
13	Ammonium nitrate	2	0	1.5	1	1	1	1	1	1
14	1,3,5-Trinitrobenzene	1	2	2	1	1	1	1	1	1
15	Trinitrotoluene	1.67	2.33	2	1	1	1	1	1	1
16	Diazomethane	1	0.5	0	1	1	1	1	1	1
17	2,4,6-Trinitrophenol	1	2	2.33	1	1	1	1	1	1
18	2, 4, 6-Trinitrobenzoic Acid	1	2.33	2.67	1	1	1	1	1	1
19	2, 4, 6-Trinitroresorcinol	1	2	2.67	1	1	1	1	1	1
20	Trinitronaphthalene	1.67	3.33	2	1	1	1	1	2	1
21	Nitrostarch	2.33	2	3.67	1	1	1	1	1	1
22	Ethylene Nitrate	2	1	3	1	1	1	1	1	1
23	2,3,5- Trinitroanisole	1.67	2.33	2.33	1	1	1	1	1	1
24	2, 4, 6-Trinitroxylene	2.33	2.67	2	1	1	1	1	1	1
25	Methyl nitrate	3	1	3	1	1	1	1	1	1

26	Pentaerythrite tetranitrate	2	1.25	3	1	1	1	1	1	1
27	Tetranitronaphthalene	1	2.5	2	1	1	1	1	1	1
28	2,3,5-Trinitro-p-xylene	2.33	2.67	2	1	1	1	1	1	1
29	Nitroglycerin	1.67	1	3	1	1	1	1	2	1
30	2,4, 6-Trinitrophenylmethylnitramine	1	1.4	1.6	1	1	1	1	1	1
31	Collodion wool	4	3	4.5	1	1	1	1	2	1
32	Octogen	1	0.5	1	1	1	1	1	1	1
33	Tetraitro-1-naphthylamine	1	2	1.6	1	1	1	1	1	1
34	Diazobenzene Nitrate	1.67	2	1	1	1	1	1	1	1
35	Nylon-6	11	6	1	2	2	2	2	2	2
36	Nylon-11	11	11	1	2	2	2	2	2	2
37	Nylon-12	23	12	1	2	2	2	2	2	2
38	Peanut	14	10	3	2	2	2	2	2	2
39	Leguminoase	13	10	8	2	2	2	2	2	2
40	Melamine plastic	1	0.5	0	2	1	1	1	1	1
41	Polyimide fibre	5	11	2.5	2	2	2	2	2	2
42	Leucine	13	6	2	2	2	2	2	2	2
43	Glycine	5	2	2	2	2	2	2	2	2
44	Alanine	7	3	2	2	2	2	2	2	2
45	Polyacrylonitrile	3	3	0	2	2	2	2	2	2
46	Sheep's wool	4.8	3.3	1.1	2	2	2	2	2	2
47	Silk	4.5	3	1.2	2	2	2	2	2	1
48	Leather	4.8	3.1	1.3	2	2	2	2	2	2
49	Orlon	3	3	0	2	2	2	2	2	2

[*]The meaning of T_{SVC} class and P_{SVC} class are the class assigned by SVC in training and prediction respectively, T_F and P_F are the class assigned by Fisher method in training and prediction respectively, KNN class is the class assigned by KNN method.

Since the prediction ability of support vector machine is dependent on the selection of kernels and the parameter C. The rate of correctness of computerized prediction tested by LOO cross-validation method has been used as the criterion of the optimization of method of SVC computation. Four kinds of kernels (linear kernel, polynomial kernel of second degree, Gaussian kernel and sigmoid kernel functions) with $10 < C < 200$ are used to test for the best combination of kernel function and value of C parameter. The result of computation is expressed in Fig. 13.2. It can be seen that the use of linear kernel with C = 100 can give satisfactory results of rate of prediction (P_A).

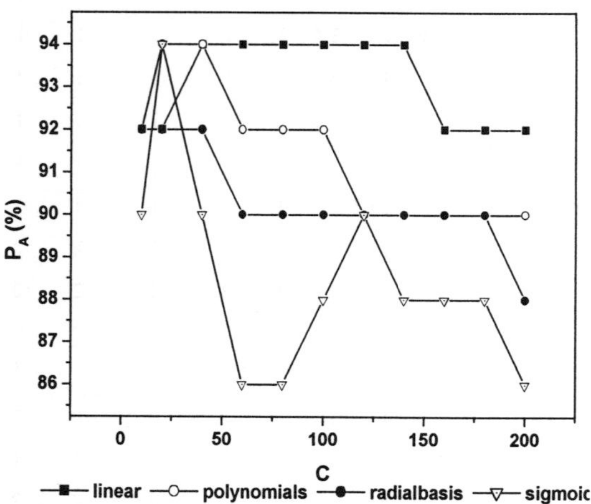

Fig. 13.2   The P_A versus C on validation set.

By using linear kernel function and C=100, the criterion to differentiate explosives from ordinary materials is found as follows:

$$1.48 - 17.8\,[\,H/N\,] - 2.00\,[\,C/N\,] + 5.66\,[\,O/N\,] > 0 \qquad (13.1)$$

the samples satisfying the above criterion are discriminated as class "1".

Table 13.7 illustrates the comparison of different methods. It can be seen that the rate of correctness of the support vector classification is better than those of two other methods.

Table 13.7   Prediction accuracy of LOO test by using different algorithms.

Algorithm	Fisher	KNN	SVM
Rate of correctness in training	98%	94%	98%
Rate of correctness in prediction	90%	94%	98%

So it can be concluded that support vector machine should be most suitable for the hidden explosive detection.

Chapter 14

SVM Applied to

Chemical and Metallurgical Technology

14.1 Physico-Chemical Basis of Modeling of Chemical Processes

It is well-known that many industrial production processes involving heat transfer, mass transfer and fluid flow can be described by a few dimensionless numbers, and therefore dimensional analysis has been widely applied to investigate these processes.

If there are chemical reactions involved in these processes, however, the application of dimensional analysis is much more difficult. Although some authors, like Damkohler and Giaconov, have been doing investigations in this direction, the progress is rather slow.

One of the difficulties in the application of dimensional analysis to chemical processes is that it is necessary to use a large number of dimensionless numbers to describe these processes. For example, according to the conclusion of Giaconov, a complicated system or process involving chemical reactions has to be described by some functions of many characteristic parameters [30; 56]:

$$X = f\left[\mathrm{Re},\ \mathrm{Pr},\ \mathrm{Pr'},\ (qC_i)/(C_p\rho\theta),\ K \cdots \right]$$

where X is the parameter describing the state of the system, K is the equilibrium constant of chemical reaction, q is the thermal effect of the process, C_i is the concentration of i-th component, C_p is the heat capacity, ρ is the density of the system , and θ is the temperature of the system. Pr

271

and Pr' are the Prandtl numbers describing heat transfer and mass transfer respectively. Re is the Reynold number describing fluid flow. Since there are many dimensionless numbers involved, it can be solved only by computerized data processing as a multivariate problem.

If we are dealing with the problem of industrial optimization or fault analysis of an existing plant, the flowsheet and equipment of this industrial process have been specified, the physico-chemical parameters such as the equilibrium constants of chemical reactions or the heat capacity of materials have been all specified, the process can be described by another series of dimensionless numbers such as the relative pressure (P/P_0), relative temperature (T/T_0), relative volume (V/V_0), and so on. Based on the above-mentioned concept, we can see that the theoretical basis of the data processing methods for industrial optimization and fault diagnosis is relevant to dimensional analysis in this respect.

14.2   Characteristics of Data Processing for Industrial Process Modeling

Industrial production of chemical and metallurgical processes accumulates a large amount of data every day. Useful information in these data sets can be extracted by data processing for twofold purposes: (1) to provide the mathematical models for the optimal control of the industrial processes, in order to realize the energy saving, yield increasing, pollution control, and production cost reduction; (2) to find the mathematical model of the fault in production for the fault diagnosis, in order to find the cause of the fault, or to find the operation condition to avoid the fault.

Compared with the more accurate laboratory data, the data in industrial records have their special characteristics [22]: (1) The data in industrial records usually have a higher noise/signal ratio. Even in a modern factory, it is still inevitable to meet many uncontrollable impacts affecting the production processes, such as the fluctuation of the composition of raw materials, the change of requirements of products, the fluctuation of environments of production processes, the impact of some accidents in production processes, and the influence of the unsteady state in the starting step and the transition processes in the process of changing from one mode to the other due to the change of production requirements; (2) The data in industrial records usually have a very non-uniform distribution, since most of the data points are concentrated within the region according to the conditions required by the operation rules, and the data points outside of this region are thinly scattered; (3) The features or variables of the data sets in industrial records are usually not independent of each other, but more or less relevant to each other. Therefore, though the number of the data sets in industrial records may be very large, they are usually low quality data sets for data processing.

In many cases, the number of industrial data may be very large, but in some special cases we can only have small data sets for modeling. For example, in fault analysis, the data set about some accident cases may be small, because accidents in production process are rare. Another example of problem of small sample size happens in petroleum refinery plants. Since crude oil is a very complex mixture, the crude oil from different oil fields or even different parts of one oil field has different composition. A large petroleum refinery consumes a batch of crude oil carried by an oil tanker within one or two weeks, and different batches of crude oil have somewhat different composition. It means that the composition of raw materials of the petroleum refinery changes every week or every two weeks. The change of raw materials will make the optimal conditions of each step (especially the upper stream unit process such as the operation of crude oil distillation tower) change in a relatively short period. If it is necessary to build a mathematical model in the early stage of these one or two weeks for the optimal control in the rest part of this period, this is just a problem of small sample size. Still another problem of small sample size is the production of diesel oil. Since the requirements of the freezing point of diesel oil are different in different seasons, the condition of operation has to be changed in every season. SVM, as a method especially useful for problems of small sample size, is suitable for the data processing tasks of this type in chemical technology.

Since it is very dangerous to put a wrong mathematical model in practice for chemical industry, we have to be particularly cautious in mathematical modeling for chemical industrial processes. On the other hand, the high noise/signal ratio sometimes may lead to some wrong conclusion in modeling process. We have to obey the following rules in our modeling work for optimization or fault diagnosis based on data processing:

(1) In order to avoid the influence of uncertainty induced by noise, it is necessary to use the *knowledge of domain experts* in this data processing. The domain experts can judge whether the mathematical model is reasonable or not, or whether it is dangerous or not. According to our experience, the creative knowledge fusion between the domain expert knowledge and the data processing results is absolutely necessary for this purpose. If we want to use data processing to solve a concrete problem in some factory, it is absolutely necessary that the domain expert must know the details of the operation and equipment of this plant.

(2) In order to depress the influence of uncertainty induced by noise, it is necessary to use *all available methods of data processing* and then make *knowledge fusion* based on the overall data processing results. As we have mentioned before, various linear projection techniques such as Fisher method, PCA and PLS methods are very useful, because linear projections can provide simpler relationships between target and features. At the same time, SVM should also be used to assure the reliability of the mathematical model obtained. If a mathematical model not only fits well the data of training set, but also gives good prediction results by leave-one-out (LOO) cross-validation test, this mathematical model should be considered as more reliable one.

(3) For the industrial optimization, the task is to find an optimal zone in the high dimensional space spanned by influencing parameters. It is not necessary to include all good sample points into the optimal zone used for optimal control, but the optimal zone must be large enough to make the control practice feasible. The optimal zone should locate far away from the distribution region of bad sample points, in order to make the optimal control more reliable. For fault diagnosis, it is also necessary to find a reliable zone to avoid the fault. Therefore, the optimal zone for industrial optimization or the safe zone to avoid fault should be selected from some subspace occupied by good sample points far away from the

bad sample region or the hyperplane of classification. So *one of the methods to make the optimal zone model obtained by SVC more reliable is to exclude the support vectors of good sample points from the optimal zone.*

(4) In industrial production, sometimes certain rough indices are used in practice, because it is impossible to make any accurate quantitative evaluation by other methods. For example, sensory evaluation plays a very important role in food science and technology. The scores obtained from panel evaluation cannot be considered as accurate data. It is reasonable to think that support vector regression with ε-insensitive loss function is just the method suitable to treat this kind of data. And if support vector regression with suitable ε-insensitive loss function can indeed give rise to a rough linear relationship between the calculated values and the actual target values of the problem involved, the upper points should be depart far away from the lower ones along the straight line in the feature space. It means that the optimal zone should be the zone occupied by some sample points near the extreme of the straight line. So *one of the methods to make the optimal zone model obtained by SVR more reliable is to exclude the good sample points far away from the extreme of the straight line obtained by SVR.*

(5) Sometimes the geometrical form of the optimal region is too complicated for data fitting, so that the difficulty happens in the modeling of industrial data. In these cases, it is usually helpful to divide the hyperspace into some subspaces, so that the geometry of optimal region will become simpler and modeling work will become easier. In our previous work, we called this strategy *local view technique.* In Vapnik's statistical learning theory, this strategy is called *local risk minimization model* [133]. This strategy can be demonstrated by Fig. 14.1. Fig. 14.1a shows that the fitting is not very satisfactory by using the polynomial function and it is necessary to use the polynomial of still higher degree to fit it (this will induce overfitting). However, if we divide this curve into two sections, the fitting will become much easier and the results will be better [133].

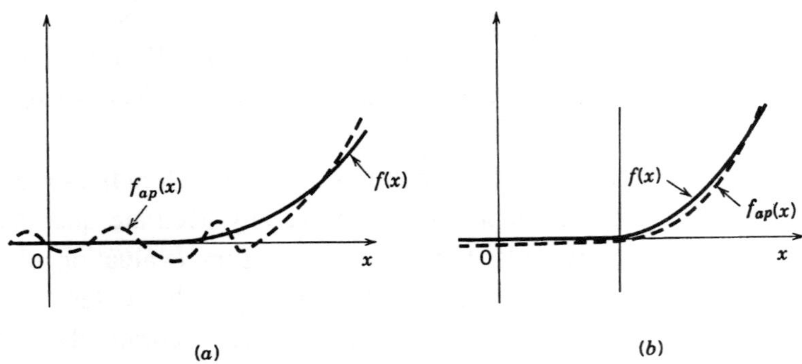

Fig. 14.1   An example to show the principle of local risk minimization.

(a) To approximate function well on interval (0, 1), one needs a polynomial of high degree.

(b) To approximate the same function on the two semi-intervals, one need only a low degree polynomials.

An interesting example can be quoted here: We were dealing with a steel making project. The task was to make a mathematical model describing the carbon content in liquid metal during the steel making process. Although the data of this industrial record were rather accurate and reliable, we still found it was rather difficult to make data modeling. But after we divided the data set into two parts: one part was the data with carbon contents higher than 0.25%, and the other part lower than 0.25%. It was found that the modeling work became much easier. Later, we have realized that the oxidation reaction in steel making

$$2C + O_2 \rightarrow 2CO \uparrow$$

exhibits different mechanism in different stages of oxidation: When the carbon content of liquid metal is higher than 0.25%, the chemical kinetics of the oxidation reaction of carbon is kinetics controlled, while after carbon content is reduced below 0.25%, the reaction becomes diffusion controlled. According to the principle of physical chemistry, different kinetics should be described by different mathematical models. And the correct point of division is just at carbon content equal to 0.25%.

This is an interesting example of the accordance between the empirical data processing and theoretical research of physical chemistry [30].

(6) Since the data sets in industrial technical records usually have the higher noise/signal ratio, sometimes the elimination of outliers is necessary. In the field of data processing, the definition of outlier is a confused concept. Some authors defined all sample points deviated from linear relation as outliers. This is of course not suitable for the data processing of the nonlinear data sets. A more reasonable method for the outlier elimination of complicated data set is based on KNN method. If the class of a sample point is different from the class predicted by its nearest neighbors, it will be considered as an outlier.

Another more reliable method of outlier elimination is based on SVM. If a sample point is misclassified in LOO cross-validation test by using several kinds of kernel functions, it can be eliminated to improve the classification. Figure 14.2 shows an example of the result of outlier elimination by using this method. In this example, a data file about the recovery of propylene in a petrochemical factory is used for the optimization of propylene production. The classification of samples of two classes becomes clear-cut after the elimination of the sample points misclassified in LOO cross-validation test with several kinds of kernel functions in computation.

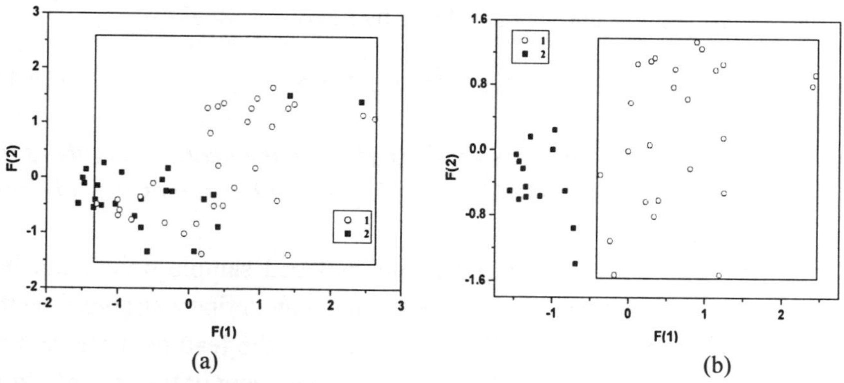

Fig. 14.2　Result of outlier elimination by support vector classification.

(a) Projection map of data structure before outlier elimination.

(b) Projection map of data structure after outlier elimination.

14.3 Optimal Zone: Strategy of Large Margin Search

The purpose of industrial optimization is to improve the production process by optimal control, that is to achieve good product quality, high rate of recovery, low energy and raw materials consumption, low pollution and low production cost, etc. Since these targets are usually determined by many factors simultaneously, multivariate analysis has to be used to make mathematic modeling of an *optimal zone* in hyperspace spanned by operation parameters.

One of the purposes of fault diagnosis is also to find an optimal zone in the high-dimensional space spanned by operation parameters in order to avoid the occurrence of fault. This is also usually a multivariate problem.

If the "good sample points" and the "bad sample points" can be separated by an optimal hyperplane determined by SVC in feature space described by kernel function, the good sample points should be divided into two categories: the support vectors and the others. Since the support vectors and the small number of misclassified sample points are relatively close to the region of bad sample points, it is reasonable to exclude them from the optimal zone to keep the bad sample points far away. And *the good sample points distributed behind the support vectors having larger margins γ_i should be considered as the members in the optimal zone*. Here γ_i can be calculated by (2.6):

$$\gamma_i = y_i \left(\left\langle \mathbf{w} \cdot \mathbf{x}_i \right\rangle + b \right) \tag{14.1}$$

If some sample points with large values of γ_i are the nearest neighbors to each other in a unified zone, this unified zone can be used as the zone for optimal control.

Similarly, if the good sample points and bad sample points can be mapped into a feature space and SVR method can define a straight line to describe the target values of all sample points in the feature space, and *if the good points located near the upper (or lower) end of the straight line are the nearest neighbors to each other in a unified zone of the input space, this zone can also be used as the zone for optimal control.*

So we have two methods to find the optimal zone from operation data records. These methods can be illustrated in Fig. 14.3 and Fig. 14.4.

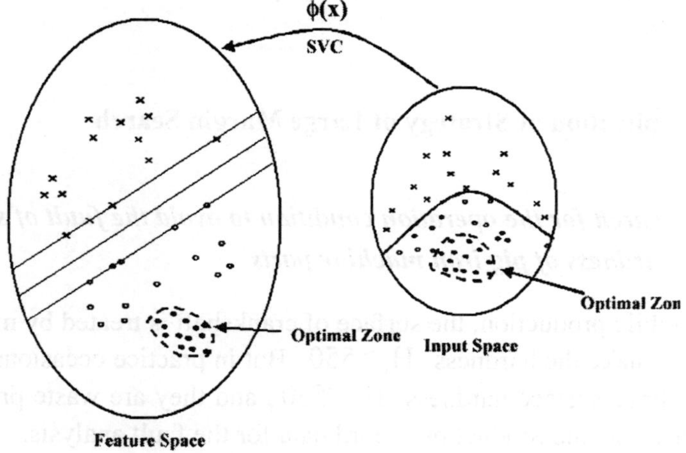

Fig. 14.3　Strategy for searching optimal zone by support vector classification.

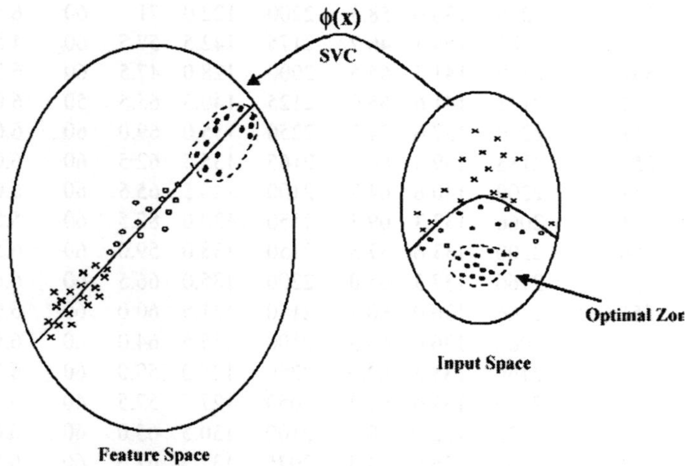

Fig. 14.4　Strategy for searching optimal zone by support vector regression.

We call these methods *"strategy of large margin search" for optimal zone modeling*. As we will see in the following paragraphs, these strategies are rather useful for industrial optimization or fault analysis in production.

14.4 Application of Strategy of Large Margin Search

14.4.1 *Search for the operation condition to avoid the fault of surface hardness of pig iron machine parts*

In automobile production, the surface of crankshaft is treated by nitriding process to make the hardness $H_v > 550$. But in practice occasionally the products have surface hardness $H_v < 550$, and they are waste products. Table 14.1 lists the production record data for the fault analysis.

Table 14.1   Records of nitriding production practice.

No	Hardness	X_1^*	X_2^*	X_3^*	X_4^*	X_5^*	X_6^*	X_7^*	X_8^*
1	550	1900	113.3	55.0	1950	146.0	50	40	5.70
2	551	2250	136.6	58.3	2200	122.0	71	60	6.50
3	556	2183	169.3	46.7	2175	142.5	57.5	60	4.50
4	551	2150	141.3	65.6	2000	128.0	47.5	60	6.30
5	551	2133	141.6	65.0	2125	140.5	65.5	50	6.00
6	551	2250	137.6	74.7	2250	135.0	69.0	60	6.00
7	551	2183	149.3	60.7	2165	135.5	62.5	60	6.00
8	551	2200	140.6	64.7	2100	140.5	65.5	60	6.00
9	551	2066	133.3	69.3	2150	130.0	87.5	60	5.80
10	556	2200	133.6	57.3	2150	138.0	59.0	60	6.50
11	551	2166	137.3	65.0	2200	135.0	66.5	60	6.00
12	551	2166	136.0	60.7	2150	131.5	60.0	60	6.00
13	551	2066	136.0	63.3	2100	135.5	64.0	60	6.00
14	551	2166	135.6	62.0	2250	125.0	59.0	60	6.00
15	534	2100	131.6	61.7	2050	127.5	57.5	40	5.70
16	551	2133	132.0	60.7	2100	130.5	65.0	60	6.00
17	551	2066	136.0	64.3	2075	137.5	62.5	60	6.50
18	550	2183	130.0	56.7	2100	131.0	60.0	60	4.50
19	551	2083	133.3	66.7	2100	135.0	65.0	60	6.00
20	551	2216	135.6	66.7	2100	135.5	67.5	60	6.00
21	551	2050	128.3	58.3	2100	137.5	67.5	60	6.00
22	551	2233	126.6	67.3	2200	127.5	57.5	60	6.00

23	500	2000	130.0	71.0	2100	127.5	60.0	60	4.50
24	551	2066	126.7	61.7	1950	122.5	62.5	60	6.00
25	551	2133	137.7	63.3	2050	130.0	50.0	60	6.00
26	508	2033	131.7	45.0	2050	137.5	52.5	60	6.00
27	534	2516	137.0	63.0	2475	134.0	62.0	60	4.80
28	551	2033	136.7	63.0	2050	137.5	64.0	60	6.00
29	551	2333	136.0	60.7	2250	131.5	55.0	60	6.00
30	551	2166	136.0	64.3	2075	132.5	62.0	60	4.50
31	515	2100	140.7	75.0	2125	137.5	67.5	60	4.60
32	551	2166	123.3	66.7	2100	130.0	65.0	60	6.00
33	500	2033	135.0	63.3	2075	135.0	62.0	60	6.00
34	551	2133	137.0	62.3	2150	122.5	62.5	60	6.00
35	551	2466	130.3	76.7	2475	131.0	70.0	60	4.50
36	525	2033	139.3	68.0	2100	135.5	65.5	60	5.30
37	561	2000	136.0	67.3	2125	129.0	63.5	60	4.50
38	551	2150	128.6	69.3	2075	129.0	67.5	60	6.00
39	515	2083	128.3	68.0	2150	130.5	66.5	60	4.50
40	551	2233	136.0	67.7	2150	135.5	65.0	60	6.00

*In this table, X_1 and X_2 denote the quantities of ammonia and alcohol consumed in the first stage of operation respectively, X_3 denotes the averaged value of furnace pressure in the first stage of operation. X_4 and X_5 denote the quantities of ammonia and alcohol in the second stage of operation respectively, X_6 denotes the averaged furnace pressure in the second stage of operation. X_7 is the time of ventilation, and X_8 is the total time of operation.

By using SVC and kernel function of second degree, the samples with $H_v > 550$ and those with $H_v < 550$ can be separated completely. After elimination of the support vectors in the "good samples", it can be found that a connected region of good samples has the following range of operation parameters:

$$2130 < X_1 < 2250$$
$$128 < X_2 < 150$$
$$61 < X_3 < 73$$
$$2100 < X_4 < 2260$$
$$128 < X_5 < 135$$
$$X_6 > 58$$
$$X_7 = 60$$
$$5.7 < X_8 < 6.0$$

The rate of waste products has been greatly depressed by keeping this condition of production.

14.4.2 *Searching the operation conditions to avoid the fault of quenching stain in tin-plate production*

Tinplate is produced by the following process: At the first step tin is electroplated onto the surface of steel plate; then the electroplated steel plate is heated by electric heating up to the melting point of tin. After molten tin becomes a uniform coating on the surface of the steel plate. The steel plate is quenched into a water bath. Then the tin-coated steel plate (tinplate) is obtained. One of the defects of tinplate is so-called *quench stain*. It looks like some dark dirty stain on the surface of the tinplate. We have cooperated with Bao Steel Co., the largest tinplate producer in China, to find the mechanism of quench stain formation and the method to avoid the quench stain formation on tinplate. In this work, both the experimental study and the data processing (including the use of SVM) have been carried out. In experimental work, the morphology of quenching stains of tinplates has been studied by scanning electron microscope. It has been found that the quenching stains are assemblies of spots of 0.01–0.1 mm size with uneven tin surfaces. It is the light scattering on uneven surfaces that makes the dark appearance of quench stains. By modeling experiments with high speed photography, it has been found that the strong disturbance induced by water boiling with unsteady transition mode causes the solidified tin surface to be uneven. According to the theory of heat transmission from a hot surface to the boiling liquids, there are two steady modes of heat transmission: with small temperature difference the boiling is initiated by vapor bubble nucleation, while with large temperature difference the hot surface is completely covered by the film of vapor and the heat is transferred through the vapor film by conduction and radiation. Between these two steady modes there is a transition mode in which the vapor film is not stable. The formation and collapse of the vapor film can induce small vigorous disturbance or oscillation. It is clear that how to avoid the unsteady mode of water boiling on the surface of molten tin in quenching process is crucial for the elimination of quench stain.

In order to find the concrete condition to avoid the formation of quench stain, SVR has been used to process the data of production process. In production practice, the tinplate products with quenching stains are classified to five classes according to the seriousness by inspection of the appearance. This is only a semi-quantitative index to describe the degree of fault of products. But it is understandable that such kind of data set may be treated by SVR with ε-insensitive loss function with good results. Figure 14.6 illustrates the results of computation. It can be seen the regularity is rather clear.

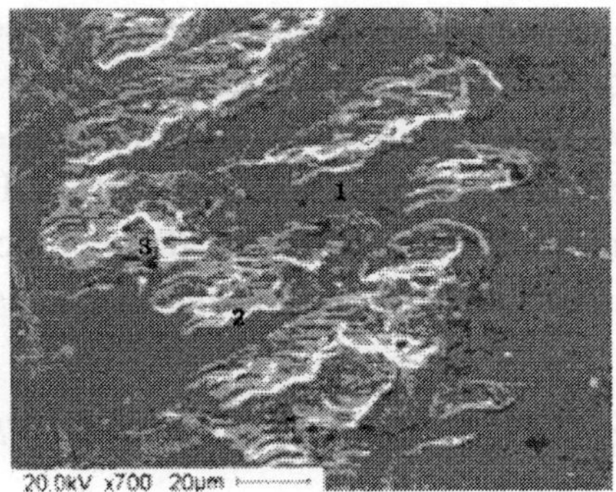

Fig. 14.5　Photograph of quenching stain ($\times$3500) by scanning electron microscope.

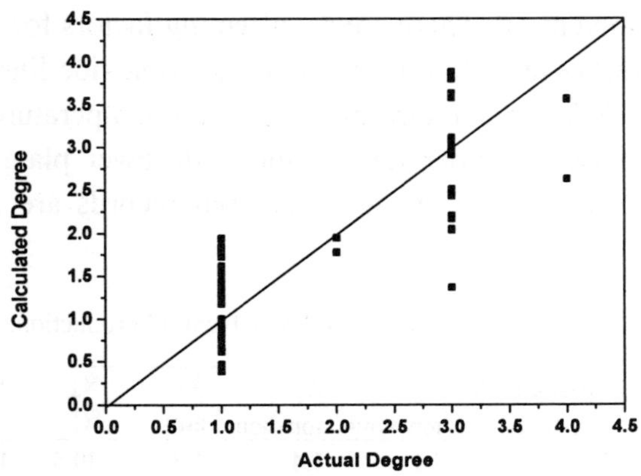

Fig. 14.6　Result of support vector regression of the industrial data of quenching stain formation. Kernel used: Gausian type with $\varepsilon = 0.15$.

By SVR with linear kernel, it is possible to find the chief factors affecting the quenching stain formation from multiple correlation coefficients. It has been found (under the concrete condition of this production process) that to depress the temperature of water in quenching bath or to increase the speed of movement of steel plate in production can avoid the formation of quenching stain. Under these conditions, the formation of quenching stain indeed can be completely avoided in production practice.

14.5 Optimal Control for Target Maximization or Minimization

In many cases of industrial optimization, it is required to control the target value of optimization to be maximized value (for yield of recovery, product quality, productivity, etc.) or minimized value (for energy consumption, pollution, production cost, etc.). In order to find the mathematical model of optimal zone for these purposes, the *large margin search* strategy, mentioned above for fault diagnosis, is also useful.

14.5.1 *SVM applied to optimization of solvent oil recovery*

Solvent oil is a by-product of the platinum reforming plant of petrochemical factories. The target of optimization is to increase the recovery of solvent oil. There are six affecting factors for the recovery of solvent oil: reflux rate (X_1), flow rate of the first side line (X_2), pressure at tower top (X_3), temperature of reflux (X_4), temperature at solvent oil tower bottom (X_5), temperature of the 35th-tower plate of solvent oil tower (X_6). Some data from the industrial records are listed in Table 14.2.

Table 14. 2   Data record of solvent oil production.

No.	Class	Productivity ton·hour^{-1}	X_1 ton·hour^{-1}	X_2 ton·hour^{-1}	X_3 atm	X_4 °C	X_5 °C	X_6 °C
1	1	3.6	7.1	1.4	1.3	30.5	181.5	91
2	1	3.8	7.5	1.4	1.2	31	183.5	90
3	1	3.8	7.3	1.4	1.2	25	182	93

4	1	3.8	7.2	1.4	1.2	25.5	180	95.5
5	1	3.7	7.5	1.4	1.2	25.5	179	96
6	1	3.7	7.6	1.4	1.2	26	171	94
7	1	3.6	7	1.4	1.2	23	173.5	93.5
8	1	3.6	7	1.4	1.2	23	177.5	95.5
9	1	3.6	7	1.4	1.2	23	179	96
10	1	3.6	6	1.4	1.3	26	186	95.5
11	1	3.8	7.1	1.4	1.3	37	170.5	101
12	1	3.6	6.4	1.4	1.3	34.5	175.5	101.5
13	1	3.6	6.4	1.4	1.3	33.5	176.5	101.5
14	1	3.6	6.4	1.4	1.3	33.5	175	101
15	1	3.6	7	1.4	1.3	33.5	170	98
16	2	3.4	8.2	1.3	1.2	33.5	158.5	97
17	2	3.4	8.4	1.3	1.2	33.5	158	96.5
18	2	3.4	8.4	1.3	1.2	36.5	156	97
19	2	3.4	8.3	1.3	1.2	37	154	97.5
20	2	3.4	8.2	1.3	1.2	37.5	157	97
21	2	3.4	8.2	1.3	1.2	35.5	158.5	97.5
22	2	3.4	8.5	1.3	1.2	34.5	157.5	97.5
23	2	3.4	8.1	1.3	1.2	34.5	158	96.5
24	2	3.4	7.6	1.3	1.3	34.5	157	96.5
25	2	3.4	7	1.3	1.3	36	156.5	99.5
26	2	3.4	7.8	1.4	1.4	26	176.5	94
27	2	3.4	7.9	1.4	1.3	25.5	176	94.5
28	2	3.4	8	1.4	1.4	25.5	174	94.5
29	2	3.4	8	1.4	1.4	26.5	174.5	97
30	1	3.8	8	1.4	1.3	28.5	160.5	101
31	2	3.2	7	1.3	1.2	38	154	98.5
32	2	3.2	7	1.3	1.2	36	152.5	98.5
33	2	3.2	7.8	1.3	1.2	33	152	98.5
34	2	3.2	8	1.3	1.2	34	151	98
35	2	3.2	7.8	1.3	1.2	34.5	151.5	97
36	2	3.2	8	1.3	1.2	36.5	154	98
37	2	3.2	8.1	1.3	1.2	35.5	157.5	98
38	2	3.2	8.3	1.3	1.2	35	158	98.5
39	2	3.2	8.8	1.3	1.3	36	157.5	98.5
40	2	3.2	8.8	1.3	1.3	35.5	158	99
41	2	3.2	7.3	1.3	1.3	34.5	158.5	98.5
42	2	3.2	6.3	1.3	1.2	33.5	159	96.5

43	2	3.2	7.2	1.3	1.3	35	159	68.5
44	2	3.2	6.8	1.3	1.3	33.5	158.5	96
45	2	3.2	7.8	1.3	1.2	34	158.5	96
46	1	3.5	8	1.3	1.3	36	160	99
47	1	3.5	7.6	1.3	1.3	35.5	158	99
48	1	3.5	7.3	1.3	1.3	36.5	157	98
49	2	3.3	7.1	1.3	1.2	36.5	162	99
50	2	3.3	7.2	1.3	1.2	36.5	160	98
51	2	3.3	7.1	1.3	1.3	40	160	101
52	2	3.3	7.2	1.3	1.3	41	160	101
53	2	3.3	8.1	1.3	1.3	40.5	156	100
54	2	3.3	7.9	1.3	1.5	37.5	157	100
55	1	3.5	7.8	1.3	1.4	38	158.5	100.5
56	1	3.5	7.5	1.3	1.4	38	159	100
57	1	3.5	7.6	1.3	1.4	38	159	100
58	1	3.5	6.9	1.3	1.3	38	163	103
59	2	3.3	7.8	1.3	1.4	40.5	161.5	98.5
60	2	3.3	7.9	1.3	1.3	40.5	159.5	100.5
61	2	3.3	7.5	1.3	1.3	40	158	100
62	2	3.3	7.6	1.3	1.3	40	158	100
63	2	3.3	8	1.3	1.3	42	153.5	98.5
64	2	3.3	6.7	1.3	1.2	37	156	100.5
65	2	3.3	6.6	1.3	1.2	35	158	102
66	2	3.3	6.9	1.3	1.3	34.5	158	103
67	2	3.3	6.9	1.3	1.5	35	158	102.5
68	2	3.3	7.5	1.3	1.4	36	157	100
69	2	3.4	7.4	1.4	1.2	26.5	169	94
70	1	3.5	7	1.4	1.2	25	166.5	92
71	1	3.6	8.5	1.4	1.3	26	178	90.5
72	1	3.6	6.4	1.4	1.3	26	182	92
73	1	3.6	6.3	1.4	1.3	26	181.5	92

In the cases listed in this table, the feed rate is kept unchanged, so the productivity of solvent oil is proportional to the recovery of solvent oil.

By SVR-based feature selection, it can be shown that the most important factors are X_5, X_6 and X_2. Since X_2 is determined by the productivity requirement of other products, the temperature control of lower part of solvent oil tower should be the chief factor affecting the recovery of solvent oil. At the same time, it has been found that the relationship between the productivity of solvent oil and the influencing factors can be expressed by SVR with a kernel function of second degree. By SVR the calculated values and actual values of productivity are compared in Fig. 14.7.

At the upper end of the straight line in Fig. 14.7 there are sample points of high recovery. We have found by KNN method that many sample points of this kind are nearest neighbors in a unified zone with the following boundaries:

$$6.4 \leq X_1 \leq 8.5$$
$$1.2 \leq X_3 \leq 1.3$$
$$25 \leq X_4 \leq 34.5$$
$$171 \leq X_5 \leq 183.5$$
$$90 \leq X_6 \leq 101.5 \quad \text{with } X_2 \text{ near } 1.4$$

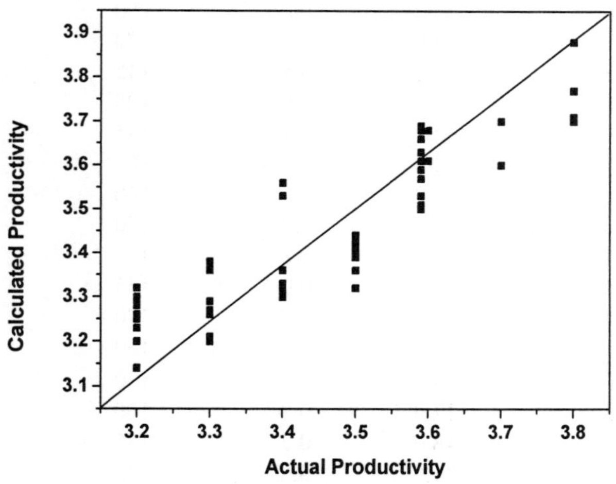

Fig. 14.7   The comparison between calculated and actual values of productivity of solvent oil ($\mathrm{t \cdot hr^{-1}}$).

By keeping the operation conditions within this range, the productivity and recovery of solvent oil can be increased significantly.

14.5.2 *SVM applied to chromium electroplating*

One of the difficulties of chromium electroplating of machine parts with complex shapes is the incomplete coverage of the surface by chromium layer. The rate of coverage is dependent on the operation conditions of electroplating. A standard method for determination of the optimal conditions is to make electroplating on a bended cathode and measure the uncovered area of electroplating result. Table 14.3 lists some data of the experimental results for the electroplating on a standard bended cathode, for the comparison of the covering ability of electroplating by different conditions.

Table 14.3   Data set for mathematical modeling of chromium plating.

No. of sample	X_1 (Temperature) °C	X_2 (Current density) $A \cdot dm^{-2}$	X_3 (H_2SO_4 concentration) $G \cdot L^{-1}$	X_4 (CrO_3 concentration) $G \cdot L^{-1}$	Area not covered by chromium layer*
1	18	6.0	0.6	160	63
2	18	12.0	0.7	170	14
3	18	18.0	0.8	145	32
4	18	21.6	0.9	185	28
5	26	6.0	0.7	145	78
6	26	12.0	0.6	185	43
7	26	18.0	0.9	160	37
8	26	21.6	0.8	170	40
9	34	6.0	0.8	185	120
10	34	12.0	0.9	145	74
11	34	18.0	0.6	170	42
12	34	21.6	0.7	160	120
13	42	6.0	0.9	170	120
14	42	12.0	0.8	160	41
15	42	18.0	0.7	185	51
16	42	21.6	0.6	145	44
17	18	12.0	0.1	160	1
18	18	15.0	0.3	175	11
19	18	18.0	0.5	190	1
20	23	12.0	0.3	190	14
21	23	15.0	0.5	160	2
22	23	18.0	0.1	175	1
23	28	12.0	0.5	175	26
24	28	15.0	0.1	190	5
25	28	18.0	0.3	160	17

26	18	12.0	0.6	160	28
27	18	15.0	0.8	175	10
28	18	18.0	1.0	190	14
29	23	12.0	0.8	190	29
30	23	15.0	1.0	160	24
31	23	18.0	0.6	175	15
32	28	12.0	1.0	175	31
33	28	15.0	0.6	190	21
34	28	18.0	0.8	160	24

*The uncovered area is counted by the units of 5×5 mm squares on a 20×100mm bended copper foil used as cathode.

If we assign the samples with target values smaller than 30 as class "1", and the rest of the sample points as class "2". By SVM computation , it can be found that the samples of number 17, 18, 19, 20, 22 and 24 are located in the region of "good samples" far away from the optimal hyperplane. And these sample points are located in an optimal zone defined by the following inequalities in the hyperspace spanned by the four features listed in Table 14.3:

$$18 \leq X_1 \leq 28$$
$$12 \leq X_2 \leq 18$$
$$0.1 \leq X_3 \leq 0.5$$
$$160 \leq X_4 \leq 190$$
$$\text{and } X_4 > 3.344X_1 + 89.7$$

14.6   Optimal Control for Problem of Restricted Response

Besides the optimal problems mentioned above, there is another type of problems of industrial optimization: the problems of restricted response. In these problems, it is necessary to control the target values within certain narrow range. Both the samples having too high or too low target values are considered as bad samples. For example, in steel production the chemical compositions (such as carbon content) of steel product must be controlled within some narrow range. In some machine part production, the surface hardness should not be kept too low. Too high surface hardness is also not good since it would be difficult for machining. Since the control of target values within a narrow range is usually a multivariate problem, it is necessary to make some mathematical model to solve this problem.

Theoretically speaking, the region with too high target values and that with too low target values should be separated by some curved hypersurface in the high dimensional space representing the working states of production process, and the sample points near this hypersurface are just the optimal sample points. Because the conditions of the production processes are usually bounded within a narrow range, sometimes this curved hypersurface can be approximated as a hyperplane. So we can use a *hyperplane model* for the optimal control. In order to build this hyperplane model, we have several methods. One method is based on a modified PCA method. We call it "method of PCA projection with good sample points". The other method is to make mathematical model by SVR technique. Here two examples of application are used to demonstrate these two methods.

14.6.1 *SVM applied to optimization of ML values in butadiene rubber production*

ML value is an index of the mechanical property of synthetic rubber. It is related to the averaged molecular weight of synthetic rubber products. In butadiene rubber production, the ML value should be controlled within 43 to 47. Too high ML values correspond to insufficiency of elasticity, while too low ML values correspond to low tensile strength of products. Table 14.4 illustrates some data of production records about the factors affecting ML values in production. It has been found that there are five chief factors influencing the ML values in production: X_1 (feed rate of butadiene monomer), X_2 (feed rate of solvent oil), X_3 (feed rate of catalyst), X_4 (temperature of feed), X_5 (temperature of the lower part of the first reactor).

Table 14.4   The chief factors affecting ML value of butadiene rubber.

Sample No.	X_1	X_2	X_3	X_4	X_5	ML value
1	67	83	22.5	6	86	46.3
2	67.5	83	22	2	83	46.5
3	66.5	84	24.5	3	88	44.7
4	67	84	21	6	83	45.2
5	67	85	21	6	84	46.5
6	52	64	19	-2	86	43.5
7	67.5	64	23	4	85	44.5
8	50	64.5	18	2	86	45.8
9	68	85	23	0	83	43.3
10	68	84	24	0	84	44.8
11	68	84	22	-1	81	44.1
12	67	85	26.5	5	90	43.7
13	67	87	28	1	91	46.2
14	67	87	24	3	85	45.6
15	50	87	19.5	-1	82	44.3
16	75	90	31	1	96	46
17	75	92	30	1	90	45.3
18	70	85	25	-1	83	44.2
19	70	84	28	2	90	43.4
20	70	85	25.5	-1	86	46.1
21	70	84.5	25	-2	84	44.8
22	67	80	26	0	90	45.0
23	67	80	24	0	87	43.7
24	50	59	16	-1	82	44.6
25	50	59.5	17.5	-3	84	43.3
26	50	59.5	18	-2	86	46.0
27	65	75	24	-2	88	46.7
28	67	76.5	26	-2	87	43.5
29	67	77.5	21.5	0	83	45.3
30	67	77	22	0	84	45.8
31	58	72	18	4	82	44.2
32	58	77	18.5	0	81	43.4
33	75	94.5	24.5	4	83	46
34	50.0	59.0	27.0	4	82	49.8
35	70.0	96.0	12.0	-2	98	41.0
36	50.9	67.6	30.9	5	90.8	54.6
37	50.7	70.3	26.8	0	87.8	54.0
38	51.3	74.6	16.2	6	88.1	42.4
39	52.4	89.3	16.7	4	88.6	42.4
40	54.0	98.0	12.0	1	97.0	47.0
41	52.9	88.7	20.0	3	88.4	41.3
42	72.0	98.6	11.7	-3	99.8	40.7

43	46.9	65.8	36.0	7	91.3	48.0
44	47.8	66.0	36.6	7	90.8	51.0
45	72.0	99.9	9.8	0	97.0	42.8
46	71.0	98.8	9.9	0	96.8	38.9
47	71.6	99.0	9.8	1	98.0	39.2
48	73.0	97.2	8.9	1	99.5	38.5
49	76.8	87.9	11.0	7	96.8	38.9
50	46.8	64.0	38.8	8	90.6	53.9
51	47.9	67.0	38.2	7	87.6	54.2
52	42.0	58.8	40.2	7	90.5	56.0
53	40.9	59.8	36.0	7	90.5	55.2
54	42.0	49.0	40.4	7	90.0	54.4
55	41.2	51.2	40.3	7	91.3	53.8
56	41.1	53.2	40.2	7	89.9	50.2
57	40.9	51.1	39.8	0	91.4	56.0
58	70.9	99.0	8.8	0	99.3	38.8
59	71.9	99.3	8.7	1	98.6	38.9
60	72.3	98.6	8.9	1	99.0	38.4

By the method of PCA projection with good samples, we classify the good samples (with 43<ML values <47) as class "1", and assign the samples having ML values ≥ 47 or ML values ≤ 43 (as "bad" samples) to be class "0" (in our software, it means that these sample points are not used as training points). Then we make KL transformation, and use some principle components as the coordinates of the projection maps. One of the maps is shown in Fig. 14.8. It can be seen that the "good" sample points chiefly distributed in a narrow band in this map. It means that this narrow band is a projection of a thin "plate" in the hyperspace, and a hyperplane expressed by the following equation in the middle of thin plate can be used as the mathematical model for the optimal control of ML value in production:

$$11.388 - 0.0652X_1 - 0.0086X_2 + 0.2061X_3 + 0.0497X_4 - 0.132X_5 = 0$$

Although the result of the example mentioned above is rather satisfactory, the PCA projection method, as a linear projection method, cannot be used for the problems having significant nonlinearity. The following section will consider an example of such kind.

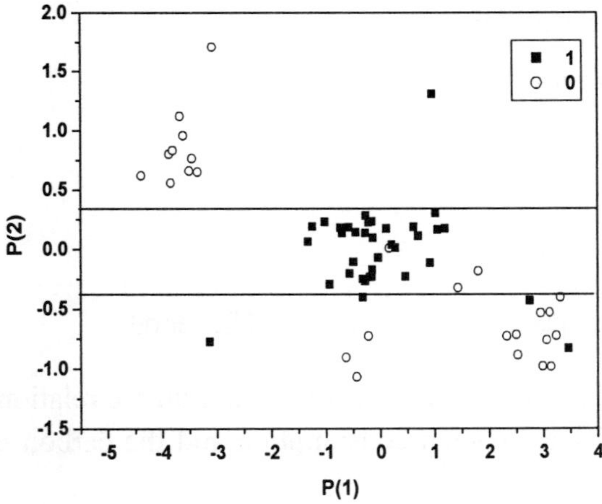

Fig. 14.8   Optimal region in projection map by principle component analysis.

14.6.2   *SVM applied to modeling for the carbon content control in steel making*

Carbon content control is crucial for the optimal control in steel making. The traditional method uses quick chemical analysis to determine the end point of a batch in steel making. This method is necessary to delay the time of one batch of steel making. In order to save the delayed time and make the steel making process smooth, it is desirable to use a mathematical model for the end point prediction in steel making.

Table 14.5 lists a data set of steel making. The factors influencing the carbon content of final products are the following 13 features:

X_1 life of oxygen blower

X_2 height of oxygen blower

X_3 oxygen consumption before the middle period of steel making

X_4 carbon content of liquid iron before the middle period of steel making

X_5 temperature of liquid iron at the middle period

X_6 quantity of lime added

X_7 quantity of magnesium oxide

294   *Support Vector Machine in Chemistry*

X_8 total weight of materials

X_9 quantity of pig iron

X_{10} quantity of iron ore

X_{11} quantity of scrap

X_{12} life of ladle

X_{13} quantity of oxygen consumption after middle period

The task of mathematical modeling is to find the relationship between the total quantity of oxygen consumption and the carbon content in the final products.

Table 14.5   Factors affecting the carbon content in liquid metal of steel making.

Kind	C^*	X_1	X_2	X_3	X_4	X_5	X_6	X_7	X_8	X_9	X_{10}	X_{11}	X_{12}	X_{13}
2	64.8	108	12.06	114.6	101.5	1635	15	2.9	10	317.7	292.2	25.5	292	450
2	64.4	134	11.9	108.6	96.7	1584	19.1	3.2	13.8	308	289.8	18.2	424	457.5
2	60.8	190	11.7	116.6	88.6	1645	14.1	3	8.5	317.4	291.6	25.8	374	457.5
2	59.3	113	11.83	113	89.2	1628	16.1	4.1	8.4	308.2	283.4	24.8	403	452.5
2	57.2	126	11.93	112.6	109.7	1627	19.1	3.8	9.4	319.8	293.8	26	310	483.8
2	51.4	113	11.8	114	90.5	1631	17	4	6.4	315.7	285.6	30.1	785	448.8
2	50.4	193	11.84	117.4	71.5	1603	16.9	4	9.5	309.8	284.9	24.9	865	447.5
2	47.6	157	11.92	115.8	82.7	1638	16.1	4.1	9.8	317.1	291.4	25.7	105	448.8
2	45.4	198	12.02	117	98.5	1646	19	4	12.7	316.4	297.6	18.8	146	490
2	43.9	151	11.8	118	62.5	1656	14.7	5.7	12.9	308.1	283.4	24.7	606	456.3
2	42.8	173	11.7	131.4	46.2	1619	13.9	5	9.6	315.2	287.2	28	628	478.8
2	40.2	110	12	117.8	85.5	1612	14.9	5	6	310	285.2	24.8	565	451.3
2	36.8	123	11.85	119.8	66	1642	16.7	4	9	311	286.3	24.7	578	477.5
2	32.2	153	11.93	126.2	53	1632	18	5.3	11	314.8	289.2	25.6	48	452.5
2	28	118	11.86	115.8	57.2	1589	14.8	3.1	15.8	312.7	284.2	28.5	302	478.8
2	27.7	124	11.93	119.8	46.6	1608	14.9	5	10.4	309	280.9	28.1	308	487.5
2	27.6	141	11.82	123	59.5	1640	17.9	3.9	8.5	315.2	286.8	28.4	325	475
2	25.2	153	11.98	122.2	46.8	1594	13.9	2.9	14.8	317.2	288.5	28.7	337	490
2	25	112	11.99	120.4	72.4	1601	16.1	5	8.4	308.4	289.4	19	60	450

2	24.8	118	11.85	120.2	54.9	1581	15	4.6	12.2	315.8	287	28.8	573	477.5
2	23.1	153	11.73	115.8	67.1	1617	13.9	3	13.6	310	278.6	31.4	443	452.5
2	22.3	186	12.09	119.2	64.3	1583	16	2.9	10.6	312.3	284.2	28.1	21	486.3
2	22	148	12.08	116.2	63.8	1612	15	4.9	14.9	310.6	282.8	27.8	332	448.8
2	21.4	165	11.97	121.4	45.7	1608	14.4	3	11.5	313.7	285.3	28.4	349	460
2	21.2	127	11.93	121.8	59.6	1614	16	4.5	10.5	315.4	286.4	29	311	483.8
2	20.7	139	11.82	116.6	59.5	1585	15.9	6.4	13	310.9	276.5	34.4	323	461.3
2	20.6	143	11.82	115.8	47.4	1594	14.9	5	20.4	339.2	310.8	28.4	327	481.3
2	20	197	11.77	130	64	1685	15	7	6.8	312.1	283.9	28.2	652	471.3
2	19.8	196	12.02	118.2	48.3	1610	14.9	7	17.1	311	282.7	28.3	144	456.3
2	19.2	161	11.8	116.2	76.3	1572	12.9	4	12.2	311.5	283.7	27.8	616	482.5
2	19	187	12.09	122.2	50	1606	16.1	2.9	11	320.7	291.6	29.1	22	486.3
2	18.5	159	11.97	119.8	41.4	1598	17.5	2.9	13.1	309.4	281.5	27.9	343	485
2	17.7	136	11.84	119.2	53	1577	15.5	4.9	15.2	312.4	283.9	28.5	808	498.8
2	17.6	173	11.99	112.6	91.6	1579	14.9	5.9	7.6	316.4	291.1	25.3	357	450
2	17.5	199	12	117.4	60.2	1626	13.5	5	14.5	312.7	284.3	28.4	34	490
2	17	144	11.92	122	49.4	1615	17	3	18.4	343.9	324.3	19.6	328	447.5
2	16.9	186	11.74	125.2	41.2	1594	14.3	4.9	14.5	313.9	285.6	28.3	858	478.8
2	16.8	107	11.86	119.4	47.5	1604	15.3	4.9	17.3	312	292.9	19.1	291	468.8
2	16.4	180	11.77	119.6	56	1574	13.4	6.1	16.4	315.5	287.5	28	635	481.3
2	16	100	11.98	120	49.9	1594	15.2	5.5	13.2	309.5	280.3	29.2	48	453.8
2	15.6	131	12.02	119	53	1578	14	4.9	15.8	309.1	281	28.1	79	482.5
2	15.5	181	11.69	125.2	45.3	1642	18	2.9	14.7	314.6	289.3	25.3	16	456.3
2	15.1	128	11.8	121.4	41.7	1562	15.8	2.9	16.9	312.4	284.1	28.3	800	485
2	14.8	119	11.74	118.8	44.6	1605	13.4	5	15.3	310.8	282.5	28.3	409	491.3
2	14.8	138	11.82	119.4	43.8	1599	15.6	5.2	10.4	310.2	276	34.2	322	463.8
2	14.6	117	11.91	122.2	42	1602	13	4	11.8	312.6	283.9	28.7	65	475
2	14.5	150	11.92	116.8	43.7	1600	15.1	4	19.9	310.7	282.4	28.3	98	468.8
2	14.4	125	11.9	122.6	48.1	1592	15.1	8	13.9	312.1	284.2	27.9	797	488.8
2	14.3	113	11.86	116.8	43.4	1594	13.9	2.9	14.3	309.5	281.2	28.3	297	488.8
2	14	116	11.86	112.6	53.6	1575	17	3	17.6	309.4	281.5	27.9	300	481.3
2	13.7	139	11.72	118	56.4	1607	12.1	5	13.5	310.5	282.7	27.8	594	491.3
2	13.5	174	11.74	124	54.7	1587	16.3	5.7	9.1	312.4	282.5	29.9	846	468.8
2	13.3	195	11.37	119.8	67.3	1600	13.3	9.3	13.7	309.3	282	27.3	650	466.3

2	13.1	181	11.89	122	43.1	1620	13.1	5	15.4	312.7	283.9	28.8	129	488.8
2	13.1	111	11.86	113.8	55.1	1594	14.9	4.9	15.5	309.3	281.3	28	295	468.8
1	12.9	163	11.89	121.4	61.6	1610	13.8	5.6	12.8	310.5	282.4	28.1	111	488.8
1	12.8	102	11.77	119.6	58.1	1576	16	4	13.8	315.6	287	28.6	286	483.8
1	12.7	109	11.86	114.4	54.5	1584	15.9	3	16.6	314.5	285.6	28.9	293	452.5
1	12.6	147	11.94	122	45	1604	13.9	5	13.9	310	279	31	42	457.5
1	12.6	150	11.63	117	50.8	1608	12	5	17.7	310.5	282.4	28.1	440	468.8
1	12.6	159	11.89	113.8	53.4	1586	13.6	5.7	19.5	309.1	281.2	27.9	107	448.8
1	12.4	128	11.53	117	68.8	1598	15.9	5	13.6	308.8	280.8	28	312	456.3
1	12.4	145	11.79	123.8	48.9	1632	14.2	3.5	13	313.4	287.4	26	435	457.5
1	12.3	103	11.97	122.4	42.7	1622	15	2.9	9.8	309	281.2	27.8	287	482.5
1	12.2	149	11.73	121.2	52.4	1626	13.2	5	14.9	310.6	282.3	28.3	439	495.3
1	12.1	135	11.82	118.4	47.5	1627	14	3.5	13.1	301.5	277.5	24	319	472.5
1	12	197	12.11	124.6	48.3	1666	15	2.9	9.9	302.6	278.5	24.1	381	483.8
1	11.8	164	11.97	117.8	43	1566	14	2.9	17.4	309.4	281.5	27.9	348	480
1	11.8	132	12.02	130.4	40.2	1626	15.9	3.9	10.8	318.7	289.9	28.8	80	490
1	11.6	143	11.72	120.2	49.8	1619	12	5	12.6	310.9	282.7	28.2	598	490
1	11.6	193	11.77	121.8	46.9	1607	14.6	5	13.1	310.9	282.5	28.4	648	476.3
1	11.5	186	11.62	122	51.5	1624	14.7	6.1	14.1	309.4	281.6	27.8	134	480
1	11.4	184	11.74	122.2	51.5	1610	14.6	5	14.1	312.4	285.9	26.5	856	472.5
1	11.3	126	11.74	120.6	55.8	1600	14.4	4.6	14	314.1	285.5	28.6	21	487.5
1	10.7	173	11.89	125.4	42	1642	14	5	13.4	310.7	282.7	28	121	487.5
1	10.5	162	11.78	127.2	42.2	1614	14.9	5.4	14.4	313.8	285.4	28.4	834	461.3
1	10.4	177	11.74	129.2	48.7	1658	18	5.8	11.1	320	294.3	25.7	849	472.5
1	10.4	128	11.8	120.2	69	1604	13.1	4.5	15.7	311	283.3	27.7	418	485
1	10.4	177	12.1	126.8	46	1651	16.6	3	16.8	305	280.3	24.7	12	469.8
1	10.2	144	11.72	119.6	44.4	1597	14	5	14.6	311	282.5	28.5	599	490
1	9.8	117	11.96	121.8	42.8	1598	13.8	5.9	16.9	310.5	282.8	27.7	12	486.3
1	9.8	194	12.02	127.2	71.2	1652	19	3	15.4	316.7	298.3	18.4	142	456.3
1	9.7	116	11.72	121.8	41	1623	13.9	4	14	318.5	293.1	25.4	406	471.3
1	9.6	137	12.04	123.6	48	1612	13.9	3.8	10.6	312.1	283.6	28.5	32	487.5
1	9.4	156	12.04	130.4	48.6	1607	15.9	3	13.1	310.6	282.7	27.9	51	465
1	9.3	177	11.89	124.4	46.3	1635	11.4	5.1	12.6	312.6	284.4	28.2	125	476.3
1	9.3	162	11.89	130.6	55	1652	14.4	5.4	13.7	316.8	291.1	25.7	110	470

1	9.2	149	11.94	124.8	58.6	1619	13.8	7	7.4	308.6	280	28.6	44	463.8
1	9.2	183	11.77	120.8	45.2	1630	14.8	5.2	14	305.3	280.8	24.5	638	465
1	8.9	112	11.69	121.4	44.7	1621	13.1	6	14.9	311.2	282.7	28.5	402	468.8
1	8.7	150	12.08	116.2	50	1610	15	3.9	18.7	310.8	282.6	28.2	334	447.5
1	8.7	185	11.74	129.6	49.1	1640	15	3.4	13.4	323.8	303.8	20	857	452.5
1	8.4	143	11.8	124.2	47	1634	14.5	5	12.8	309.5	281.7	27.8	433	487.5
1	8.3	150	11.8	123.2	42.5	1580	14.7	5.8	14.7	315.3	286	29.3	605	483.8
1	8.3	173	12	124.4	41.9	1616	15.8	4.9	17.5	315.1	286.2	28.9	8	466.3
1	8.2	184	11.89	122.4	51.5	1619	13.9	3	13.8	309.1	281.8	27.3	132	486.3
1	8.1	179	11.77	127.2	43.7	1645	14.1	5.3	15.1	320.9	295	25.9	634	483.8
1	7.8	123	11.9	123.6	49	1576	16.6	5.4	14.1	315.7	287.2	28.5	795	491.3
1	7.7	185	11.89	125.2	44.1	1612	12.5	3	12.9	312.8	285.2	27.6	133	485
1	7.5	101	11.85	125.2	54.5	1625	15.9	5	9.1	320.7	288.6	32.1	556	480
1	7.5	191	11.74	123.8	42.8	1612	14.1	5	13.5	313.8	285.8	28	863	467.5
1	7.1	119	11.85	121	43.3	1622	16.3	4.6	10	311.2	286.3	24.9	574	448.8
1	6.8	126	11.8	121.4	43.1	1619	13.5	4.7	15.4	313.7	286	27.7	416	482.5
1	6.8	116	11.69	120	41.2	1605	14	5	14.3	314.8	286.7	28.1	571	490
1	6.8	159	11.75	118.2	61.5	1600	14.8	0.7	17.7	310.5	283.1	27.4	449	456.3
1	6.7	127	11.55	120.8	53.6	1583	15.4	4	12.6	310.4	282.7	27.7	582	488.8
1	6.3	199	11.74	124.2	44.3	1578	14	7	12.5	315	287.1	27.9	654	472.5
1	6.2	186	11.67	119.6	50.1	1574	13.1	5.9	15.9	311	282.6	28.4	641	480
1	6	146	11.94	122.6	43.3	1601	13	5	11.6	313.9	279.5	34.4	41	468.8
1	5.9	191	11.77	121.4	54.2	1597	13.3	7	15.7	314.8	286.6	28.2	646	470
1	5.8	179	12.1	120.2	44.2	1562	21.7	4.9	19.7	313.5	284.9	28.6	14	488.8
1	5.7	113	11.74	125.4	47.1	1614	14.6	6.9	9.8	314.7	286.5	28.2	568	491.3
1	5.2	120	12.06	126.2	40	1616	13.8	7	15.3	314.7	286.6	28.1	15	488.8
1	3.5	127	11.6	122.2	41	1621	15.5	4.1	13.1	308.5	281.2	27.3	417	471.3

*The unit of carbon content is $10^{-2} \times C\%$.

Figure 14.9 shows an example of the result of the PCA projection of good samples. Here the carbon content of the final product is required to be controlled within the range of 0.15±0.05%. From Fig. 14.9 it can be seen that the separation of the "good" points from "bad" points is not very satisfactory. This is due to the nonlinear nature of the relationship between the features and target in computation.

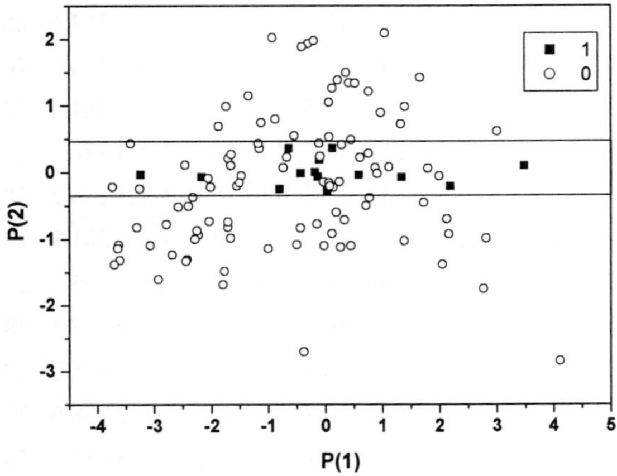

Fig. 14.9   Modeling of carbon content by PCA projection.

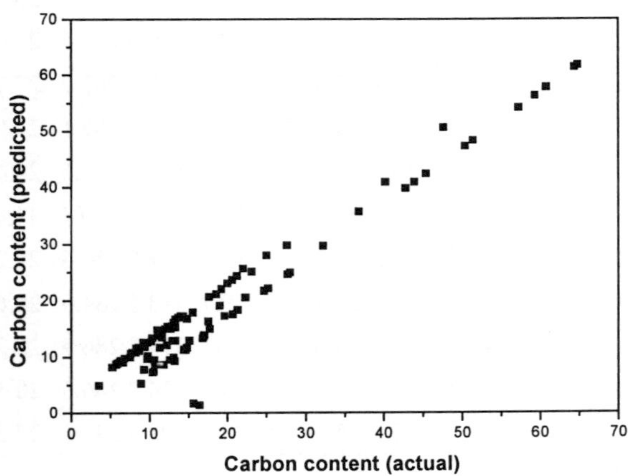

Fig. 14.10   Modeling of carbon content by SVR.

Figure 14.10 shows the results of support vector regression (using Gaussian kernel, and $\varepsilon = 0.05$) of the data listed in Table 14.5. It can be seen that the regularity is rather good, and the sample points with carbon content within the range of 0.15±0.05% distribute within a narrow range in the middle part of the straight line. 91.3% of the samples calculated to be in this range are actually having carbon content with the range of 0.15±0.05%. So it is possible to use this relationship for the control of carbon content in the steel production.

14.7   Materials Properties Estimation for Production Process

The physico-chemical properties of slags are very important basic information for the investigation of pyrometallurgical process. It is desirable to make the mathematical model of the physico-chemical properties of slags for various practical purposes. Figure 14.11 illustrates the comparison between the experimental data of viscosities of four-component liquid slags at 1650°C [46] and the calculated values by SVR and kernel function of second degree.

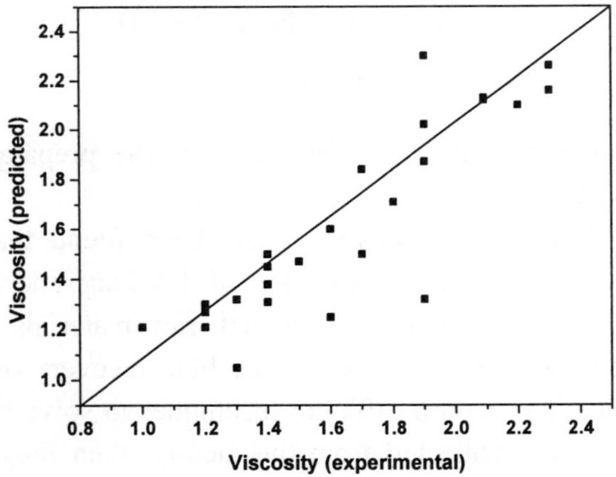

Fig. 14.11 Comparison between actual and predicted values of viscosity of four component slags (the unit of viscosity is centpoise).

14.8   Comprehensive Strategy for Industrial Optimization

Although SVM is indeed a useful tool of modeling for the industrial optimization and fault analysis, its results will be more believable if we can find the physical meaning of this mathematical model by some experimental research work. In this section we will use an example to show how to do research work of this type.

Soda-lime sintering process is one of the chief methods for alumina production. In this process, sodium carbonate and limestone are mixed with bauxite or red mud (the sludge of Bayer process) and sintered in rotary kiln, so that the following reactions take place:

$$Na_2CO_3 + Al_2O_3 \rightarrow Na_2O \cdot Al_2O_3 + CO_2$$

$$2CaCO_3 + SiO_2 \rightarrow Ca_2SiO_4 + CO_2$$

$$\text{or } Na_2O \cdot Al_2O_3 \cdot 2SiO_2 + 4CaCO_3 \rightarrow Na_2O \cdot Al_2O_3 + 2Ca_2SiO_4 + CO_2$$

According to the traditional view-point of alumina production, the best composition of the raw materials should be in accordance to the following relations:

$$Na_2O/(Al_2O_3 + Fe_2O_3) = 1.0$$

$$CaO/SiO_2 = 2.0$$

This is so-called "saturation formula" for the preparation of raw materials for sintering process.

In industrial practice, however, it has been found that saturation formula cannot assure the highest rate of leaching and recovery of alumina, especially for the iron oxide rich raw materials. In order to make a mathematical model for achieving high recovery of alumina in industrial production, we use different techniques to solve this problem. At first raw data are collected from the factory, then feature selection methods are used to find the chief factors affecting the recovery of alumina, and the mathematical model assuring high recovery is found by the computation based on these chief parameters. Finally, the physico-chemical basis of the mathematical model is investigated by experimental method.

Feature selection techniques are used for finding the chief factors influencing the target. Preliminary statistical analysis indicates that the data set has inclusive structure, so we can use two methods for this feature selection work: one is based on KNN method, and the other is based on SVR using Gaussian kernel function. Fortunately both methods give the same result. The data file after feature selection is shown in Table 14.6.

Table 14.6   Factors affecting recovery of alumina.

No. of sample	Class	Recovery %	Fe/Al	$Na_2O/(Al_2O_3+Fe_2O_3)$	CaO/SiO_2
1	1	93	0.1	1	2.05
2	1	96	0.2	1	2
3	1	97	0.1	1	2.04
4	1	97	0.1	0.97	2.03
5	2	80	0.1	0.96	2.02
6	2	80	0.1	0.8	2
7	2	75	0.2	0.75	2
8	2	70	0.1	0.7	2.05
9	2	65	0.2	0.65	1.95
10	2	60	0.1	0.6	1.95
11	2	75	0.1	1.2	2.2
12	2	70	0.2	1.3	2.3
13	2	70	0.1	1.4	2.3
14	2	65	0.2	1.2	2.3
15	2	60	0.1	1.2	2.4
16	2	75	0.3	1.2	1.8
17	2	70	0.4	1.3	2.7
18	2	70	0.3	1.4	2.7
19	2	65	0.4	1.3	2.6
20	2	60	0.3	1.4	2.64
21	1	95	1	0.7	2.4
22	1	94	1	0.7	2.35
23	1	93	1	0.7	2.45
24	1	95	1	0.7	2.5
25	1	96	1	0.7	2.31
26	2	75	1	1	2

27	2	70	1	1	2
28	2	65	1	1	2
29	2	60	1	0.4	2.4
30	2	60	1	0.5	2.3
31	1	93	0.1	1	2
32	1	95	0.2	0.95	1.95
33	1	92	0.1	1.02	2.02
34	1	91	0.2	1.04	2.05
35	1	94	0.1	0.98	2.03
36	2	70	0.1	0.8	1.95
37	2	75	0.2	0.85	2
38	2	66	0.1	0.75	2.05
39	2	68	0.2	0.75	2.02
40	2	70	0.2	0.73	1.95
41	2	70	0.2	0.75	2.04
42	2	65	0.2	0.75	2.04
43	2	70	0.2	0.75	2.06
44	2	65	0.3	0.7	2.03
45	2	70	0.3	0.75	2.06
46	1	93	0.5	0.85	2.2
47	1	94	0.5	0.88	2.3
48	1	96	0.4	0.9	2.15
49	1	92	0.4	0.9	2.2
50	1	94	0.8	0.8	2.3
51	1	93	0.7	0.82	2.4
52	2	75	0.7	0.82	2.6
53	2	70	0.6	0.9	2.65
54	2	65	0.7	0.6	2
55	1	90	0.7	0.8	2.35
56	2	80	0.7	0.8	2

57	2	60	0.9	0.8	1.9
58	2	70	0.8	0.8	2
59	2	50	0.9	0.7	1.8
60	2	60	0.9	0.8	1.9
61	1	90	1	0.9	2.4
62	1	92	0.9	0.92	2.3
63	1	93	1	0.85	2.35
64	1	92	1.1	0.85	2.35
65	1	94	1.1	0.9	2.4
66	1	95	1.2	0.84	2.5
67	2	70	1.1	0.6	2
68	2	75	1.1	0.85	2.7
69	2	70	1.1	0.7	2.15
70	2	65	0.1	1.1	1.5
71	1	91	0.2	1	2
72	2	80	0.3	0.9	2.5

By SVR with Gaussian kernel, and relative large ε-insensitive loss function ($\varepsilon = 0.15$ for the data processing of normalized data set), it is found that the data set listed in Table 14.6 can be roughly described by a linear relationship in the feature space described by Gaussian kernel function (Fig. 14.12), and the sample points of high recovery are located near the upper end of the straight line. By using KNN method, it can be found that these "good points" form several clusters in the regions with different ranges of Fe/Al ratio. The good points corresponding to raw materials of low iron content just obey the "saturation formula" of composition, while the good sample points with the Fe/Al ratio around 1.0 are located in the following range in the input space:

$$0.7 \leq Na_2O/(Al_2O_3 + Fe_2O_3) \leq 0.9$$
$$2.3 \leq CaO/SiO_2 \leq 2.5$$

In other words, it should decrease the quantity of soda and increase the quantity of calcium oxide in the raw materials for alumina production when the iron content of raw materials is high.

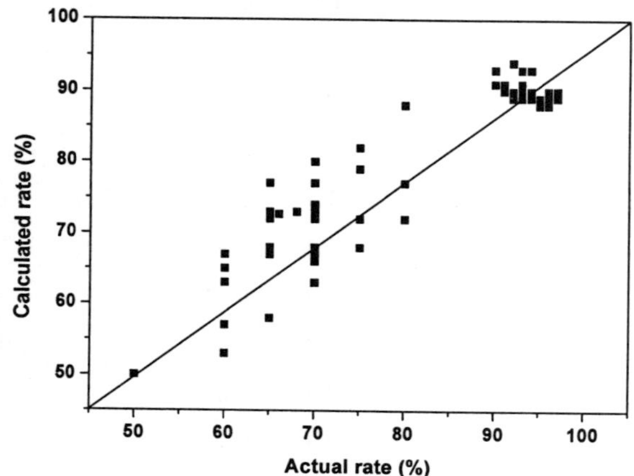

Fig. 14.12   Rate of leaching can be described by SVR with Gaussian kernel.

The empirical regularity found is in agreement with the results of our physico-chemical research work. By X-ray diffraction and infrared spectroscopic studies, it has been found that the following reaction does occur in the sintering process [147]:

$$5Ca_2SiO_4 + 4NaFeO_2 \rightarrow Na_4Ca_8(SiO_4)_5 + 2CaFe_2O_4$$

And in the leaching process, $Na_4Ca_8(SiO_4)_5$ can react with sodium aluminate to form $Ca_3[Al(OH)_6]_2$ precipitate and make alumina leaching rate (and the recovery) decreased. A better method to increase the leaching rate of alumina is to add more CaO in raw materials to make iron oxide not form $NaFeO_2$ but $CaFe_2O_4$. This is just the reason why we should add more CaO and less soda in high iron raw materials for sintering.

Appendix A

The Implementation of SVM

A.1 Introduction

From this book, we have known that, support vector machine (SVM) has generated a great interest in the community of machine learning, chemistry and other related fields due to its excellent generalization performance in a wide variety of learning problems, especially for the data processing of small data training sets. In this appendix, we will discuss the training of support vector machine.

The SVM training tasks involve optimization of a quadratic Lagrangian. Some applicable QP packages, such as MINOS and LOQO, can be used to train the SVM rapidly but they have the disadvantage that the kernel matrix is stored in memory and the matrix is so large that it will exhaust the memory.

Currently there are several algorithms for SVM, such as Chunking, Decomposition and sequential minimum optimization (SMO), which makes a great progress in the training of SVM. Many implementations of SVM are available on the internet. Some are implemented by C/C++, or Java, some others are by MatLab.

Firstly, we provide an overview of the typical SVM training algorithms such as Chunking and Decomposition. Secondly, we introduce SMO. Then some well known implementations of SVM are introduced. At last, we present our software of SVM, ChemSVM.

A.2   Chunking and Decomposition

For completeness, the QP problem to train an SVM is shown below:

$$\max_{\alpha} W(\alpha) = \sum_{i=1}^{\ell} \alpha_i - \frac{1}{2}\sum_{i=1}^{\ell}\sum_{j=1}^{\ell} y_i y_j K(\mathbf{x}_i, \mathbf{x}_j)\alpha_i\alpha_j, \qquad (A.1)$$

$$0 \le \alpha_i \le C, \forall i,$$

$$\sum_{i=1}^{\ell} y_i \alpha_i = 0.$$

A number of optimization techniques can be directly applied to QP, such as Newton method, conjugate gradient, and primal dual interior-point method. But in fact, those methods are very hard to use, so they are not widely used in SVM.

For large scale problems, there exists a so-called active set or working set method in optimization: if one knew in advance which constrain was active, it would be possible to discard all the inactive constrains and simplify the problem. This leads to several strategies, all based on how to guess the active set, the simplest heuristic is known as Chunking. A more advanced algorithm is Decomposition, which only updates a fixed size of subset.

A.2.1   *Vapnik's chunking method*

Vapnik first suggested the decomposition approach in a method known as *Chunking*. The algorithm sometimes is called decomposition PCG (projected conjugate gradient). It starts with an arbitrary subset or *Chunk* of the data, and trains an SVM using a generic optimizer on that portion of the data. The algorithm retains the support vectors from the *chunk* and discards the other points. Then it uses the hypothesis found to test the points in the remaining part of the data. The M points that most violate the KKT conditions are added to the support vectors of the previous problem, to form a new *chunk*, where M is a parameter of the system. The procedure is iterated, initiating α for each new sub-problem with the output of values from the previous stage, finally halting when some stopping criterion is satisfied.

The chunk of data being optimized at a particular stage is sometimes referred to the working set. Typically the working set grows, sometimes it will decrease, until in the last iteration the machine is trained on the set of support vectors representing the active constraints.

Pseudo-code is described as below:

Given training set S

$$\alpha \leftarrow 0$$

Select an arbitrary working set $\hat{S} \subset S$
Repeat
  Solve optimization problem on $\hat{S}$
  Select new working set from data not satisfying KKT conditions
Until stopping criterion satisfied
Return α

A.2.2 *Osuna's decomposition method*

In 1997, Edgar Osuna [105] and his colleagues proposed a theorem that suggests a whole new set of quadratic program algorithms for SVM. The theorem proves that the large quadratic program problem can be broken down into a series of smaller sub-problems of quadratic program. Osuna and his colleagues proposed to keep a matrix of constant size for every sub-problem of quadratic program. So every times a new point can be added to the working set, another point has to be removed. Using a matrix of constant size will allow the training on arbitrarily sized data sets.

Here is the Pseudo-code:
Given an training set
Select an arbitrary working set B of parameters
The set N of parameters
While KKT violated (there exists some $i \in N$, such that)

$$\alpha_i = 0, \qquad f(\mathbf{x}_i)y_i < 1$$
$$\alpha_i = C, \qquad f(\mathbf{x}_i)y_i > 1$$
$$0 < \alpha_i < C, \qquad f(\mathbf{x}_i)y_i = 1$$

Select new set B by replacing any $\alpha_i, i \in B,$
Solve optimization problem on B
Return $\boldsymbol{\alpha}$

A.3   Platt's SMO Algorithm

The sequential minimal optimization (SMO) algorithm is derived from the idea of the decomposition method to its extreme and the optimization for a minimal subset of just two points at each iteration. It was first devised by Platt [110], and applied to text categorization problems. SMO is a simple algorithm that can quickly solve the SVM QP problem without any extra matrix storage and without using numerical QP optimization steps at all. SMO decomposes the overall QP problem into QP sub-problems, using Osuna's theorem to ensure convergence.

SMO chooses to solve the smallest possible optimization problem at every step. For the standard SVM QP problem, the smallest possible optimization problem involves only two Lagrange multipliers. So at every step, SMO chooses two Lagrange multipliers to jointly optimize, finds the optimal values for the multipliers, and updates the SVM to reflect the new optimal values.

The advantage of SMO lies in the fact that solving for two Lagrange multipliers can be done analytically. Thus, an entire inner iteration due to numerical QP optimization is avoided. The inner loop of the algorithm can be expressed in a small amount of C code, rather than invoking an entire iterative QP library routine. Even though more optimization sub-problems are solved in the course of the algorithm, each sub-problem is so fast that the overall QP problem can be solved quickly. In addition, SMO does not require extra matrix storage because manipulation of large matrices is avoided. SMO can be less susceptible to numerical precision problems.

There are three components of SMO:

1) An analytic method to solve optimization of two Lagrange multipliers.

2) A heuristic for choosing which multipliers to optimize in next step.

3) A method for computing b.

A.3.1 *Solving for two lagrange multipliers*

Because there are only two multipliers, without loss of generality, we assume that the two chosen elements are α_1, α_2. From

$$\sum_{i=1}^{n} y_i \alpha_i = 0,$$

we can get the formula:

$$y_1\alpha_1 + y_2\alpha_2 = Const = y_1\alpha_1^{old} + y_2\alpha_2^{old}$$

In (α_1, α_2) space and in the box defined by $0 \le \alpha_1, \alpha_2 \le C$. The one-dimensional problem results from the restriction of the objective function to such a line that can be solved analytically. The constraints can be displayed in two dimensions as follows:

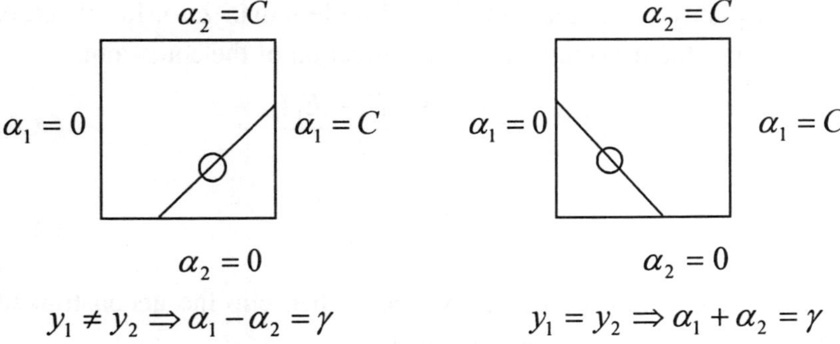

The two Lagrange multipliers must fulfill all of the constraints of the full problem. The inequality constraints cause the Lagrange multipliers to lie in the box. The linear equality constraint causes them to lie on a diagonal line. Therefore, one step of SMO must find an optimum of the objective function on a diagonal line segment.

In order to solve for the two Lagrange multipliers, SMO first computes the constraints on these multipliers and then solves for the constrained maximum.

$$U = \max(0, \alpha_2^{old} - \alpha_1^{old}), V = \min(C, C + \alpha_2^{old} - \alpha_1^{old}). \qquad y_1 \neq y_2$$

$$U = \max(0, \alpha_1^{old} + \alpha_2^{old} - C), V = \min(C, \alpha_1^{old} + \alpha_2^{old}). \qquad y_1 = y_2 \qquad \text{(A.2)}$$

The second derivative of the objective function along the diagonal line can be expressed as

$$k = 2K(\mathbf{x}_1, \mathbf{x}_2) - K(\mathbf{x}_1, \mathbf{x}_1) - K(\mathbf{x}_2, \mathbf{x}_2) \qquad \text{(A.3)}$$

Always the equation can be simplified as

$$k = 2K_{12} - K_{11} - K_{22}$$

The next step of SMO is to compute the location of the constrained maximum of the objective function in the following equation while allowing only two Lagrange multipliers to change. Under normal circumstances ($k \neq 0$), there will be a maximum along the direction of the linear equality constrain, and k will be less than zero, In this case, SMO computes the maximum along the direction of the constraint.

$$\alpha_2^{new} = \alpha_2^{old} - \frac{y_2(E_2 - E_1)}{k} \qquad \text{(A.4)}$$

$$E_i = f^{old}(\mathbf{x}_i) - y_i \qquad \text{(A.5)}$$

E_i is the error on the *i*th training example. Then clip the unconstrained maximum to the ends of the line segment

$$\alpha_2^{new,clipped} = \begin{cases} V, & if(\alpha_2^{new} \geq V) \\ \alpha_2^{new}, & if(U \leq \alpha_2^{new} \leq V) \\ U, & if(\alpha_2^{new} \leq U) \end{cases}$$

$$\alpha_1^{new} = \alpha_1^{old} + y_1 y_2 (\alpha_2^{old} - \alpha_2^{new,clipped}) \qquad (A.6)$$

Under unusual circumstances, k will not be negative. A zero k can occur if more than one training example has the same input vector $\mathbf{x}$. In any event, SMO will work even when k is not negative, in which case the objective function W should be evaluated at each end of the line segment.

To summarize, given α_1, α_2 (and the corresponding y_1, y_2, K_{11}, K_{12}, K_{22}, $E_2^{old} - E_1^{old}$), we can optimize the two α's by the following procedure:

1. $k = 2K_{12} - K_{11} - K_{22}$.

2. if $k < 0$

$$\Delta \alpha_2 = \frac{y_2(E_2^{old} - E_1^{old})}{k}$$

and clip the solution within the feasible region. Then

$$\Delta \alpha_1 = -s \Delta \alpha_2$$

3. If $k = 0$, we need to evaluate the objective function at the two endpoints, and set α_2^{new} to be the one with larger objective function value.

The objective function is:

$$L_D = \frac{1}{2} k \alpha_2^2 + (y_2(E_1^{old} - E_2^{old}) - k\alpha_2^{old})\alpha_2 + Const$$

A.3.2　*Heuristics for choosing which multipliers to optimize*

Unlike the common decomposition method, SMO will always alter only two Lagrange multipliers to move uphill in the objective function projected into the one-dimensional feasible subspace. In order to speed convergence, SMO uses heuristics to choose which two Lagrange multipliers to jointly optimize.

There are two separate choice heuristics, one for the first Lagrange multiplier and the other for the second. The choice of the first heuristic provides the outer loop of the SMO algorithm. The outer loop first iterates over the entire training set, determining whether each example violates the KKT conditions. If an example violates the KKT conditions, it is then eligible for immediate optimization. Once the first Lagrange multiplier is chosen, SMO chooses the second Lagrange multiplier to maximize the size of the step taken during joint optimization. Evaluating the kernel function K is time consuming, so SMO approximates the step size by the absolute value of the numerator in equation $\left| E_2 - E_1 \right|$, where E_1 is defined in formula (A.5).

First Choice Heuristic　The first point $\mathbf{x}_1$ is chosen from among those points that violate the KKT conditions. The outer loop of the algorithm goes through the training set looking for points that violate KKT conditions and selects any it finds for update. When one such point is found, the second heuristic is used to select the second point.

Second Choice Heuristic　The second point $\mathbf{x}_2$ must be chosen in such a way that updating on the pair α_1, α_2 causes a large change, which should result in a large increase of the dual objective. In order to find a good point without performing too much computation, a quick heuristic is to choose $\mathbf{x}_2$ to maximize the quantity $\left| E_2 - E_1 \right|$.

1) If E_1 is positive, SMO chooses an example $\mathbf{x}$ with minimum error E_2. While E_1 is negative, SMO maximizes the error E_2. A cached list of errors for every non-bound point in the training set is kept to further reduce the computation.

2) If this choice fails to deliver a significant increase in dual objective, SMO tries each non-bound point in turn.

3) If there is still no significant progress, SMO goes through the entire training set for a suitable point. The loops go through the

non-bound points and the whole training set, and start from random locations in the respective lists, so that no bias is introduced towards the examples occurring at the beginning of them.

After the two multipliers are chosen and jointly optimized, the SVM is then updated using these two new multiplier values, and the outer loop resumes looking for KKT violators. In order to increase the chances of finding KKT violations, the outer loop goes through the points which the corresponding parameter α_i satisfies $0 < \alpha_i < C$ implying that its value is not on the boundary of the feasible region, and only when all such points satisfy the KKT conditions to the specified tolerance level is a complete loop through all the training set again undertaken.

A.3.3 *Threshold and error cache*

Training the SVM need to solve the Lagrange multipliers $\boldsymbol{\alpha}$ and the threshold b. After solving the Lagrange multipliers $\boldsymbol{\alpha}$, b must be computed separately.

The threshold b is re-computed after each step, so that the KKT conditions are fulfilled for both optimized examples. The threshold b_1 is valid when the new α_1 is not at the bounds, because it forces the output of the SVM to be y_1 when the input is $\mathbf{x}_1$:

$$b_1 = E_1 + y_1(\alpha_1^{new} - \alpha_1^{old})K(\mathbf{x}_1,\mathbf{x}_1) + y_2(\alpha_2^{new,clipped} - \alpha_2^{old})K(\mathbf{x}_1,\mathbf{x}_2) + b^{old}$$

The following threshold b_2 is valid when the new α_2 is not at bounds, because it forces the output of the SVM to be y_2 when the input is $\mathbf{x}_2$:

$$b_2 = E_2 + y_1(\alpha_1^{new} - \alpha_1^{old})K(\mathbf{x}_1,\mathbf{x}_2) + y_2(\alpha_2^{new,clipped} - \alpha_1^{old})K(\mathbf{x}_2,\mathbf{x}_2) + b^{old}$$

When both b_1 and b_2 are valid, they are equal. When both new Lagrange multipliers are at bound and if U is not equal to V, then the interval between b_1 and b_2 are all thresholds that are consistent with the KKT conditions. SMO chooses the threshold to be halfway in between b_1 and b_2.

$$b^{new} = \frac{b_1 + b_2}{2}$$

To improve the training rate, a cached error value E is kept for every example whose Lagrange multiplier is neither zero nor C. When a Lagrange multiplier is non-bound and involved in a joint optimization, its cached error is set to zero. Whenever a joint optimization occurs, the stored errors for all non-bound multipliers α_k that are not involved in the optimization are updated according to

$$E_k^{new} = E_k^{old} + y_1(\alpha_1^{new} - \alpha_1^{old})K(\mathbf{x}_1,\mathbf{x}_k) + y_2(\alpha_2^{new,clipped} - \alpha_2^{old})K(\mathbf{x}_2,\mathbf{x}_k) + b^{old} - b^{new}$$

When an error E is required by SMO, it will look up the error in the error cache if the corresponding Lagrange multiplier is not at bound. Otherwise, it will evaluate the current SVM decision function based on the current $\boldsymbol{\alpha}$ vector.

A.4   Overview of Well-known Implementations

Now, many SVM implementations and packages are available, including C, C++, Fortran and MATLAB codes. Some of them can handle not only binary classification but also multiple classification.

Some of them are list below:

Optimization packages:

—CPLEX: Barrier/QP Solver

—LOQO: Linear and Quadratic Optimization Package by Robert Vanderbei

—MINOS: Linear and Quadratic Solver

C/C++, Fortran implementations

—BSVM: A decomposition method for bound-constrained SVM formulations

—HeroSvm2.0: A high performance library for training SVM on a very large training set efficiently

—LIBSVM: An SVM library with a graphical interface

—LOOMS: A leave-one-out model selection for SVM

—Least Squares - Support Vector Machines: MATLAB /C Toolbox

—M-SVM: Multi-class SVM for very large scale problems

—mySVM: SVM implementation for pattern recognition and regression

—Nearest Point Algorithm by Sathiya Keerthi (in FORTRAN)

—SMOBR: An implementation of the original SMO proposed by Platt, written in C++

—SVM^{light} by Thorsten Joachims (written in C++)

MATLAB implementations

—Matlab SVM Toolbox by Steve Gunn

—Matlab SVM Toolbox, MATLAB implementation in the style of SVMlight, can train 1-norm and 2-norm SVM

—OSU SVM Classifier Matlab Toolbox, A MATLAB toolbox with a C++ MEX core to perform fast

—SVM Object Oriented MATLAB toolbox, including C++ MEX implementation of the SMO by Gavin Cawley

—SVM Tools MATLAB implementation by Glenn Fung and Olvi Managsarian

More details are available at *http://www.kernel-machines.org/*

Among them, LIBSVM [20] performs better. It is the winner of EUNITE worldwide competition on electricity load prediction. The latest version is v2.5. The basic algorithm is a mixed version of both SMO by Platt and SVM^{Light} by Joachims [74]. It is also a simplification of the modification of SMO by Keerthi et al. The executable program and the source code can be downloaded from:

http://www.csie.ntu.edu.tw/~cjlin/libsvm/.

LIBSVM 2.5 includes such functions as scale, train, and predict. It supplies Support Vector Classification and Regression, such as C-SVC, ν-SVC, One-class SVC, C-SVR and ν-SVR. It also provides four kinds of kernel functions. As far as the multi-class problem is concerned, LIBSVM supplies some multi-class methods such as one to one, one-to-rest and SVMDAG. The FAQ on the web gives more detailed information about how to use the LIBSVM.

A.5   Introduction to ChemSVM

In this section, the software package ChemSVM is introduced. ChemSVM is a convenient SVM implementation for chemists. The program was developed at the laboratory of chemical data mining, Shanghai University in China. In order to analyze the chemical data more conveniently, the program integrates support vector classification and support vector regression. The demo version ChemSVM of executable program in Windows operating system can be downloaded from http://www.seawallsoft.com.

The main modules of ChemSVM are listed as follow:

- Preprocessing of data file
- Processing the data set using support vector classification (SVC)
- Leave-one-out cross-validation based on SVC
- Processing the data set using support vector regression (SVR)
- Leave-one-out cross-validation based on SVR
- Visualization of results available

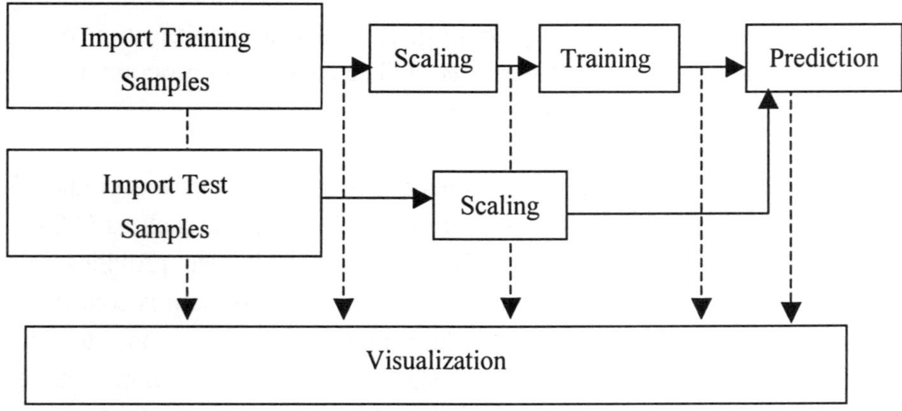

The training data set and the test data set should be scaled before processing. The sample will be standardized to the range of 0.0 to 1.0 using an affine function.

The Visualization module can display the information of the training samples and the test samples by 3D graph. Both training and test results can be output in ASCII text file or 2D graph mode. The main interface of

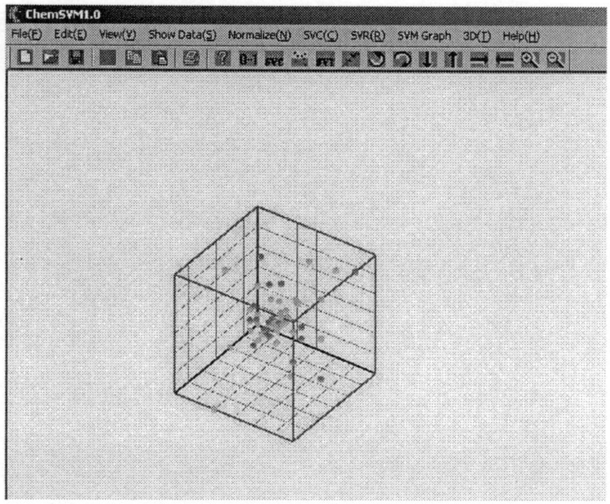

Fig. A.1   The main interface of ChemSVM

ChemSVM is presented in Fig. A.1.

The implementation of SVM is still a hot spot of machine learning. New algorithms are coming forth all along. The latest news and updates of software can be referred to http:// www.kernel-machines.org/.

Bibliography

[1] Adachi, M. and Nakamura, S. (1991). Comparison of the INDO/S and the CNDO/S method for the absorption wave length calculation of organic dyes, *Dyes and pigments*, 17, pp. 287–296.

[2] Alberty, R. A. and Reif, A. K. (1988). Standard chemical thermodynamic properties of polycyclic aromatic hydrocarbons and their isomer groups, I, benzene series, *Journal of physical and chemical reference data*, 17, pp. 241–253.

[3] Alberty, R. A., Chung, M. B. and Reif, A. K. (1989). Standard chemical thermodynamic properties of polycyclic aromatic hydrocarbons and their isomer groups, II. pyrene series, naphthaopyrene series and coronene series, *Journal of physical and chemical reference data*, 18, pp. 77–108.

[4] Alex J. S. and Schölkopf, B. (1998). A Tutorial on Support Vector Regression, NeuroCOLT Technical Report NC-TR-98-030, Royal Holloway College, University of London, UK.

[5] Alves de Lima Ribeiro, F. and Ferreira, F. M. C. (2003). QSPR models of boiling point, octanol-water partition coefficient and retention index of polycyclic aromatic hydrocarbons, *J. Mol. Struct. (Theochem)*, 663, pp. 109–126.

[6] Balaban, A. T. (1997). From chemical topology to three dimensional geometry, Plenum Press, New York.

[7] Bao, X. H., Lu, W. C. and Chen, N. Y. (2002). Support vector machine applied to intelligent data base of phase diagrams of molten salt systems, *Computer and applied chemistry*, 19, pp. 723–725.

[8] Bao, X. H., Lu, W. C., Liu, L. and Chen, N. Y. (2003). Hyper-polyhedron model applied to computer-aided molecular screening of N-(3-Oxo-3,4-dihydro-2H-benzo[1,4]oxazine-6-carbonyl) guanidines as Na/H exchange inhibitors, *Acta pharmacologica sinica*, 24, pp. 472.

[9] Bao, X. H., Pan, Q. Y. and Chen, N. Y. (2002). Support vector regression model for controlling the thickness of semiconductor In_2O_3 film, *Computer and applied chemistry*, 19, pp. 733–736.

[10] Basanov, S. S. (1962). Electronegativity of elements and chemical bond, Publishers of Siberia Branch of Academy of Sciences of USSR, Novosibirsk.

[11] Basanov, S. S. (1981). On the temperature of formation of solid solutions of alkali halides, *Journal of inorganic chemistry (Russia)*, 26, pp. 2145–2146.

[12] Bellman, R. E. (1961). Adaptive Control Processes, Princeton University Press, Princeton, NJ.

[13] Bienfait, B. (1994). Applications of high resolution self-organizing maps to retrosynthetic and QSAR analysis. *J. Chem. Inf. Comput. Sci.*, 34, pp. 890–898.

[14] Blander, M. (1964). Molten salt chemistry, Interscience Publishers, New York.

[15] Boser, B. E., Guyon, I. M. and Vapnik, V. (1992). A training algorithm for optimal margin classifiers, Fifth Annual Workshop on Computational Learning Theory, ACM, Pittsburgh.

[16] Burbidge, R., Trotter, M., Buxton, B. and Holden, S. (2001). Drug design by machine learning: support vector machines for pharmaceutical data analysis, *Computer and chemistry*, 26, pp. 5–14.

[17] Cai, Y. D., Feng, K. Y., Li, Y. X. and Chou, K. C. (2003). Support vector machine for predicting alpha-turn types. *Peptides*. 24, pp. 629–630.

[18] Caruana, R. (1997). Multitask learning, *Machine learning*, 28, pp. 41–75.

[19] Caruana, R. and Virginia de Sa, R. (2003). Benefiting from the variables that variable selection discards, *Journal of machine learning research*, 3, pp. 1245–1264.

[20] Chang, C. C. and Lin, C. J. (2001). LIBSVM, A library for support vector machines, Software available at http://www.csie.ntu.edu.tw/~cjlin/libsvm.

[21] Chen, N. Y. (1976). Parameters of chemical bond and their application, Science Press, Beijing.

[22] Chen, N. Y. (1997). Industrial optimization by pattern recognition techniques and its application, Chinese Petrochemical Industry Press, Beijing.

[23] Chen, N. Y., Chen, R. L., Lu, W. C., Li, C. H. and Villars P. (1999). Regularities of formation of ternary intermetallic compounds Part 4. Ternary intermetallic compounds between two nontransition elements and one transition element, *Journal of alloys and compounds*, 292, pp. 129–133.

[24] Chen, N. Y., Ding, Y. P., Li, G. Z., Ye, C. Z. and Wu, Q. S. (2002). Support vector regression applied to simultaneous determination of Pb, Cd, Zn by spectrophotometric method, *Computer and applied chemistry*, 19, pp. 717–718.

[25] Chen, N. Y., Liu G., Li, C. H., Qin, P. and Liu, H. L. (1997). Regularities of melting points and melting types of simple and complex ionic solids, *Journal of physics and chemistry of solids*, 58, pp. 731–735.

[26] Chen, N. Y., Lu, W. C., Chen, R. L., Qin, P. and Villars, P. (1999). Regularities of formation of ternary intermetallic compounds, *Journal of alloys and compounds*, part 1, 289, pp. 120–125; part 2, pp. 126–130; part 3, pp. 131–134.

[27] Chen, N. Y., Lu, W. C., Chen, R. L., Ye, C. Z. and Li, G. Z. (2002). Support vector machine applied to differentiation of oolong tea from black tea or green tea, *Computer and applied chemistry*, 19, pp. 719–720.

[28] Chen, N. Y., Lu, W. C., Liu, X., Ye, C. Z. and Li, G. Z. (2002). On the relationships between the geometrical parameters of polycyclic aromatic hydrocarbons and their environmental properties, *Computer and applied chemistry*, 19, pp. 749–751.

[29] Chen, N. Y., Lu, W. C., Wu, H. S. and Xu, X. H. (2002). Support vector machine applied to modeling preparation process of AlN film, *Computer and applied chemistry,* 19, pp. 726–728.

[30] Chen, N. Y., Qin, P., Chen, R. L. and Lu, W. C. (2000). Pattern recognition applied to chemistry and chemical technology, Science Press, Beijing.

[31] Chen, N. Y., Qin, P., Chen, R. L. and Lu, W. C. (2002). Pattern recognition applied to chemistry and chemical technology (2nd ed.), Science Press, Beijing.

[32] Chen, N. Y., Yan, L. C., Lu, W. C., Bao, X. H., Fang, J. H. and Ding, Y. M. (2001). Computerized prediction of thermodynamic properties and intelligent data base of phase diagrams of molten salt systems, *Proceedings of 6th international symposium on molten salt chemistry and technology,* Shanghai, China, pp. 1–8.

[33] Chen, R. L., Lu, W. C. and Chen, N. Y. (2003). Support vector machine applied to relationship of trace element contents and hypertension disease, *Computer and applied chemistry,* 20, pp. 567–570.

[34] Chen, Y. Q. *et al.* (1999). Quantitative structure–activity relationships study of herbicides using neural networks and different statistical methods, *Chemometrics and intelligent laboratory systems,* 45, pp. 267–276.

[35] Cheng, H. S., He, W. Q., Yang, F. J. and Zhou, F. T. (2002). Study on Ru-porcelain of Song dynasty, *Proceedings of international symposium of ancient ceramics '02,* Shanghai, China.

[36] Chinese Central Station of Environmental Monitoring. (1990). The background values of soil of China, Publisher of Chinese Environmental Science, Beijing.

[37] Chiristopher, J. C. and Buges, C. J. C. (1998). A Tutorial on Support Vector Machines for Pattern Recognition, *Data mining and knowledge discovery,* 2, pp. 121–167.

[38] Chukanov, V. N. and Chikanov, N. D. (2000). Interaction in binary bromide systems, *Journal of inorganic chemistry (Russia),* 45, pp. 1221–1224.

[39] Chung, K. M., Kao, W. C., Sun, C. L., Wang, L. L. and Lin, C. J. (2003). Radius margin Bounds for Support vector Machines with the RBF Kernel, *Neural computation,* 15, pp. 2643–2681.

[40] Cibas, T., Soulie, F. and Gallinari, P. (1996). Variable selection with neural networks. *Neurocomputing,* 12, pp. 223–248.

[41] Cortes, C. and Vapnik, V. (1995). Support Vector Networks. *Machine learning,* 20, pp. 273–297.

[42] Cristianini, N. and Taylor, J. S. (2000). An Introduction to Support Vector Machines and Other Kernel-based Learning Methods, Cambridge University Press, Cambridge.

[43] Cun, Y. L., Denker, J. S. and Solla, S. A. (1990). Optimal brain damage, Edited by Touret-zky, D., Advances in Neural Information Processing Systems, Morgan Kaufmann, Inc., pp. 598–605.

[44] Dai, Q. H. (1980). Research on chemical carcinogen and mechanism of chemical carcinogenesis, *Scientia sinica,* 23, pp. 453–470.

[45] Dash, M. and Liu, H. (1997). Feature selection for classification. *Intelligent data analysis,* 1, pp. 131–156.

[46] Department of Alloy steel smelting of Japanese iron and steel society. (1979). The properties of slags of electroslag remelting process, Japanese iron and steel society, Tokyo.

[47] Ding, Y. M., Chi, L. and Chen, N. Y. (2002). Computerized prediction and experimental confirmation of the phase diagram of $CsF-CaF_2$ system, *Computer and applied chemistry,* 19, pp. 721–722.

[48] Ding, Y. P., Chen, N. Y., Wu, Q. S., Li, G. Z. and Yang, J. (2002). Derivative spectrum simultaneous determination of NO_3--NO_2- by SVR method, *Computer and applied chemistry,* 19, pp. 752–754.

[49] Dunn, W. J., Wold, S., Edlund, U., Hellberg, S., *et al.* (1984). *J. Quant. Struct.-Act. Relat.,* 3, pp. 131–137.

[50] Elena, M., Marisa, L., Broccia, F., Di Renzo, *et al.* (2001). Antifungal triazoles induce malformations *in vitro, Reproductive toxicology,* 15, pp. 421–427.

[51] Feng, Z. L., Huang, J. C. and Li, Z. X. (1987). Modern research on trace elements, Environmental Science Press of China, Beijing.

[52] Ferreira, M. M. C. (2001). Polycyclic aromatic hydrocarbons: a QSPR study, *Chemosphere,* 44, pp. 125–146.

[53] Fletcher, R. (1987). Practical Methods of Optimization. John Wiley and Sons, Inc. (2nd ed.).

[54] Galasso, F. S. (1990). Perovskites and high Tc superconductors, Wiley, New York.

[55] Galvez, J. (2003). Prediction of molecular volume and surface of alkanes by molecular topology, *Journal of chemical information and computer science,* 43, pp. 1231–1239.

[56] Giaconov, G. K. (1956). Problems of theory of similarity in physico-chemical processes, Publisher of Academy of Sciences of USSR, Moscow.

[57] Goldberg, D.E. (1989). Genetic Algorithms in Search, Optimization and Machine Learning, Addison-Wesley, MA.

[58] Grodzins L. (1991). Nuclear technologies for finding clandestine explosives, Edited by Vouvopoulos, G. and Paradellis, T., *International Conference on the Applications of Nuclear Techniques,* Crete, Greece, June, 1990. World Scientific Press, pp. 338–360.

[59] Guermeur, Y. (2002). A Simple Unifying Theory of Multi-class Support Vector Machines, Technical Report RR-4669, INRIA.

[60] Gunn, S. R. (1998). Support Vector Machines for Classification and Regression, Technical Report, University of Southampton.

[61] Gunn, S. R., Brown, M. and Bossley, K. M. (1997). Network Performance Assessment for Neurofuzzy Data Modelling, *Lecture notes in computer science,* 1280, pp. 313–323.

[62] Guo Q. T. and Kleppa, O. J. (2001). The standard enthalpies of the compounds of early transition metals with late transition metals and with noble metals as determined by Kleppa and coworkers at the University of Chicago-A review, *Journal of alloys and compounds,* 321, pp. 169–182.

[63] Guo, J. K., Su, H., Li, C. H., Qin, P. and Chen, N. Y. (1995). Pattern recognition applied to the ancient Jun glaze, *Proceedings of the international symposium ancient ceramics '95,* Shanghai, China.

[64] Guyon, I., Weston, J., Barnhill, S. and Vapnik, V. (2002). Gene selection for cancer classification using support vector machines. *Machine learning,* 46, pp. 389–422.

[65] Hadamard, J. (1923). Lectures on the Cauchy Problem in Linear Partial Differential Equations, Yale University Press, New Haven.

[66] Han, C. Z., Zhao, X. W., Jing, J. X., Guo, J. G., Zhu, Q. J. and Cai, J. W. (1999). Evaluation of serum copper zinc levels and copper/zinc ratio in the diagnosis of malignant tumor, *Chinese tumor science,* 8, pp. 572–573.

[67] Hansch, C., Muir, R. M., Fujita, T., Maloney, P. P., Geiger, F. and Streich, M. (1963). The Correlation of Biological Activity of Plant Growth Regulators and Chloromycetin Derivatives with Hammett Constants and Partition Coefficients, *J. Am. Chem. Soc.,* 85, pp. 2817–2824.

[68] Hawkins, D. M. (2004). The problem of overfitting, *Journal of chemical information and computer science,* 44, pp. 1–12.

[69] Haykin, S. (1999). Neural Networks (A Comprehensive Foundation), Prentice Hall.

[70] Holla, P. (1995). Detecting hidden explosives, *Analyt. Chem.,* 67, pp. 184A–189A.

[71] Hsu, C. W. and Lin, C. J. (2001). A Comparison of Methods for Multi-class Support Vector Machines, Technical Report, 19, Department of Computer Science and Information Engineering, National Taiwan University.

[72] Hultgren, R., Orr, R. L., Anderson, P. D. and Kelley, K. K. (1963). Selected values of thermodynamic properties of metals and alloys, John-Wiley&Sons Inc, Berkley.

[73] Jain, A. and Zongker, D. (1997). Feature selection: Evaluation, application, and small sample performance. *IEEE transactions on pattern analysis and machine intelligence,* 19, pp. 153–158.

[74] Joachims, T. (1998). Making large-scale SVM learning practical, edited by Schölkopf, B., Burges, C. J. C. and Smola, A. J., Advances in Kernel Methods-Support Vector Learning, MIT Press, pp. 169–184.

[75] Keerthi, S. S. and Lin, C. J. (2003). Asymptotic Behaviors of Support vector Machines with Gaussian Kernel, *Neural computation,* 15, pp. 1667–1689.

[76] Kier, L. B. and Hall, L. H. (1986). Molecular connectivity in structure-activity analysis, Research Studies Press-Wiley, New York.

[77] Kirino, O., Takayama, C. and Mine, A. (1986). Quantitative structure relationships of herbicidal N-1-methyl-1-phenylethyi phenylacetamides, *Journal pesticide science,* 11, pp. 611–617.

[78] Kochikarov, Z. A., Lokiyaev, S. M., Otrova, I. A. and Gasanaliev, A. M. (2001). Four-component reciprocal system $Na,K|F,CO_3,MoO_4$, *Journal of inorganic chemistry,* 46, pp. 335–343.

[79] Kohavi, R. and George, J. (1997). Wrappers for feature subset selection. *Artificial intelligence,* 97, pp. 273–324.

[80] Korobov, M. V and Smith, A. L. (2000). Fullerenes, chemistry, physics and technology, John-Wiley, New York.

[81] Kraus, R. A. (1996). Nuclear resonance absorption (NRA) explosive detection proto-type demonstration, Federal Aviation Administration, DOT/FAA/AR-9/85-1.

[82] Kubo, Y., Yoshida, K., Adachi, M., Nakamura, S. and Maeda, S. (1991). Experimental and theoretical study of near-infrared absorbing naphthoquinone

methide dyes with a nonlinear geometry, *Journal of American chemical society,* 113, pp. 2868–2873.

[83] Kumar, K. C. H. and Wollants, P. (2001). Some guidelines for thermodynamic optimization of phase diagrams, *Journal of alloys and compounds,* 320, pp. 189–198.

[84] Li, G. Z., Wang, Z. X., Yang, J., Yao, L. X. and Chen, N. Y. (2002). A SVM-based feature selection method and its applications. *Computer and applied chemistry,* 19, pp. 703–705.

[85] Li, J. Z. (1998). The history of Science and Technology of China, *History of Chinese ceramics,* 22, Science Press, Beijing (in Chinese).

[86] Li, J. Z., Deng, Z. Q., Wu, J., Du, Z. X. and Ma, D. F. (2003). Studies on trace element compositions of Laohudong guan kiln ware and Ru guan ware, *Journal of building materials (China),* 6, pp. 118–122.

[87] Li, J. Z., Zhang, Z. G., Deng, Z. Q. and Chen, S. P. (1997). Investigation on the porcelain chips, kiln tools collected at south part of Ancient Middle River of Hangzhou for inferring the technique and location of the porcelain kiln, *Journal of palace museum of Beijing,* 4, pp. 66–77.

[88] Lin, H. T. and Lin, C. J. (2003). A Study on Sigmoid kernels for SVM and the Training of Non-PSD Kernels by SMO-type Methods, Technical Report, Department of Computer Science and Information Engineering, National Taiwan University.

[89] Liu, H. X., Zhang, R. S., Luan, F., Yao, X. J., Lin, M. C., Hu, Z. D. and Fan, B. T. (2003). Diagnosing breast cancer based on support vector machines, *Journal of chemical information and computer science,* 43, pp. 900–907.

[90] Liu, K. M. (1981). Investigation of Jun glaze, Shandong Ceramics, 1, pp. 40–57 (in Chinese).

[91] Liu, L., Lu, W. C. and Chen, N. Y. (2004). Regularities of formation and lattice distortion of perovskite-type complex halides, *Journal of physics and chemistry of solids,* 65, pp. 855–860.

[92] Liu, X., Lu, W. C., Chen, N. Y., Yao, L. X. and Ye, C. Z. (2002). Support vector machine applied to composition design of cathode materials of Ni/H battery, *Computer and applied chemistry,* 19, pp. 731–732.

[93] Livingstone, D. J. (1996). Multivariate data display using neural networks, Chapter 7 in Neural Networks in QSAR and Drug Design, Edited by Devillars, *J., Academic Press,* London, pp. 157–176.

[94] Lu, W. C., Chen, N. Y., Ye, C. Z., Li, G. Z. and Zhu, D. P. (2002). Support vector machine applied to detection of explosives for aviation security examination, *Computer and applied chemistry,* 19, 709–711.

[95] Lupaliev, V. V., Kozmincheva, G. M. and Heliebov, E. P. (2000). Crystal-chemical analysis of perovskite-like phases of 1222 type and problem of superconductivity, *Perspective materials,* 3, pp. 10–21.

[96] Masamoto, A., Kiyoshi, H., and Kimito, F. (2003). Novel Alignment Method of Small Molecules Using the Hopfield Neural Network, *J. Chem. Inf. Comput. Sci,* 43, pp. 1390–1395.

[97] Mercer, J. (1909). Functions of positive and negative type and their connection with the theory of integral equations, Philosophical Transactions of the Royal Society, London, A 209, pp. 415–446.

[98] Miedema, A. R. and Nissen, A. K. (1983). The enthalpy of solution for solid binary alloys of two 4-d transition metals, *CALPHAD*, 7, pp. 27–36.

[99] Miedema, A. R., Boom, R. and De Boer, F. R. (1975). On the heat of formation of solid alloys, *Journal of the less-common metals*, 41, pp. 283–298.

[100] Minoux, M. (1986). Mathematical Programming: Theroy and Algorithms, John Wiley and Sons.

[101] Muller, K. R., Mika, S., Ratsch, G., Tsuda, K. and Schölkopf, B. (2001). An Introduction to kernel-based learning Algorithms. IEEE Neural Networks, 12, pp. 181–201.

[102] Muller, O. and Roy, R. (1974). The major ternary structural families, Springer-Verlag, Berlin.

[103] Nguyen, M. N. and Rajapakse, J. C. (2003). Multi-class Support Vector Machines for Protein Secondary Structure Prediction, *Genome informatics*, 14, pp. 218–227.

[104] Niessen, A. K., De Boer, F. R., Boom, R., De Chatel, Mattens, W. C. M. and Miedema, A. R. (1983). Model prediction for the enthalpy of transition metal alloys, *CALPHAD*, 7, pp. 51–70.

[105] Osuna, E., Freund, R. and Girosi, F. (1997). Improved training algorithm for support vector machines. Proc. IEEE Neural Networks in Signal Processing'97.

[106] Pasipaiko, V. E. and Alekseeva, E. A. (1977). Diagrams of fusibility of salt systems, Metallurgy Press, Moscow.

[107] Pauling, L. (1960). Nature of the chemical bond, Cornell Univrsity Press, Ithaca.

[108] Pearson, W. B. (1972). The crystal chemistry and physics of metals and alloy, Wiley-Interscience, New York.

[109] Pitzer, K. S. (1991). Activity coefficients in electrolyte solutions (2nd ed.), CRC Press, Boca Raton.

[110] Platt, J. (1998). Fast training of support vector machines using sequential minimal optimization, edited by Schölkopf, B., Burges, C. and Smola, A., Advances in kernel methods: support vector learning. MIT Press.

[111] Poggio, T., Torre, V. and Koch, C. (1985). Computational Vision and Regularization Theory, *Nature*, 317, pp. 314–319.

[112] Pudil, P., Novovicova, J. and Kittler, J. (1994). Floating search methods in feature selection. *Pattern recognition letters*, 15, pp. 1119–1125.

[113] Rakotomamonjy, A. (2003). Variable selection using SVM-based criteria. *Journal of machine learning research*, 3, pp. 1357–1370.

[114] Randic, M. (1975). On characterization of molecular branching, *Journal of American chemical society*, 97, pp. 6609–6615.

[115] Rawlings, J. O. (1988). Applied Regression Analysis: A Research Tool, Wadsworth & Brooks.

[116] Reiss, H. (1961). Theory of corresponding state of fused salts, *Journal of chemical physics*, 35, pp. 820–826.

[117] Ren, S. J. (2003). Phenol mechanism of toxic action classification and prediction: a decision tree approach, *Toxicology letters*, 144, pp. 313.

[118] Sandersen, B. T. (1976). Chemical Bonds and bond energy, Academic Press, New York.

[119] Sangster, J. and Pelton, A. D. (1987). Phase diagrams and thermodynamic properties of the 70 binary alkali halide systems having common ions, *Journal of physico-chemical reference data,* 16, pp. 509–561.

[120] Schölkopf, B. and Smola, A. J. (2001). Learning with Kernels, MIT Press, Cambridge, MA.

[121] Schölkopf, B. Bartlett, P. L. Smola, A. J. and Williamson, R. C. (1998). Shrinking the tube: a New Support Vector Regression Algorithm, edited by Kearns, M. S., Solla, S. A. and Cohn, D. A., *Advances in neural information processing systems,* 11, MIT Press.

[122] Schölkopf, B., Burges, C. J. C. and Vapnik, V. (1995). Extracting support data for a given task, edited by Fayyad, U. M. and Uthurusamy, R., Proc. First International Conference on Konwledge Discovery and Data Mining, AAAI Press, Menlo Park, CA.

[123] Sementosova, D. V., Bukhanova, G. A. and Mataiko, Z. A. (1967). System of CsF-CaF2, *Journal of inorganic chemistry,* 12, pp. 1645–1649.

[124] Shannon, R. D. and Preweitt, C. T. (1969). Effective ionic radii in oxides and fluorides, *Acta crystallographica,* B25, pp. 925–945.

[125] Shannon, R. D. and Preweitt, C. T. (1976). Effective ionic radii in oxides and fluorides, *Acta crystallographica,* A32, pp. 751–767.

[126] Smola, A. J., Bartlett, P. J. *et al.* (2000). Advances in Large Margin Classifiers, MIT Press.

[127] Suykens, J. A. K., Gestel, T. V., Brabanter, J. D., De Moor, B. and Vandewalle, J. (2002). Least squares support vector machines, World Scientific Publishing Co., Singapore.

[128] Tashiro-Itoh, T., Ichida, T. and Matsuda, Y. (1997). Metallothionein expression and concentration of copper and zinc are associated with tumar differentiation in hepatocellular carcinoma, *Liver,* 17, pp. 300–306.

[129] Timofei, S., Kurunczi, L., Suzuki, T., Fabian, W. M. F. and Muresan, S. (1997). Multiple linear regression and neural network calculation of some diazo dye absorption on cellulose, *Dyes and pigments,* 34, pp. 181–193.

[130] Topliss, J. G. and Edwards, R. P. (1979). Chance Factors in Studies of Quantitative Structure -Activity Relationships, *J. Med. Chem.,* 22, pp. 1238–1244.

[131] Vapnik, V. (1982). Estimation of Dependencies Based on Empirical Data, Springer-Verlag, New York.

[132] Vapnik, V. (1995). The Nature of Statistical Learning Theory, Springer-Verlag, New York.

[133] Vapnik, V. (1998). Statistical Learning Theory, John Wiley and Sons, Inc. New York.

[134] Vapnik, V. (1999). An Overview of Statistical Learning Theory, IEEE Trans, On Neural Networks, 10, pp. 988–999.

[135] Vapnik, V. (2000). The nature of statistical learning theory, Spring-Verlag, New York.

[136] Wang, K., Lu, W. C., Chen, N. Y., Li, G. Z. and Yao, L. X. Investigation of Ru-porcelain of Song dynasty by trace element-SVM method, *Computer and applied chemistry.* (accepted).

[137] Weston, J., Elissee, A., Bakir, G. and Sinz, F. (2004). The spider. http://www.kyb.tuebingen.mpg.de/bs/people/spider/index.html, Feb.

[138] Wiener, H. (1947). Structural determination of paraffin boiling points, *Journal of American chemical society,* 69, pp. 17–20.

[139] Wu, J. P., Leungm, Z., Li, M. J. and Stokes, M. J. (2002). EDXRF studies on Chinese yue ware, *X-ray spectrometry,* 31, pp. 408–413.

[140] Wu, J., Li, J. Z., Liang, B. L. (2002). Study on Chinese blue and white porcelain from different production places, *Proceedings of international symposium of ancient ceramics,* Shanghai, China.

[141] Xu, H. B., Zhu, Z. L., Chen, N. Y. and Jiang, N. X. (1983). Computerized pattern recognition (NIM method) applied to investigation of antagonistic action of selenium, *Journal of molecular science and chemical research,* 3, pp. 139–142.

[142] Yamamoto, T., Hori, M., Watanabe, I., Tsutsui, H., Harada, K., Ikeda, S. *et al.* (1998). Synthesis and quantitative structure-activity relationships of N-(3-Oxo-3,4-dihydro-2H-benzo[1,4]oxazine-6-carbonyl) guanidines as Na/H exchange inhibitors. *Chem Pharm Bull,* 46, pp. 1716–1723.

[143] Yashimirskii, K. B. (1951). Thermochemstry of complex compounds, Publisher of Academy of Sciences of USSR, Moscow.

[144] Ye, C. Z., Li, G. Z., Yao, L. X., Chen, N. Y. and Lu, W. C. (2002). Support vector regression applied to thermodynamic analysis of solubility of C_{60} in different solvents, *Computer and applied chemistry,* 19, pp.729–730.

[145] Ye, C. Z., Yang, J., Yao, L. X. and Chen, N. Y. (2001). Regularities of the formation and lattice distortion of perovskite-type compounds, *Chinese science bulletin,* 46, pp. 1951–1953.

[146] Zelinskii, O. Y., Pavlyuk, V. V. and Zelinskii, A. V. (2002). *Journal of alloys and compounds,* 333, pp. 81–83.

[147] Zhan, Q. B., Li, M., Zhou, J. C. and Chen, N. Y. (1993). Reaction between Ca_2SiO_4 and $NaFeO_2$ and composition optimization of aluminate sinters, *Acta metallurgica sinica,* 29B, pp. 49.

[148] Zhang, D. Z., Zhou, T. S., Wu, Y. J., Liu, C. M. *et al.* (1997). Synthesis and Antifungal Activity of 1-(1H-1,2,4-Triazole-1-Yl)-2-(2,4-Difluorophenyl)-3-Substituted-2-Pr-opanols, *Acta pharmaceutica sinica,* 32, pp. 943–949.

[149] Zhang, L. M., Lu W. C., Liu X., Chen N. Y. and Yao, L. X. (2003). Support vector regression applied to estimation of activity coefficients of some concentrated electrolytic solutions, *Computer and applied chemistry,* 20, pp.745–749.

[150] Zhao, H., Lu, W. C., Liu, L. and Song, H. F. (2004). Support vector machine applied to prediction of absorption maximum wavelength of azo dyestuff, *Acta chemica sinica,* pp. 649–656.

[151] Zhong, H. L. (1987). The epidemiology and prevention of cancer, Guangdong Branch of Publishing Company for Science Propagation, Guangzhou.

Index